INTEGRATED WASTEWATER MANAGEMENT AND VALORIZATION USING ALGAL CULTURES

INTEGRATED WASTEWATER MANAGEMENT AND VALORIZATION USING ALGAL CULTURES

Edited by

GOKSEL N. DEMIRER
School of Engineering and Technology & Institute for Great Lakes Research, Central Michigan University, Mount Pleasant, MI, United States

SIBEL ULUDAG-DEMIRER
Biosystems and Agricultural Engineering Department & The Anaerobic Digestion Research and Education Center (ADREC), Michigan State University, East Lansing, MI, United States

ELSEVIER

Elsevier
Radarweg 29, PO Box 211, 1000 AE Amsterdam, Netherlands
The Boulevard, Langford Lane, Kidlington, Oxford OX5 1GB, United Kingdom
50 Hampshire Street, 5th Floor, Cambridge, MA 02139, United States

Copyright © 2022 Elsevier Inc. All rights reserved.

No part of this publication may be reproduced or transmitted in any form or by any means, electronic or mechanical, including photocopying, recording, or any information storage and retrieval system, without permission in writing from the publisher. Details on how to seek permission, further information about the Publisher's permissions policies and our arrangements with organizations such as the Copyright Clearance Center and the Copyright Licensing Agency, can be found at our website: www.elsevier.com/permissions.

This book and the individual contributions contained in it are protected under copyright by the Publisher (other than as may be noted herein).

Notices

Knowledge and best practice in this field are constantly changing. As new research and experience broaden our understanding, changes in research methods, professional practices, or medical treatment may become necessary.

Practitioners and researchers must always rely on their own experience and knowledge in evaluating and using any information, methods, compounds, or experiments described herein. In using such information or methods they should be mindful of their own safety and the safety of others, including parties for whom they have a professional responsibility.

To the fullest extent of the law, neither the Publisher nor the authors, contributors, or editors, assume any liability for any injury and/or damage to persons or property as a matter of products liability, negligence or otherwise, or from any use or operation of any methods, products, instructions, or ideas contained in the material herein.

ISBN: 978-0-323-85859-5

For information on all Elsevier publications visit our website at https://www.elsevier.com/books-and-journals

Publisher: Charlotte Cockle
Acquisitions Editor: Peter Adamson
Editorial Project Manager: Aleksandra Packowska
Production Project Manager: Sojan P. Pazhayattil
Cover Designer: Matthew Limbert

Typeset by TNQ Technologies

Contents

Contributors	*ix*
Preface	*xiii*

1. Role of microalgae in circular economy   1
Ozgul Calicioglu and Goksel N. Demirer

1.	Background: circular economy and waste valorization	1
2.	Conceptual framework: potential role of microalgae in the circular economy	3
3.	Techno-economic feasibility: scale-up potential and limitations of integrated wastewater-derived microalgal biorefineries	7
4.	Conclusions	8
	Acknowledgments	9
	References	9
	Further reading	12

2. Recent advancements in algae—bacteria consortia for the treatment of domestic and industrial wastewater   13
Duygu Ozcelik, F. Koray Sakarya, Ulas Tezel and Berat Z. Haznedaroglu

1.	Introduction	13
2.	Algae in biological treatment of wastewater	14
3.	Mechanism of symbiosis between algae and bacteria	27
4.	Examples of wastewater treatment by algae—bacteria consortia	32
5.	Circular economy, process design, and modeling aspects of algae—bacteria based wastewater treatment	38
6.	Conclusions	42
	References	43

3. Integrated algal-based sewage treatment and resource recovery system   51
N. Nirmalakhandan, I.S.A. Abeysiriwardana-Arachchige, S.P. Munasinghe-Arachchige and H.M.K. Delanka-Pedige

1.	Introduction	51
2.	Details of the STaRR system	60
3.	Performance of the STaRR system	63
4.	Outlook	74
	Acknowledgments	75
	References	75
	Further reading	80

v

vi   Contents

4. Microalgae-based technologies for circular wastewater treatment   81
Tânia V. Fernandes, Lukas M. Trebuch and René H. Wijffels

1. Introduction to circular wastewater treatment	81
2. Microalgae use for circularity	84
3. Microalgae-based technologies for wastewater treatment	95
4. Conclusions and perspectives	102
References	104

5. Treatment of anaerobic digestion effluents by microalgal cultures   113
Nilüfer Ülgüdür, Tuba Hande Ergüder-Bayramoğlu and Göksel N. Demirer

1. Introduction	113
2. Treatment of digestates by microalgal cultures	114
3. Microalgal growth in digestates	129
4. Challenges and potential remedies for digestate treatment by microalgae	131
5. Pilot scale plants	135
6. Outcomes of techno-economic and life cycle assessment analysis	140
7. Conclusions	142
References	142

6. Techno-economic analysis and life cycle assessment of algal cultivation on liquid anaerobic digestion effluent for algal biomass production and wastewater treatment   149
Sibel Uludag-Demirer, Mauricio Bustamante, Yan Liu and Wei Liao

1. Introduction	149
2. Material and methods	150
3. Results and discussion	153
4. Conclusions	162
Acknowledgments	163
References	163

7. Biomethane production from algae biomass cultivated in wastewater   165
Esteban Serrano, Maikel Fernandez, Raul Cano, Enrique Lara, Zouhayr Arbib and Frank Rogalla

1. Introduction	165
2. Material and methods	166
3. Results and discussion	171
4. Conclusions	180
References	180

Contents vii

8. Anaerobic digester biogas upgrading using microalgae — 183

Kaushik Venkiteshwaran, Tonghui Xie, Matthew Seib, Vaibhav P. Tale and
Daniel Zitomer

1. Introduction	183
2. Raw biogas characteristics	183
3. Required gas quality for heat and power equipment	184
4. Current commercial biogas conditioning technologies	187
5. Novel microalgae biogas conditioning technology	194
6. Benefits of algae biogas upgrading	199
7. Limitations of algae biogas upgrading systems	200
8. Potential carbon and nutrient sources for algae	200
9. Algal bioreactor configurations and operations	201
10. Biogas constituents removed	204
11. Algal reactor products utilization and management	207
12. Conclusions and recommendations for future research	208
References	208

9. Large-scale demonstration of microalgae-based wastewater biorefineries — 215

Zouhayr Arbib, David Marín, Raúl Cano, Carlos Saúco, Maikel Fernandez,
Enrique Lara and Frank Rogalla

1. Introduction	215
2. Wastewater versus seawater as culture medium	216
3. Large-scale raceway ponds construction	217
4. Microalgae consortium	223
5. Harvesting process	223
6. Potential WWT using microalgae at large scale	225
7. Conclusion	231
References	232

10. Cultivation of microalgae on agricultural wastewater for recycling energy, water, and fertilizer nutrients — 235

Lijun Wang and Bo Zhang

1. Introduction	235
2. Microalgae for agricultural wastewater treatment	235
3. Microalgae for feeds, fuels, and fertilizers	240
4. Microalgae cultivation systems	246
5. Challenges of resource recovery using microalgae cultivation in wastewater	250

viii Contents

6. Enhancement of microalgae cultivation in wastewater		253
7. Conclusions		256
Acknowledgment		257
References		257

Index   *265*

Contributors

I.S.A. Abeysiriwardana-Arachchige
Civil Engineering Department, New Mexico State University, Las Cruces, NM, United States

Zouhayr Arbib
FCC Aqualia S.A. Innovation and Technology Department, Madrid, Spain

Mauricio Bustamante
Biosystems Engineering, University of Costa Rica, San Pedro, San José, Costa Rica

Ozgul Calicioglu
The World Bank, Environment, Natural Resources and Blue Economy Global Practice, Washington, DC, United States

Raúl Cano
FCC Aqualia S.A. Innovation and Technology Department, Madrid, Spain

H.M.K. Delanka-Pedige
Civil Engineering Department, New Mexico State University, Las Cruces, NM, United States

Göksel N. Demirer
School of Engineering and Technology & Institute for Great Lakes Research, Central Michigan University, Mount Pleasant, MI, United States

Tuba Hande Ergüder-Bayramoğlu
Department of Environmental Engineering, Middle East Technical University, Ankara, Turkey

Tânia V. Fernandes
Department of Aquatic Ecology, Netherlands Institute of Ecology (NIOO-KNAW), Wageningen, the Netherlands

Maikel Fernandez
FCC Aqualia S.A. Innovation and Technology Department, Madrid, Spain

Berat Z. Haznedaroglu
Institute of Environmental Sciences, Bogazici University, Istanbul, Turkey

F. Koray Sakarya
Institute of Environmental Sciences, Bogazici University, Istanbul, Turkey

Enrique Lara
FCC Aqualia S.A. Innovation and Technology Department, Madrid, Spain

Wei Liao
Biosystems Engineering, Michigan State University, East Lansing, MI, United States

Yan Liu
Biosystems Engineering, Michigan State University, East Lansing, MI, United States

David Marín
FCC Aqualia S.A. Innovation and Technology Department, Madrid, Spain

S.P. Munasinghe-Arachchige
Civil Engineering Department, New Mexico State University, Las Cruces, NM, United States

N. Nirmalakhandan
Civil Engineering Department, New Mexico State University, Las Cruces, NM, United States

Duygu Ozcelik
Institute of Environmental Sciences, Bogazici University, Istanbul, Turkey

Frank Rogalla
FCC Aqualia S.A. Innovation and Technology Department, Madrid, Spain

Carlos Saúco
FCC Aqualia S.A. Innovation and Technology Department, Madrid, Spain

Matthew Seib
Madison Metropolitan Sewerage District, Madison, WI, United States

Esteban Serrano
FCC Aqualia S.A. Innovation and Technology Department, Madrid, Spain

Vaibhav P. Tale
Chemtron Riverbend Water, Saint Charles, MO, United States

Ulas Tezel
Institute of Environmental Sciences, Bogazici University, Istanbul, Turkey

Lukas M. Trebuch
Department of Aquatic Ecology, Netherlands Institute of Ecology (NIOO-KNAW), Wageningen, the Netherlands; Bioprocess Engineering, AlgaePARC, Wageningen University, Wageningen, the Netherlands

Nilüfer Ülgüdür
Department of Environmental Engineering, Düzce University, Düzce, Turkey

Sibel Uludag-Demirer
Biosystems and Agricultural Engineering Department & The Anaerobic Digestion Research and Education Center (ADREC), Michigan State University, East Lansing, MI, United States

Kaushik Venkiteshwaran
Department of Civil, Coastal and Environmental Engineering, University of South Alabama, Mobile, AL, United States

Lijun Wang
Department of Natural Resources and Environmental Design, North Carolina Agricultural and Technical State University, Greensboro, NC, United States

René H. Wijffels
Bioprocess Engineering, AlgaePARC, Wageningen University, Wageningen, the Netherlands; Faculty of Biosciences and Aquaculture, Nord University, Bodø, Norway

Tonghui Xie
School of Chemical Engineering, Sichuan University, Chengdu, Sichuan, China

Bo Zhang
Department of Natural Resources and Environmental Design, North Carolina Agricultural and Technical State University, Greensboro, NC, United States

Daniel Zitomer
Department of Civil, Construction and Environmental Engineering, Marquette University, Milwaukee, WI, United States

Preface

Intensive development efforts and linear economic activities since the last century resulted in excessive natural resource use as well as environmental and health problems. When the cost of waste management and health care as well as declining levels of several important natural resources are considered, it is evident that this also represents major economic and social challenges. This new paradigm brought along integrating the concept of sustainability with all anthropogenic activities to create more sustainable means of producing, processing, and consuming the natural resources along with sustainable waste management practices.

Among others, circular economy aims at accomplishing a closed-loop system to maximize the recovery of raw materials derived from the waste at end-of-life. The wastes should be considered as "misplaced renewable resources" that can be used again to generate valuable and marketable products, replacing the fossil-based resources. Reducing the carbon footprint of waste management activities, recycling and reuse of valuable materials, wastewater reuse, and bioproduct and biofuel generation from wastes characterize the main features of sustainable waste management. Thus, it is not only an integral part of circular economy but also offers a solid framework to alleviate sustainability and resource efficiency—related problems.

Algal cultures have been used for nutrient removal from wastewaters for a long time. However, the research on algal biotechnology has been accelerating recently since it can integrate several processes with environmental and economic benefits, such as CO_2 sequestration via photosynthesis, nutrient removal and recovery from wastewaters, and production of valuable products such as biofuels, human food, cosmetics, pharmaceuticals, animal, and aquaculture feed and fertilizers.

This book revisits algal biotechnology and its potential contribution to sustainable wastewater management, resource efficiency, circular economy, and the United Nation's Sustainable Development Goals. We would like to express our sincere appreciation to the diverse group of 35 experts from 7 different countries and 16 institutions who shared their experience, work, and vision.

Goksel N. Demirer
Sibel Uludag-Demirer
Okemos, MI, USA
December 2021

CHAPTER 1

Role of microalgae in circular economy

Ozgul Calicioglu[1] and Göksel N. Demirer[2]

[1]The World Bank, Environment, Natural Resources and Blue Economy Global Practice, Washington, DC, United States;
[2]School of Engineering and Technology & Institute for Great Lakes Research, Central Michigan University, Mount Pleasant, MI, United States

1. Background: circular economy and waste valorization

The concept of Circular Economy (CE) was introduced by the British economists Pearce and Turner (1989), but the most widely adopted definition was presented by the Ellen MacArthur Foundation as "*An industrial economy that is restorative or regenerative by intention and design.*" With this approach, the concept of waste disappears, since its components return to form part of natural or industrial cycles with a minimal consumption of energy. The components of the waste that are of organic origin will be biodegraded, while those that are of technological or industrial origin will be reused in a straightforward way and with low energy cost (Potocnik, 2013). The essence is closing the life cycle of products, i.e., moving away from a linear model of the economy (produce, use, and discard) to a model that is circular, as occurs in nature. The current linear model based on increased production, consumption, and economic growth seems to be coming to an end (Ghisellini et al., 2016).

In contrast to linear economy, progress toward CE is not only an opportunity, but also is a necessity for long-term economic, environmental, and social sustainability. For instance, a study conducted by European Environment Agency concluded that fostering CE would bring advantages including: (1) Reduced demand for primary raw materials, which in turn would improve resource security by decreasing the dependence on imported materials; (2) Reduced greenhouse gas emissions and decrease in the overall environmental impact of the anthropogenic activities (environmental benefits); (3) Enhanced and new opportunities for economic growth and technological innovation, along with benefits from enhanced resource efficiency (economic benefits); and (4) Increased job creation among all skill levels and enhanced health and safety practices among the consumers due to advancements in the consumer behavior (social benefits) (Reichel et al., 2016).

A CE strives for designing out waste and utilizing renewable resources (Allesch and Brunner, 2014) and allows an effective prevention, minimization, and valorization of wastes (Zorpas et al., 2014; EC, 2015). Kalmykovaa et al. (2018) clearly demonstrated the relationship between the CE and relevant concepts such as eco-efficiency, waste prevention, recycling, reuse, and industrial symbiosis. The study indicated that waste

---

Integrated Wastewater Management and Valorization using Algal Cultures
ISBN 978-0-323-85859-5, https://doi.org/10.1016/B978-0-323-85859-5.00003-8

© 2022 Elsevier Inc.
All rights reserved.

avoidance principles and tools are fundamental elements of CE. For instance, it has been estimated that eco-design, waste prevention, and reuse can bring net savings for the European Union (EU) businesses of up to EUR 600 billion, while reducing greenhouse gas emissions. Moreover, the additional measures to increase resource productivity by 30% by 2030 could boost the gross domestic product by nearly 1% and also create 2 million additional jobs in the EU (EC, 2015). In 2020, The European Commission adopted a new Circular Economy Action Plan, one of the main blocks of the European Green Deal, as part of Europe's new agenda for sustainable growth. The new Action Plan announces initiatives along the entire life cycle of products, targeting, for example, their design, promoting CE processes, fostering sustainable consumption, and aiming to ensure that the resources used are kept in the EU economy for as long as possible (https://ec.europa.eu/commission/presscorner/detail/en/ip_20_420).

CE deployment requires collaboration among stakeholders throughout the entire value chain of the economic sectors, including the production, consumption, as well as waste collection and recycling stages. Often, the responsibilities for waste collection and recycling are attributed to the governments and usually allocated at the municipality level. Yet, from a perspective of CE, the urban and regional systems have also started exploring the options for better collaboration and increased producer responsibility for the supply of high-quality recycling services, as well as enhanced biological treatment of waste streams such as biorefining and anaerobic digestion. For example, a CE of the local food systems would not only comprise of local production, distribution, communication, and promotion and local consumption by informed citizens, but also the management of organic waste in a systematic manner by the consumers, which could then be utilized as a raw material, or could be valorized in another form such as biogas (Bačová et al., 2016).

Similarly, wastewater treatment has traditionally been the responsibility of municipalities, and CE entails obtaining the highest benefit at the highest efficiency from the *resource* compared to conventional treatment technologies. In this respect, as a wastewater processing alternative, microalgae could capture nutrients before they reach water bodies. These systems would not only enable the upcycling of nutrients but also of atmospheric carbon dioxide (CO_2) into microalgal biomass, a precursor for value-added products (e.g., biofuels, biochemicals, and proteins). For instance, microalgal biomass could be used as a substrate for biogas production or supplemented to already-existing biodigesters of municipal solid waste to balance their nutrient contents (Calicioglu and Demirer, 2019). When integrated, the nutrient removal, CO_2 sequestration, and biofuel production processes could address the challenges of linear economy around waste management and fossil-based resource dependence. This approach would bring circularity to urban and agro-industrial systems by (1) "disassembling" and "reconstructing" the wastewater nutrients in an available form (i.e., as a fertilizer or protein precursor) and (2) producing renewable energy for other activities of the CE.

The following sections aim to discuss the conceptual framework and technical feasibility of integrated nutrient removal and biogas production using microalgal and anaerobic microbial cultures, emphasizing the potential of waste valorization as a significant activity of CE. To this purpose, the potential role of microalgae in CE; municipal waste valorization through integrated microalgal and anaerobic bioprocesses in a biorefinery concept; as well as the technical feasibility, scale-up potential and limitations are presented.

2. Conceptual framework: potential role of microalgae in the circular economy

Production of microalgae is not necessarily circular per se. Indeed, the linear production of microalgae using fertilizer and freshwater is proven to be not sustainable from a life cycle perspective (Murphy and Allen, 2011). Nevertheless, microalgae are suitable agents for upcycling waste nutrients and CO_2 into various bioproducts through biorefining, as an integrated, closed loop process (Venkata Mohan et al., 2020). Fig. 1.1 conceptually depicts this integrated, closed loop process.

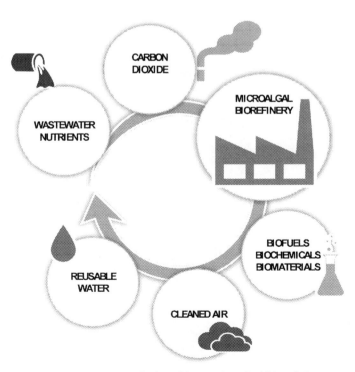

Figure 1.1 Integrated, closed loop microalgal biorefining.

2.1 Wastewater treatment by microalgal cultures

Microalgal cultures have been widely used for the treatment of a variety of wastewaters including those from municipal, agricultural, and industrial origin (Mata et al., 2010; Calicioglu and Demirer, 2016; Dogan-Subasi and Demirer, 2016) and have been proven to be effective in the removal of nitrogen, phosphorus, and metals (Mata et al., 2010; Mussgnug et al., 2010). Significant quantities of nutrient removal and biomass production reported in these studies demonstrated the feasibility of coupled microalgae cultivation and wastewater treatment processes (Cai et al., 2013). In particular, *Chlorella* species (including *Chlorella vulgaris*) have been successfully used for nitrogen, phosphorus, and chemical oxygen demand removal from wastewaters with a wide range of operating conditions (Wang et al., 2010; Calicioglu and Demirer, 2015, 2016, 2019).

Wastewater treatment using algae has many advantages. The process offers the potential to recycle these nutrients into algae biomass as a fertilizer and thus can offset treatment costs. After treatment using algae, an oxygen-rich effluent is released into water bodies, which prevents eutrophic conditions (Becker, 2004). Removing nitrogen and carbon from water, microalgae can help reduce the eutrophication in the aquatic environment (Mata et al., 2010) and contribute to biodiversity preservation by eliminating harmful impacts of sewage effluents and nitrogenous industrial wastewaters. Wastewater rich in CO_2 provides a conducive growth medium for microalgae because the CO_2 balances the Redfield ratio (molecular ratio of carbon, nitrogen, and phosphorus in marine organic matter, C:N:P $= 106{:}16{:}1$) of the wastewater allowing for faster production rates, reduced nutrient levels in the treated wastewater, decreased harvesting costs, and increased lipid production (Brennan and Owende, 2010).

2.2 Biosequestration of CO_2 emissions by microalgal cultures

Sequestration of CO_2 through photosynthesis has attracted attention as an alternative strategy over chemical methods, since the former is comparatively less capital and energy intensive. Microalgae efficiently capture CO_2 from high-CO_2 streams such as flue gases and flaring gases (CO_2 content 5%$-$15%) (Hsueh et al., 2007) in comparison with terrestrial plants, which typically absorb only 0.03%$-$0.06% CO_2 from the atmosphere. Microalgae can typically be used to capture CO_2 from three different sources like atmospheric CO_2, CO_2 emission from power plants and industrial processes, and CO_2 from soluble carbonate. Capture of atmospheric CO_2 is probably the most basic method to sink carbon and relies on the mass transfer from the air to the microalgae in their aquatic growth environments during photosynthesis (Singh and Ahluwalia, 2013). However, according to Brennan and Owende (2010) due to low CO_2 concentration in air (around 360 ppm), the potential yield from the atmosphere is limited. On the contrary, because of the higher CO_2 concentration of flue gas, much higher recovery levels can be achieved (Bilanovic et al., 2009).

Most importantly, the sequestered CO_2 as microalgal biomass from the atmosphere would be available as organic carbon, which can be considered as a chemical energy carrier, or a raw material for the synthesis of other biochemicals.

2.3 Microalgal biorefineries

The concept of biorefinery is analogous to traditional petroleum refinery which produces multiple fuels and products from petroleum, such that in biorefineries biomass is converted into marketable chemicals, fuels, and products (Pérez et al., 2017). The main difference between biorefinery and petroleum refinery is in terms of the raw materials and the technology employed. The biorefinery concept encompasses a large array of technologies for the production of building blocks (i.e., carbohydrates, proteins, lipids) from resources of biological origin such as wood, grasses, corn, different wastes, etc. These building blocks could be further processed into a variety of value-added products, including fuels and chemicals, as a substitute for the petroleum-based alternatives. In a biorefinery, the processes which convert biomass into an array of products are integrated in order to yield a variety of end products at various values, qualities, and quantities, including transportation fuels, power, and chemicals from biomass (Cherubini, 2010).

Developments in technology have enabled envisioning the derivation of materials and products from renewable biomass as an alternative to finite fossil-based resource consumption. These advancements fostered the transition to bioeconomy in developed countries, not only from a research and innovation perspective but also in terms of formulating supporting policies to develop a market pull, and accompanying consumer demand (Bracco et al., 2020). Yet, despite the availability of technology, biomass resources are very scarce in some developed countries (Bracco et al., 2018). In this respect, waste biorefineries could be a feasible option for the valorization of waste streams in these countries. Particularly, the concept of waste biorefinery is pertinent and vital not only in developed but also in developing countries, since the conventional waste disposal activities and fulfilling the increasing energy and raw material demands are both environmentally and economically challenging (Nizami et al., 2017). On the other hand, the estimates point out that the world market could gain a total value of up to $ 410 billion only by the recovery and recycling of municipal waste for beneficial purposes (Ismail and Nizami, 2016). Yet, only one quarter of this potential has been utilized (Guerrero et al., 2013; Hossain et al., 2014).

Microalgae contain high amounts of proteins, lipids, and carbohydrates which could be the feedstock for different products. For example, extracted microalgal lipids can be utilized as a potential feedstock for biodiesel production, while microalgal carbohydrates can be used as a carbon source in fermentation industries to replace conventional carbohydrate sources like simple sugars or treated lignocellulosic biomass. Furthermore, long-chain fatty acids found in microalgae have important functions as health food supplements, while proteins and pigments found in microalgae exhibit properties desired in the pharmaceutical industries to treat certain diseases (Yen et al., 2013). The important role of microalgae in the production of biofuels and bio-based chemicals makes it a promising feedstock to be considered as an alternative to many natural components and sources (Chew et al., 2017).

Due to high lipid contents, microalgal biomass has become particularly of interest as a biodiesel feedstock (Slade and Bauen, 2013). Microalgae offer many superiorities over traditional oil crops to produce biofuels and high-value chemicals, such as robust environmental adaptability, no competition with food or arable land, rapid fixation of environmental carbon, cultivation on wastewater, and year-round cultivation (Ho et al., 2014). They transform solar energy into high value biofuels and bio-based products by acting as biosolar machines which incorporate light capture and carbon sequestration mechanisms and convert these inputs into high value biomolecules/metabolites into the biomass. Due to significantly low biomass productivities of autotrophic cultivation methods, algal bioprocesses have shifted focus toward mixotrophic and heterotrophic approaches by implementing multiple-product strategies in a biorefinery approach (Yen et al., 2013).

Integrating biorefinery concept with wastewater treatment would provide efficient utilization of algae biomass, reduce overall residual waste component, and favor sustainable economics. The residual biomass can be subjected to a range of biochemical processes such as fermentation and anaerobic digestion for recovery of methane and biohydrogen (Subhash and Venkata Mohan, 2014). Thermo-chemical conversion of algae biomass can also be performed for synthesis of biooil and biochar (Agarwal et al., 2015). Moreover, microalgae that contain glucose-based carbohydrates are the most feasible feedstock for bioethanol production. The high value coproducts can be preferred for economic support of main process (Venkata Mohan et al., 2015).

Despite the interest in the utilization of microalgae as a biodiesel feedstock, the large-scale applications of this process are rather limited due to the limitations in the current harvesting and processing technologies (Vuppaladadiyam et al., 2018; Alcántara et al., 2013; Passos and Ferrer, 2014). Therefore, anaerobic conversion of the microalgae, which would not require concentration of biomass as opposed to biodiesel production, could be an energetically favorable mechanism for the valorization of the wastewater-derived biomass in the form of biomethane (Passos and Ferrer, 2014; Tan et al., 2014; Wiley et al., 2011). In addition, this process would provide a stabilized and concentrated stream of nutrients which could be utilized as fertilizers in a complete biorefinery concept (Dogan-Subasi and Demirer, 2016). However, anaerobic bioprocessing of microalgae also faces some challenges: (1) the cell wall structure of the raw microalgae is rigid and resistant to digestion by the anaerobic microbial consortia (Axelsson et al., 2012; Demuez et al., 2015); (2) the average carbon-to-nitrogen (C:N) ratio of microalgae (6:1) (Yen and Brune, 2007) may result in ammonia inhibition, since this ratio is lower than the optimum range for anaerobic digestion, 20:1–30:1 (Parkin and Owen, 1986). Therefore, prior disintegration of the cell wall structure by pretreatment and/or balancing the C:N ratio in the digester by the addition of a cosubstrate could improve the anaerobic digestibility and biomethane yields. Numerous studies have focused on methods to enhance anaerobic biodegradability of microalgae (Axelsson et al., 2012; Mendez et al., 2013; Ometto et al., 2014; Calicioglu and Demirer, 2016), by means of physical, chemical, and biological pretreatment (Demuez et al., 2015; He et al., 2016; Ometto et al., 2014). In order to balance the overall C:N ratio, researchers also codigested microalgal biomass with other substrates high in C:N ratio, such as waste paper (Yen and Brune, 2007), corn straw (Zhong et al., 2012), kitchen waste (Zhao and Ruan, 2013; Calicioglu and Demirer, 2019), and switchgrass (El-Mashad, 2013).

3. Techno-economic feasibility: scale-up potential and limitations of integrated wastewater-derived microalgal biorefineries

The potential technologies for biochemical or thermochemical processing of microalgae in large-scale biorefineries have been well established. In particular, anaerobic digestion is a robust technology which has been commercialized decades ago. Similarly, microalgae and value-added products out of its biomass have been successfully produced from laboratory to demonstration scales (Stiles et al., 2018), including cosmetics (Spolaore et al., 2006), biofuels (Suganya et al., 2016; Bose et al., 2020), human or animal feed (Becker, 2007), or as a soil treatment and slow release fertilizer (Mulbry et al., 2005). Therefore, the bottleneck of the integrated microalgal biorefineries would be the production of wastewater-derived biomass under a controlled environment with a consistent quality at a large scale.

One limitation of the microalgae production systems using wastewater as the growth media is the susceptibility to bacterial growth due to carbon sources present in the wastewater, as bacteria may outcompete the algae. In addition, the population dynamics in these microalgal–bacterial systems could result in inconsistencies of the final product. Therefore, the large-scale cultivation processes would need enhanced control (Van Den Hende et al., 2014; Silkina et al., 2017). Therefore, a closed microalgae seed system might be necessary to supplement the large-scale production in wastewater in open ponds (Benemann, 2013). The closed microalgae seed system would ensure that the desired microalgae species predominate other microorganisms in the system, whereas a sole open system would be more susceptible to the outgrowing of unwanted species.

In addition to the technical feasibility studies, in order to accelerate the large-scale application of integrated processes, market analyses should be performed for the high-value products derived from microalgae grown in wastewater. Therefore, the techno-economic analysis of microalgal biorefineries must go hand in hand with market analysis of the bioproducts (Ruiz et al., 2016). The multidisciplinary research conducted on integrated systems and technical know-how must be harmonized with the economic and regulatory information to facilitate the applications at a large scale (Stiles et al., 2018). Nevertheless, case studies and feasibility assessments at this angle exist. For instance, in a report on a potential CE business case based on production of a variety of commodity products from microalgae in Japan, it was discussed that using wastewater as nutrients source could reduce the costs of fertilizers by up to a 75%, and provide 70% reduction in the costs by recycling acetic acid and ammonia. The reductions in the cost originated from both the improvement of the existing technology and the increase in the revenue of wastewater treatment up to 50 yen per cubic meters. It was concluded that this scenario could compete with fossil oil prices (Herrador, 2016). Indeed, this alternative would not only be economical, but also environmentally and socially attractive. For instance, it has been reported that the life cycle impact of microalgae-derived biofuels is dominated by the cultivation phase, and the environmental and economic feasibility of the system could be improved by coupling biofuel production with wastewater treatment (Clarens et al., 2010; Murphy and Allen, 2011). Such a system could also perform well in terms of social wellbeing indicators, such as food security (Efroymson et al., 2017). In addition, increasing demands on all three components of the food—energy—water nexus require societies to search for more sustainable development equilibria (Calicioglu and Bogdanski, 2021). In this respect, internalizing social and environmental costs of fossil-based alternatives could also provide a fair comparison ground for microalgal biorefinery products in the market.

4. Conclusions

The development of a CE will require high-quality, secondary raw materials that can be fed back into production processes. In this sense, the waste management sector will have to become a key partner in building new business models that focus both on waste prevention and turning waste into a resource (EPRS, 2017).

Biosequestration of CO_2 emissions by microalgal cultures is a relatively new research area among the waste management applications of microalgal cultures. Nutrient removal by microalgal cultures and anaerobic digestion of waste microalgal biomass and subsequent biogas, hydrogen, and fertilizer production were investigated in the past. Integrating nutrient removal, greenhouse gas mitigation and biofuel production through microalgal and anaerobic cultures will not only tackle with waste management issues (wastewater treatment and CO_2 mitigation) but also generate renewable energy in the form of biogas.

Algae are rich in protein and lipids as well as many other compounds which have several applications in the pharmaceutical and cosmetics industries in addition to food, animal feed, and fuel industries. Thus, it is a good example of waste valorization that aligns well with both waste management and circularity objectives of the CE for several sectors. This approach, however, requires more research and development for economically feasible scaling up. Creating a market pull for the high value products is therefore essential for utilizing microalgae as a mediator of CE.

Acknowledgments

The views represented in this chapter reflect those of the authors and do not necessarily reflect the views of the World Bank Group.

References

Agarwal, M., Tardio, J., Venkata Mohan, S., 2015. Effect of pyrolysis parameters on yield and composition of gaseous products from activated sludge: towards sustainable biorefinery. Biomass Convers. Biorefin. 5 (2), 227–235.

Alcántara, C., García-Encina, P.A., Muñoz, R., 2013. Evaluation of mass and energy balances in the integrated microalgae growth-anaerobic digestion process. Chem. Eng. J. https://doi.org/10.1016/j.cej.2013.01.100.

Allesch, A., Brunner, P.H., 2014. Assessment methods for solid waste management: a literature review. Waste Manag. Res. 32 (6), 461–473.

Axelsson, L., Franzén, M., Ostwald, M., Berndes, G., Lakshmi, G., Ravindranath, N.H., 2012. Perspective: Jatropha cultivation in southern India: assessing farmers' experiences. Biofuels Bioprod. Biorefin 6 (3), 246–256.

Bačová, M., Böhme, K., Guitton, M., van Herwijnen, M., Kállay, T., Koutsomarkou, J., et al., 2016. Pathways to a Circular Economy in Cities and Regions. European Week of Regions and Cities. Retrieved from. http://europa.eu/rapid/press-release_IP-15-6203_en.htm%0Ahttps://www.rabobank.com/en/images/Pathways-to-a-circular-economy.pdf.

Becker, W., 2004. Microalgae in human and animal nutrition. In: Richmond, A. (Ed.), Handbook of Microalgal Culture. Blackwell, Oxford, U.K, pp. 312–351.

Becker, E.W., 2007. Micro-algae as a source of protein. Biotechnol. Adv. 25 (2), 207–210.

Benemann, J., 2013. Microalgae for biofuels and animal feeds. Energies 6 (11), 5869–5886.

Bilanovic, D., Andargatchew, A., Kroeger, T., Shelef, G., 2009. Freshwater and marine microalgae sequestering of CO_2 at different C and N concentrations—response surface methodology analysis. Energy Convers. Manag. 50, 262–267.

Bose, A., O'Shea, R., Lin, R., Murphy, J.D., 2020. A perspective on novel cascading algal biomethane biorefinery systems. Bioresour. Technol. 304, 123027. https://doi.org/10.1016/j.biortech.2020.123027.

Bracco, S., Calicioglu, Ö., Flammini, A., San Juan, M.G., Bogdanski, A., 2020. Analysis of standards, certifications and labels for bio-based products in the context of sustainable bioeconomy. Int. J. Stand. Res. https://doi.org/10.4018/ijsr.2019010101.

Bracco, S., Calicioglu, O., Juan, M.G.S., Flammini, A., 2018. Assessing the contribution of bioeconomy to the total economy: a review of national frameworks. Sustainability 10. https://doi.org/10.3390/su10061698.

Brennan, L., Owende, P., 2010. Biofuels from microalgae—a review of technologies for production, processing and extractions of biofuels and co-products. Renew. Sustain. Energy Rev. 14, 557–577.

Cai, T., Park, S.Y., Li, Y., 2013. Nutrient recovery from wastewater streams by microalgae: status and prospects. Renew. Sustain. Energy Rev. 19, 360–369.

Calicioglu, Ö., Bogdanski, A., 2021. Linking the bioeconomy to the 2030 sustainable development agenda: can SDG indicators be used to monitor progress towards a sustainable bioeconomy? N. Biotech. https://doi.org/10.1016/j.nbt.2020.10.010.

Calicioglu, O., Demirer, G.N., 2015. Integrated nutrient removal and biogas production by Chlorella vulgaris cultures. J. Renew. Sustain. Energy 7, 033123.

Calicioglu, O., Demirer, G.N., 2016. Biogas production from waste microalgal biomass obtained from nutrient removal of domestic wastewater. Waste Biomass Valorization 7 (6), 1397–1408.

Calicioglu, O., Demirer, G.N., 2019. Carbon-to-nitrogen and substrate-to-inoculum ratio adjustments can improve co-digestion performance of microalgal biomass obtained from domestic wastewater treatment. Environ. Technol. 40 (5), 614—624.

Cherubini, F., 2010. The biorefinery concept: using biomass instead of oil for producing energy and chemicals. Energy Convers. Manag. 51 (7), 1412—1421.

Chew, K.W., Yap, J.Y., Show, P.L., Suan, N.H., Juan, J.C., Ling, T.C., Lee, D.J., Chang, J.S., 2017. Microalgae biorefinery: high value products perspectives. Bioresour. Technol. 229, 53—62.

Clarens, A.F., Resurreccion, E.P., White, M.A., Colosi, L.M., 2010. Environmental life cycle comparison of algae to other bioenergy feedstocks. Environ. Sci. Technol. 44, 1813—1819.

Demuez, M., Mahdy, A., Tomás-Pejó, E., González-Fernández, C., Ballesteros, M., 2015. Enzymatic cell disruption of microalgae biomass in biorefinery processes. Biotechnol. Bioeng. 112 (10), 1955—1966.

Dogan-Subasi, E., Demirer, G.N., 2016. Anaerobic digestion of microalgal (Chlorella vulgaris) biomass as a source of biogas and biofertilizer. Environ. Prog. Sustain. Energy 35 (4), 936—941.

Efroymson, R.A., Dale, V.H., Langholtz, M.H., 2017. Socioeconomic indicators for sustainable design and commercial development of algal biofuel systems. In: GCB Bioenergy. https://doi.org/10.1111/gcbb.12359.

El-Mashad, H.M., 2013. Kinetics of methane production from the codigestion of switchgrass and Spirulina platensis algae. Bioresour. Technol. 132, 305—312.

European Commission, 2015. Communication from the Commission to the Parliament, the Council and the European Economic and Social Commitee and the Commitee of the Regions: Closing the Loop — an EU Action Plan for the Circular Economy. COM (2015) 614 final.

European Parliamentary Research Service, 2017. Towards a Circular Economy: Waste Management in the EU, Scientific Foresight Unit (STOA) PE 581.913, IP/G/STOA/FWC/2013-001/LOT 3/C3. September 2017.

Ghisellini, P., Cialani, C., Ulgiati, S., 2016. A review on circular economy: the expected transition to a balanced interplay of environmental and economic systems. J. Clean. Prod. 114, 11—32.

Guerrero, L., Maas, G., Hogland, W., 2013. Solid waste management challenges for cities in developing countries. Waste Manag. 33 (1), 220—232.

He, S., Fan, X., Katukuri, N.R., Yuan, X., Wang, F., Guo, R.B., 2016. Enhanced methane production from microalgal biomass by anaerobic bio-pretreatment. Bioresour. Technol. 204, 145—151.

Herrador, M., 2016. The Microalgae/Biomass Industry in Japan -an Assessment of Cooperation and Business Potential with European Companies — Tokyo. Retrieved from. https://www.eu-japan.eu/sites/default/files/publications/docs/microalgaebiomassiindustryinjapan-herrador-min16-1.pdf.

Ho, S.H., Ye, X., Hasunuma, T., Chang, J.S., Kondo, A., 2014. Perspectives on engineering strategies for improving biofuel production from microalgae — a critical review. Biotechnol. Adv. 32 (8), 1448—1459.

Hossain, H.M.Z., Hossain, Q.H., Monir, M.M., Ahmed, M.T., 2014. Municipal solid waste (MSW) as a source of renewable energy in Bangladesh: revisited. Renew. Sustain. Energy Rev. 10, 11—21.

Hsueh, H.T., Chu, H., Yu, S.T., 2007. A batch study on the bio-fixation of carbon dioxide in the absorbed solution from a chemical wet scrubber by hot spring and marine algae. Chemosphere 66, 878—886.

Ismail, I.M.I., Nizami, A.S., 2016. Waste-based biorefineries in developing countries: an imperative need of time. In: Paper Presented at the Canadian Society for Civil Engineering: 14th International Environmental Specialty Conference, in London, Ontario, Canada, June 1—4, 2016. Available online. http://ir.lib.uwo.ca/csce2016/London/Environmental/9/.

Kalmykovaaa, Y., Sadagopanb, M., Rosadoc, L., 2018. Circular economy — from review of theories and practices to development of implementation tools, Resources. Conserv. Recycl. 135, 190—201.

Mata, T.M., Martins, A.A., Caetano, N.S., 2010. Microalgae for biodiesel production and other applications: a review. Renew. Sustain. Energy Rev. 14, 217—232.

Mendez, L., Mahdy, A., Timmers, R.A., Ballesteros, M., González-Fernández, C., 2013. Enhancing methane production of Chlorella vulgaris via thermochemical pretreatments. Bioresour. Technol. 149, 136—141.

Mulbry, W., Westhead, E.K., Pizarro, C., Sikora, L., 2005. Recycling of manure nutrients: use of algal biomass from dairy manure treatment as a slow release fertilizer. Bioresour. Technol. 96 (4), 451—458.

Murphy, C.F., Allen, D.T., 2011. Energy-water nexus for mass cultivation of algae. Environ. Sci. Technol. 45, 5861—5868.

Mussgnug, J.H., Klassen, V., Schluter, A., Kruse, O., 2010. Microalgae as substrates for fermentative biogas production in a combined biorefinery concept. J. Biotechnol. 150, 51−56.

Nizami, A.S., Rehan, M., Waqas, M., Naqvi, M., Ouda, O.K.M., Shahzad, K., Miandad, R., Khan, M.Z., Syamsiro, M., Ismail, I.M.I., Pant, D., 2017. Waste biorefineries: enabling circular economies in developing countries. Bioresour. Technol. 241, 1101−1117.

Ometto, F., Quiroga, G., Pšenička, P., Whitton, R., Jefferson, B., Villa, R., 2014. Impacts of microalgae pre-treatments for improved anaerobic digestion: thermal treatment, thermal hydrolysis, ultrasound and enzymatic hydrolysis. Water Res. 65, 350−361.

Parkin, G.F., Owen, W.F., 1986. Fundamentals of anaerobic digestion of wastewater sludges. J. Environ. Eng. 112 (American Society of Civil Engineers).

Passos, F., Ferrer, I., 2014. Microalgae conversion to biogas: thermal pretreatment contribution on net energy production. Environ. Sci. Technol. 48 (12), 7171−7178.

Pearce, D.W., Turner, R.K., 1989. Economics of Natural Resources and the Environment. Johns Hopkins University Press, Baltimore, U.S.A.

Pérez, A.T.E., Camargo, M., Rincón, P.C.N., Marchant, M.A., 2017. Key challenges and requirements for sustainable and industrialized biorefinery supply chain design and management: a bibliographic analysis. Renew. Sustain. Energy Rev. 69, 350−359.

Potocnik, J., 2013. Towards the Circular Economy - Economic and Business Rationale for an Accelerated Transition. Ellen MacArthur Foundation. https://www.ellenmacarthurfoundation.org/assets/downloads/publications/Ellen-MacArthur-Foundation-Towards-the-Circular-Economy-vol.1.pdf.

Reichel, A., De Schoenmakere, M., Gillabel, J., 2016. Circular Economy in Europe, Developing the Knowledge Base. European Environment Agency, Copenhagen, Denmark.

Ruiz, J., Olivieri, G., De Vree, J., Bosma, R., Willems, P., Reith, J.H., Eppink, M.H.M., Kleinegris, D.M.M., Wijffels, R.H., Barbosa, M.J., 2016. Towards industrial products from microalgae. Energy Environ. Sci. 9 (10), 3036−3043.

Silkina, A., Zacharof, M.P., Hery, G., Nouvel, T., Lovitt, R.W., 2017. Formulation and utilisation of spent anaerobic digestate fluids for the growth and product formation of single cell algal cultures in heterotrophic and autotrophic conditions. Bioresour. Technol. 244, 1445−1455.

Singh, U.B., Ahluwalia, A.S., 2013. Microalgae: a promising tool for carbon sequestration. Mitig. Adapt. Strategies Glob. Change 18, 73−95.

Slade, R., Bauen, A., 2013. Micro-algae cultivation for biofuels: cost, energy balance, environmental impacts and future prospects. Biomass Bioenergy 53 (0), 29−38.

Spolaore, P., Joannis-Cassan, C., Duran, E., Isambert, A., 2006. Commercial applications of microalgae. J. Biosci. Bioeng. 101 (2), 87−96.

Stiles, W.A.V., Styles, D., Chapman, S.P., Esteves, S., Bywater, A., Melville, L., Silkina, A., Lupatsch, I., Grunewald F., C., Lovitt, R., Chaloner, T., Bull, A., Morris, C., Llewellyn, C.A., 2018. Using microalgae in the circular economy to valorise anaerobic digestate: challenges and Opportunities. Bioresour. Technol. 267 (18), 732−742.

Subhash, G.V., Venkata Mohan, S., 2014. Deoiled algal cake as feedstock for dark fermentative biohydrogen production: an integrated biorefinery approach. Int. J. Hydrogen Energy 39 (18), 9573−9579.

Suganya, T., Varman, M., Masjuki, H.H., Renganathan, S., 2016. Macroalgae and microalgae as a potential source for commercial applications along with biofuels production: a biorefinery approach. Renew. Sustain. Energy Rev. 55, 909−941.

Tan, C.H., Show, P.L., Chang, J.S., Ling, T.C., Lan, J.C.W., 2014. Novel approaches of producing bioenergies from microalgae: a recent review. Biotechnol. Adv. 33 (6), 1219−1227.

Van Den Hende, S., Beelen, V., Bore, G., Boon, N., Vervaeren, H., 2014. Up-scaling aquaculture wastewater treatment by microalgal bacterial flocs: from lab reactors to an outdoor raceway pond. Bioresour. Technol. 159, 342−354.

Venkata Mohan, S., Hemalatha, M., Chakraborty, D., Chatterjee, S., Ranadheer, P., Kona, R., 2020. Algal biorefinery models with self-sustainable closed loop approach: trends and prospective for bluebioeconomy. Bioresour. Technol. 295, 122128. https://doi.org/10.1016/j.biortech.2019.122128.

Venkata Mohan, S., Rohit, M., Chiranjeevi, P., Chandra, R., Navaneeth, B., 2015. Heterotrophic microalgae cultivation to synergize biodiesel production with waste remediation: progress and perspectives. Bioresour. Technol. 184, 169−178.

Vuppaladadiyam, A.K., Prinsen, P., Raheem, A., Luque, R., Zhao, M., 2018. Sustainability analysis of microalgae production systems: a review on resource with unexploited high-value reserves. Environ. Sci. Technol. 52 (24), 14031–14049.

Wang, L., Min, M., Li, Y., Chen, P., Chen, Y., Liu, Y., Wang, Y., Ruan, R., 2010. Cultivation of green algae Chlorella sp. in different wastewaters from municipal wastewater treatment plant. Appl. Biochem. Biotechnol. 162, 1174–1186.

Wiley, P.E., Campbell, J.E., McKuin, B., 2011. Production of biodiesel and biogas from algae: a review of process train options. Water Environ. Res. 83 (4), 326–338.

Yen, H.W., Brune, D.E., 2007. Anaerobic co-digestion of algal sludge and waste paper to produce methane. Bioresour. Technol. 98 (1), 130–134.

Yen, H.W., Hu, I.C., Chen, C.Y., Ho, S.H., Lee, D.J., Chang, J.S., 2013. Microalgae-based biorefinery from biofuels to natural products. Bioresour. Technol. 135, 166–174.

Zhao, M.X., Ruan, W.Q., 2013. Biogas performance from co-digestion of Taihu algae and kitchen wastes. Energy Convers. Manag. 75, 21–24.

Zhong, W., Zhang, Z., Luo, Y., Qiao, W., Xiao, M., Zhang, M., 2012. Biogas productivity by co-digesting Taihu blue algae with corn straw as an external carbon source. Bioresour. Technol. 114, 281–286.

Zorpas, A.A., Lasaridi, K., Abeliotis, K., Voukkali, I., Loizia, P., Fitiri, L., Chroni, C., Bikaki, N., 2014. Waste prevention campaign regarding the waste framework directive. Fresenius Environ. Bull. 23 (11a), 2876–2883.

Further reading

Alzate, M.E., Munoz, R., Rogalla, F., Fdz-Polanco, F., Perez-Elvira, S.I., 2012. Biochemical methane potential of microalgae: influence of substrate to inoculum ratio, biomass concentration and pretreatment. Bioresour. Technol. 123, 488–494.

Astals, S., Musenze, R.S., Bai, X., Tannock, S., Tait, S., Pratt, S., Jensen, P.D., 2015. Anaerobic co-digestion of pig manure and algae: impact of intracellular algal products recovery on co-digestion performance. Bioresour. Technol. 181, 97–104.

Callaghan, F.J., Wase, D.A.J., Thayanithy, K., Forster, C.F., 2002. Continuous co-digestion of cattle slurry with fruit and vegetable wastes and chicken manure. Biomass Bioenergy 22 (1), 71–77.

Çaylı, D., Uludag-Demirer, S., Demirer, G.N., 2018. Coupled nutrient removal from the wastewater and CO_2 biofixation from the flue gas of iron and steel manufacturing. Int. J. Glob. Warming 16 (2), 148–161.

Chynoweth, D.P., Turick, C.E., Owens, J.M., Jerger, D.E., Peck, M.W., 1993. Biochemical methane potential of biomass and waste feedstocks. Biomass Bioenergy 5 (1), 95–111.

Eskicioglu, C., Ghorbani, M., 2011. Effect of inoculum/substrate ratio on mesophilic anaerobic digestion of bioethanol plant whole stillage in batch mode. Process Biochem. 46 (8), 1682–1687.

González-Fernández, C., Sialve, B., Bernet, N., Steyer, J.P., 2012. Comparison of ultrasound and thermal pretreatment of Scenedesmus biomass on methane production. Bioresour. Technol. 110, 610–616.

Miao, H., Wang, S., Zhao, M., Huang, Z., Ren, H., Yan, Q., Ruan, W., 2014. Codigestion of Taihu blue algae with swine manure for biogas production. Energy Convers. Manag. 77, 643–649.

Passos, F., García, J., Ferrer, I., 2013. Impact of low temperature pretreatment on the anaerobic digestion of microalgal biomass. Bioresour. Technol. 138, 79–86.

Passos, F., Uggetti, E., Carrère, H., Ferrer, I., 2014. Pretreatment of microalgae to improve biogas production: a review. Bioresour. Technol. 172, 403–412.

Schwede, S., Rehman, Z.U., Gerber, M., Theiss, C., Span, R., 2013. Effects of thermal pretreatment on anaerobic digestion of Nannochloropsis salina biomass. Bioresour. Technol. 143, 505–511.

Vlyssides, A.G., Karlis, P.K., 2004. Thermal-alkaline solubilization of waste activated sludge as a pretreatment stage for anaerobic digestion. Bioresour. Technol. 91 (2), 201–206.

Wang, M., Park, C., 2015. Investigation of anaerobic digestion of Chlorella sp. and Micractinium sp. grown in high-nitrogen wastewater and their co-digestion with waste activated sludge. Biomass Bioenergy 80 (813), 30–37.

Wang, M., Sahu, A.K., Rusten, B., Park, C., 2013. Anaerobic co-digestion of microalgae Chlorella sp. and waste activated sludge. Bioresour. Technol. 142, 585–590.

CHAPTER 2

Recent advancements in algae—bacteria consortia for the treatment of domestic and industrial wastewater

Duygu Ozcelik, F. Koray Sakarya, Ulas Tezel and Berat Z. Haznedaroglu
Institute of Environmental Sciences, Bogazici University, Istanbul, Turkey

1. Introduction

As we explore a bioeconomy driven sustainable development and abandon fossil fuels and nonrenewable energy resources, algae gain more popularity as a tangible feedstock for many sectors including energy, food, health, agricultural, and environmental sectors. Relying on waste streams such as flue gas for carbon requirements, and wastewater for nutrient requirements, algae will be an indispensable tool for generating environmentally sustainable and economically feasible products and technologies. With respect to applications in wastewater treatment, algae have already been component of tertiary treatment for enhanced removal of nutrients from wastewater for many years. As we understand metagenomic dynamics and uncover metabolic requirements in complex systems such as wastewater, we are more comfortable to offer algae—bacteria based bioremediation power to energy-intensive processes of wastewater treatment. Algae—bacteria—based bioremediation can also decrease overall costs by generating additional value added products such as biofuels, biofertilizers, biosurfactants, animal, and aquaculture feed.

Despite the wide portfolio of products that can be obtained from wastewater-fed algae, several challenges remain to be addressed to improve nutrient/contaminant removal efficiencies and achieve better economical metrics for wastewater treatment. As wastewater treatment plants (WWTPs) host millions of different species, the heterogeneity of metabolic needs, biotic, and abiotic factors pose new risks and problems that need to be resolved. Although photoautotrophy is the primary form of metabolism among algae, mixotrophic growth is very likely to generate competition between algae—bacteria for similar nutrients and carbon sources. Meanwhile, high nutrient concentrations in wastewater (especially nitrogen) most commonly result in lower lipid accumulation in most green algae species suitable for wastewater treatment.

Algae—bacteria consortia in wastewater treatment offer several advantages such as decreased cost of biomass production, and their mutual growth can positively induce

---

Integrated Wastewater Management and Valorization using Algal Cultures
ISBN 978-0-323-85859-5, https://doi.org/10.1016/B978-0-323-85859-5.00002-6

© 2022 Elsevier Inc.
All rights reserved.

more robust and durable biomass against fluctuating nutrient loads and other disturbances in WWTPs. With this framework of the chapter, we present the most recent developments of algae—bacteria consortia and provide context to some of these advantages and disadvantages in wastewater treatment.

2. Algae in biological treatment of wastewater

2.1 Conventional biological treatment of wastewater: microbial ecology and function

Biological wastewater treatment (BWWT) is traditionally used for removal of suspended solids, nutrients, and dissolved organics in wastewater. For these treatment systems, numerous types of microorganisms such as protozoa, fungi, algae, and bacteria are used (Ranjit et al., 2021). Decomposition of organic matter in the wastewater is mostly ensured with aerobic metabolism of these microorganisms, where a mixture of anaerobic, facultative, and heterotrophic metabolisms are used (Ranjit et al., 2021; Peavy and Tchobanoglous, 1985). There are mainly two types of BWWT processes: "suspended culture systems" and "attached culture systems." Suspended culture systems contain single or flocs of microorganisms, and the cultures could be maintained in completely mixed reactors or plug flow reactors. Attached culture systems are composed of microbial biomass attached to inert surfaces and the wastewater passes through those surfaces (Satyanarayana et al., 2018).

2.1.1 Suspended culture systems

In urban WWTPs, conventional activated sludge (CAS) treatment process is the most common system used for the removal of carbonaceous materials and nutrients. The process combines pretreated wastewater with microorganisms for the uptake of nutrients into microbial biomass and oxidation of carbonaceous organic matter. Firstly, pretreated wastewater is mixed with aerobic bacteria in aeration tanks supplied with oxygen (Eckenfelder and Cleary, 2013). Following, the effluent is sufficiently treated; it is sent to secondary clarifier tank for a solid—liquid separation which is a physical operation in WWTPs. Then, settled activated sludge is recycled back to the aeration tanks to sustain the microbial population for the biodegradation of organic pollutants. In biological treatment systems, in addition to regular maintenance of WWTP equipment, microbial cultures should be closely monitored as the microbial culture in activated sludge system is prone to be affected from drastic changes in influent wastewater, ambient conditions, toxic metabolites or chemicals, insufficient nutrients, etc (Karna and Visvanathan, 2019). Although it is a low cost and efficient system (up to 95% suspended solids, 95% biological oxygen demand (BOD) and 98% pathogen removal), the most challenging disadvantages of CAS process are the excessive sludge formation and intensive energy requirements (Ranjit et al., 2021).

In addition to CAS system, there are other suspended culture systems such as waste stabilization ponds (WSPs) and sequencing batch reactors (SBRs). Natural suspended culture systems such as WSPs may contain a microbial culture that is either aerobic or facultative (Peavy and Tchobanoglous, 1985). Among WSPs, facultative microbial culture system is the common one due to its convenience for application (Mara, 2003). In facultative ponds, in the upper layer of the system, aerobic microbial growth is favored as there is a supply of oxygen to aerobic bacteria from algae and atmospheric interactions (Fig. 2.1) (Mara, 2003). Bacteria use the dissolved oxygen to remove the organic compounds in the system. The carbon dioxide produced by bacteria is fixed by algae. Due to a thin colloidal film formation on the surface, oxygen transfer to the lower parts of the system is hindered; thus, anaerobic conditions predominate. However, these layers are not stable, and they may create unstable ambient conditions. These conditions trigger formation of a facultative zone, where there is slightly less dissolved oxygen and sunlight than the upper part (Peavy and Tchobanoglous, 1985; Mara, 2003). It is crucial to maintain algal population in facultative ponds healthy as algae estimated to supply 80% of the dissolved oxygen in WSP systems for aerobic bacteria to use in organic carbon decomposition (Pearson, 2003). While low operating costs and simplicity of operation are advantageous for facultative ponds, they require considerable land, perform poorly in cold climates, and generate odor and insect problems in the operating area (Ranjit et al., 2021; Peavy and Tchobanoglous, 1985; Pearson, 2003).

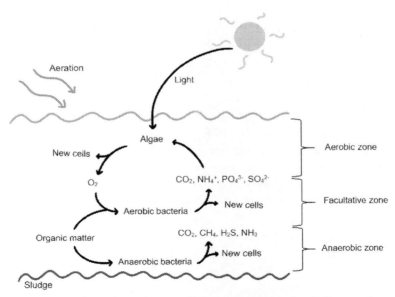

Figure 2.1 Algae—bacteria mutualistic relationship in facultative ponds.

In general, WSPs follow maturation ponds where pathogenic fecal bacteria are destroyed via pond algae (Mara, 2003). Rapid photosynthesis metabolism of algae boosts the stripping of carbon dioxide, that is generated by bacteria, from the pond aqueous medium. This stripping triggers the carbon dioxide equilibrium in the system and hydroxyl ion formation is favored Eqs. (2.1)–(2.3) (Mara, 2003). As a result, hydroxyl ions build up in the system and raise the pH of the medium to a level higher than 9, which is fatal to fecal bacteria. Increase in temperature, visible light intensity, and dissolved oxygen levels also enhance the death of fecal bacteria in the process (Curtis et al., 1992).

$$CO_{2(g)} \rightleftarrows CO_{2(aq)} + H_2O \rightleftarrows H_2CO_3 \tag{2.1}$$

$$H_2CO_3 \rightleftarrows HCO_3^- + H_3O^+ \tag{2.2}$$

$$HCO_3^- + OH^- \rightleftarrows CO_3^{2-} + H_2O \tag{2.3}$$

As a modified version of CAS system, the aeration process and sedimentation operation are carried out in the same tank in a sequencing mode, instead of separate aeration tanks and secondary clarifier tanks in the SBR process. These reactors are efficient in terms of removing nitrogen and phosphorus and the effluent has low organic carbon content (Ahuja et al., 2014).

2.1.2 Attached culture systems

Attached culture systems are the other conventional BWWT technology based on the attachment of microbial films onto surfaces (Peavy and Tchobanoglous, 1985). They are classified according to the reactor design and aerobic bacteria are used for the biodegradation of organic pollutants in the medium. In most cases, heterotrophic bacteria with dominant facultative bacteria, protozoa, and metazoa are attached onto the solid surface in attached culture systems with a similar diversity in CAS system. Microbial diversity is dependent on ambient conditions as light accessibility enables algae presence, and low levels of organic carbon indicate the presence of nitrifiers within the system. Typical attached culture systems are trickling filters, rotating biological contractors, biotowers, and packed bed reactors.

Trickling filters have stationary surfaces and wastewater is passed over the biofilm medium. Trickling filters are classified as aerobic processes. The microbial ecology of these systems is very similar to the microbial community in activated sludge systems, where heterotrophic organisms are dominant in the culture. The most abundant one being the facultative bacteria, fungi, and protozoa are also ample. Algae could be found nearby the surface where light is available. Microorganisms are attached to the surface and produce a sticky layer that allows them to retain suspended particles and colloids to solubilize those compounds (Peavy and Tchobanoglous, 1985). The removal of organic pollutants in the wastewater is highly dependent on temperature, flow rate of wastewater, organic loading rate, and diffusion rate of substrate and oxygen into the biofilm medium. Trickling filters are easy to operate, low-cost, and adaptable in hot climates. Yet, it is not possible to treat raw sewage unless primary sedimentation is applied and excess sludge generated by the process is managed properly (Ranjit et al., 2021; Satyanarayana et al., 2018).

Rotating biological contractors are frequently preferred for municipal and industrial WWT and have partially submerged discs that move through the air and the wastewater alternately (Ranjit et al., 2021). The removal of pollutants in rotating biological contractors is highly dependent on organic loading rates, submergence of the discs, rotating speed of the shafts, temperature, and retention time of wastewater (Satyanarayana et al., 2018). The process can remove 90% of BOD in the wastewater and has the advantage of low operating costs.

Apart from these traditional BWWT systems, membrane bioreactors (MBRs) gained popularity due to their higher energy efficiency (70% higher competence compared to CAS) and higher treatment capacity (50% extra capacity) in the last decade (Karna and Visvanathan, 2019). Among MBR technologies, membrane-aerated bioreactors (MABRs) eliminate the interference of microbial growth taking place from bubble aeration by introducing nonbubbling aeration on the membrane (Karna and Visvanathan, 2019). Landfill leachate, pharmaceutical wastewater, domestic wastewater, high carbon-nitrogen wastewater, volatile organic compounds, xenobiotics, and anaerobic digestion liquor are treated with MABRs (Karna and Visvanathan, 2019).

2.2 Nutrient removal

Wastewater generated by agricultural, urban, and industrial activities are the main sources of excessive nutrients and micropollutants in aquatic environments when treated inadequately (Li et al., 2019; Grandclément et al., 2017). These problems expand further as algal blooms and groundwater contamination, which pose additional risks to human health and ecosystems. Latest research investigating promising technologies for the advanced nutrient removal processes include promising physical, chemical, and biological treatment approaches. Among them, BWWT is considered as a sound option as it is a cost-effective process with high treatment efficiency, demands less energy, and has possibility of producing less sludge (Banu et al., 2009; Banu and Yeom, 2009; Ma et al., 2016). While conventional biological WWTPs deal with suspended solids and organics in wastewater, nutrient removal capacity of these WWTPs is insufficient in dealing with high nutrient loads and they are not efficient for heavy metals and micropollutant removal (Li et al., 2019; Wollmann et al., 2019; Muñoz et al., 2009; Sousa et al., 2018; Eerkes-Medrano et al., 2019). Besides being a sustainable biomass production reserve, the use of microalgae in wastewater treatment (WWT) is considered as a feasible and cost-effective way of fixing (sequestering) carbon dioxide biologically (Mohsenpour et al., 2020; Almomani et al., 2019).

Using microalgae in WWT, which has been practiced since early 1950s, has two different scopes. The first one is using microalgae for direct uptake or biodegradation of contaminants in wastewater. The second practice is using microalgae to enhance the bioremediation efficiency of bacterial BWWT systems (Section 2.1). Even though the latter is widely preferred for conventional use, there exist studies showing that microalgae could be used for bioremediation of wastewater that has high nutrient load and other contaminants as most microalgae are flexible in terms of their metabolism, namely phototrophic, mixotrophic, and heterotrophic growth (Li et al., 2019; Mohsenpour et al., 2020).

In BWWT, commonly used microalgae and cyanobacteria genera are *Chlorella*, *Scenedesmus*, *Desmodesmus*, *Botryococcus*, *Chlamydomonas*, *Nannochloropsis*, *Dunaliella*, *Arthrospira*, *Anabaena*, *Phormidium*, *Phaeodactylum*, *Arthrospira*, *Synechococcus*, and *Ettlia* (formerly *Neochloris*) (Li et al., 2019). To be able to select the right strain for efficient nutrient removal from wastewater, microalgae are screened based on growth and nutrient removal rates, resilience in multiple wastewater types and high biomass productivity. High growth rate is commonly the determinant factor for selecting algal species because it is an indicator for microalgae with higher growth rates to possess higher resilience to different wastewater media and high nutrient removal capability (Li et al., 2019).

Chlorella vulgaris is a common species pursued for the treatment of various types of wastewater such as mixed-piggery wastewater, digested piggery effluent, and secondary effluent. Nutrient removal rates of *Chlorella vulgaris* for these were reported as 19%–100%, 54% and 60%–80% for ammonia, 28%–95%, 88.4%, and 53%–80% for total phosphorus (TP), respectively (Zheng et al., 2018; Kumar et al., 2010; Ruiz-Marin et al., 2010). Other recent studies reported 61% and 99% total nitrogen (TN) and 71% and 60% TP removal for treated municipal wastewater and biological nutrient removal treatment effluent, respectively, for *Chlorella vulgaris* (Tao et al., 2017; Filippino et al., 2015). For advanced nutrient removal, *Chlorella vulgaris* was reported to remove 99% and 56% of TN and 99% and 82% of TP, respectively (Ji et al., 2013; Gao et al., 2014). Yu et al. reported 99% TN removal along with 100% of phosphorus removal during tertiary treatment in a forward osmosis membrane photobioreactor (PBR) where biomass productivity was achieved at 5 g/L (Yu et al., 2017). In a similar study, *Chlorella* sp. was cultured in piggery and winery mixed wastewater and achieved a removal rate of 89% for TN and 49% for TP (Ganeshkumar et al., 2018).

In a different wastewater stream, palm oil mill effluent (POME) collected from a facultative pond, *Chlamydomonas* sp. was reported to remove 72.9% of TN, 100% of ammonia, and 63.5% of TP from the medium (Ding et al., 2016). In other studies, POMEs were investigated with different microalgal species, *Scenedesmus dimorphus* and

Chlamydomonas incerta, and nutrient removal profiles of these microalgae were reported as 92.5% and 70% for TP, respectively (Rajkumara and Takriffab, 2015; Kamyab et al., 2017).

Chlamydomonas sp. was also cultivated recently in leachate and removed 70% and 83% of ammonia from wastewater in two different studies (Paskuliakova et al., 2018). In POME collected from a facultative pond, *Chlamydomonas* sp. was reported to remove 72.9% of TN, 100% of ammonia, and 63.5% of TP from the medium (Ding et al., 2016). *Desmodesmus* sp. were cultivated in both sewage wastewater and synthetic industrial wastewater and reported to remove 80% of TN and 38.7% of TP from sewage wastewater, and 94% of TP content of synthetic industrial wastewater (Komolafe et al., 2014; Rugnini et al., 2018). Another commonly studied microalgae species, *Scenedesmus obliquus,* showed good performance in secondary effluent of urban wastewater with a removal rate of 96.6%—100% of ammonia and 55.2%—83.3% of TP (Ruiz-Marin et al., 2010). In aquaculture wastewater with polyvinylidene fluoride, removal of nitrogen with *S. obliquus* was 86.1% and phosphorus was 82.7% (Yu et al., 2017). *Spirulina* sp., a common species utilized in both WWT and biomolecule synthesis, was reported to remove 77% of TN and 69% of TP from synthetic dairy wastewater (Sumithrabhai et al., 2016). It was also shown that *Spirulina* sp. can achieve 80% nitrate removal rate and 72% orthophosphate removal rate in dairy wastewater collected from a dairy factory (Al Hamed, 2014). *Arthrospira platensis,* one member of the *Spirulina,* was also cultivated in POME and removed 91% of TN along with 96.8% of TP (Rajkumara and Takriffab, 2015). Besides single species cultures, mixture of different species of microalgae are also experimented with various wastewater such as POME, textile wastewater, and urban wastewater (Kamyab et al., 2017; Huy et al., 2018; Marella et al., 2018). In textile wastewater, TN removal was 70% and TP removal was reported to be 100% by the tested microalgae mixture (Huy et al., 2018). Marella et al. found that another microalgal composition removed 95.1% of TN and 88.9% of TP from urban wastewater (Marella et al., 2018).

In addition to selecting which species to utilize, integrating microalgae into WWT has several challenges such as dealing with large quantities of wastewater and different types of wastewater with varying compositions. Thus, investigating the wastewater technology types for microalgal cultivation is as nearly important as experimenting with various microalgal species for nutrient removal from different wastewater. The performance of the microalgal treatment system is highly dependent on the design and operation of the bioreactor. Microalgal BWWT is categorized into two groups that are suspended culture systems and immobilized culture systems which both can be either open or closed systems. Closed systems are entitled as PBRs in general and there are unique designs of PBRs such as sequencing batch membrane photobioreactors (SB-MPBR), twin-layer PBR, air lift PBR, and multilayer PBR. Open systems are usually named as high-rate algal ponds (HRAPs) or open raceway ponds (ORPs).

2.2.1 Suspended culture systems for microalgal wastewater treatment

Open pond systems are readily available, low-cost, and low-maintenance systems for microalgal WWT because of its ease of operation and low capital costs. HRAPs can be classified as a traditional microalgae cultivation systems that are designed as raceways. Continuous mixing for aeration and prevention of biomass precipitation is ensured with paddle wheels in HRAPs. Nevertheless, evaporation of water, lower biomass production and nutrient removal rate, large footprint, susceptibility of culture to extreme weather conditions, and high costs for harvesting the biomass are some of the challenges of open ponds (Li et al., 2019).

PBRs are the other traditional suspended culture systems for microalgae cultivation and there are various designs such as airlift, column, flat panel, tubular, hybrid, and soft-column PBRs (Vo et al., 2019). PBRs, in general, have higher biomass productivity as they have better control over the culturing conditions and higher biomass productivity because of efficient light utilization due to higher surface area to volume ratios compared to open ponds. However, PBRs require relatively higher capital and operation costs (Mohsenpour et al., 2020). In addition to economic issues, PBRs have other disadvantages such as accumulation of oxygen in the reactor (limits photosynthetic activity), biofouling (limits light penetration), and seasonal variations if operated as outdoor cultivation (Mohsenpour et al., 2020). A review study reported that between 39 research studies on suspended culture PBR systems, the highest N and P removal rates were 82.9% and 87.3% in average, respectively (Mohsenpour et al., 2020).

In order to cope with the economic challenges of PBRs and low biomass productivity of open ponds, recent studies focus on new suspended culture PBR systems. A recent study compared PBR with membrane photobioreactor (MPBR) and reported that MPBR achieved nine times higher biomass productivity than conventional PBR system. It was also stated that the dilution rate of wastewater influences biomass productivity and nutrient removal in both PBRs. High nutrient removal was observed (>80%) in both PBRs when the dilution rate was low. Lower nutrient removal profiles were observed in high dilution rates, where 50% and 30% phosphorus removal rates were observed for MPBR and PBR, respectively (Marbelia et al., 2014). In another study, operation cost of MPBRs was calculated as 0.113 USD/m^3 with a treatment capacity of 5520 m^3/day which is way lower than conventional PBRs which usually have operation costs between 0.65 and 0.96 USD/m^3 (Sheng et al., 2017).

2.2.2 Immobilized systems for microalgal wastewater treatment

Immobilized systems are categorized as passive and active systems. Active immobilized systems are formed by entrapping the microalgae into gel matrices, flocculants, or chemical agents such as alginates and acrylamides. Passive immobilization systems are formed by attaching the microalgae onto a bedding material or other suitable surfaces (Mohsenpour et al., 2020; Ting et al., 2017). The microalgae biofilm is usually restricted to have only one layer, with a 0.052—2 mm optimal thickness, to ensure carbon dioxide and oxygen exchange and light penetration (Irving and Allen, 2011). In a recent study, it was reported that microalgal biofilm on the PBR wall enhanced the nutrient removal from the wastewater significantly (Su et al., 2016). Compared to suspended culture systems for microalgal WWT, immobilization systems have several advantages such as higher biomass yield and lower biomass harvesting costs as biomass is already attached to the surface. A dry cell weight of 3.3 g/L was achieved with microalgal immobilization system, whereas 1.5—1.7 g/L and 0.25—1.0 g/L were obtained with suspended tubular PBR and HRAPs, respectively (Whitton et al., 2015; Christenson and Sims, 2011). In contrast, another study suggested that *S. intermedius* and *Nannochloris* sp. removed slightly more N and P when cultivated in suspended system compared to immobilized one (Jimenez-Perez et al., 2004). Despite their efficient nutrient removal, immobilization systems are hard to implement at pilot scales due to large surface area requirements and high costs of active immobilization modules (Gonçalves et al., 2017). In addition, availability of light and fluctuations in pH and temperature of the system affect the growth of microalgae and oxygen generation in the reactor which in turn leads to reduced nutrient removal rates (Gross and Wen, 2014). Among 39 reported research studies, active immobilization systems have the highest nitrogen removal efficiency with an average of 94.4%. However, it is also indicated that microalgae is not the only contributor to removal of nitrogenous compounds; anionic and cationic interactions of ammonia and nitrates with the immobilization polymer are also associated with the nitrogen reduction (Mohsenpour et al., 2020).

2.3 Removal of heavy metals

Extensive use of metals in certain industries leads to metal contamination in industrial and urban wastewater streams. Mining and electronics industry, fertilizers in agriculture, manure treatment sludge and waste disposal, batteries, paints, and pigments are the main sources of metals dissipating into environment (Monteiro et al., 2012). If not removed, these metals undergo bioaccumulation by organisms in natural environments and cause biomagnification in the food chain leading to major adverse effects to higher organisms including humans (Sridhara Chary et al., 2008). Thus, the removal of metals at the source before discharged into the environment is critical. Conventional physico-chemical treatment methods involve adsorption, chemical precipitation/coagulation, membrane separation, filtration, or solvent extraction (Monteiro et al., 2012). Despite their effectiveness at high metal levels, those systems are not cost efficient at low metal concentrations, and may yield high sludge volumes which remains toxic and require further treatment (Nawaz and Ahsan, 2014).

Algae can be utilized for metal removal through biosorption that can be done by dead cells where only adsorption onto dead biomass occurs, or by live cells of algae where adsorption is followed by assimilation of metals. Presence of functional groups like hydroxyl (-OH), carboxyl (-COOH), amino (-NH$_2$), and sulfhydryl (-SH) on the surface of microalgal cell facilitates adsorption of positively charged metal ions onto dead biomass (Chai et al., 2021). Biosorption depends on abiotic and biotic factors such as pH, temperature, contact time of biomass to metals in the matrix, biomass concentration, and initial metal ion concentration (Zeraatkar et al., 2016). Different algae have different biosorbent capacity to varying metals and this capacity may be altered in the presence of multiple metals in the matrix (Saavedra et al., 2018). For example, Çetinkaya Dönmez et al. in their biosorption capacity study observed higher adsorption of Cu^{2+} and Zn^{2+} onto *C. vulgaris* biomass compared to Cr^{6+} (Çetinkaya Dönmez et al., 1999).

Assimilation of heavy metals by microalgae starts with absorption of heavy metals. The removal of heavy metals by microalgae occurs through the synthesis of phytochelatins (PCs), type III metallothioneins (Gekeler et al., 1988). The enzyme phytochelatin synthase (PCS) is responsible for synthesis of PCs from glutathione and its homologs (Gupton-Campolongo et al., 2013). Different microalgae species may have different response to the same heavy metal. For instance, Ahner et al. exposed several marine algae to varying cadmium (Cd) concentrations, upon which only three algae species, namely *D. tertiolecta*, *P. lutheri*, and *T. weissflogii* among eight strains preserved their growth rate at up to 30 nM Cd^{2+} concentrations (Ahner et al., 1995). Different metals have different effect on the same algae. Tsuji et al. observed enhanced biosynthesis of PCs in *D. tertiolecta* when induced with Zn^{2+} rather than Cd^{2+} (Tsuji et al., 2003). Algae are capable of biotransforming heavy metals into less toxic complex forms or into mineral form and store in their vacuoles (Perales-Vela et al., 2006). Eukaryotic algae may assimilate SO_4^{2-} with heavy metals to detoxify heavy metals in sulfide form (Edwards et al., 2013). Addition of sufficient amounts of SO_4^{2-} may reduce the Cd^{2+} toxicity to algae (Mera et al., 2014). Detoxification of Ag^+ ions via sulfide complexation and mineralization into vacuoles was confirmed by Leonardo et al. below 0.1 mM external Ag^+ concentrations (Leonardo et al., 2016). Fig. 2.2 summarizes the heavy metal assimilation processes in the algae. Detoxification of heavy metals can be accomplished by algae through biotransformation to less toxic ion form. In their study with two axenic algae species, *Chlorella* sp. and *Monoraphidium arcuatum*, Levy et al. observed that both species transformed more toxic As(V) into less toxic As(III) form in the cells (Levy et al., 2005).

2.4 Biotransformation of organic micropollutants

In 2019, 294.6 million tons of chemical substances were consumed only in Europe, and about 74% of the total amount was hazardous chemicals (EUROSTAT, 2021). Those hazardous chemicals are present in the environment at microgram or nanogram per liter concentrations but pose a risk to human and nature even at low concentrations. Those chemical substances in the environment are called micropollutants. Micropollutants are released into

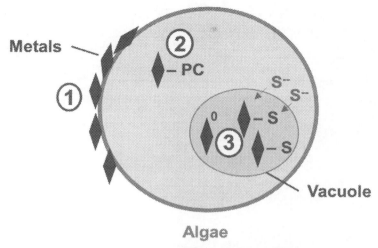

Figure 2.2 Phycoremediation of heavy metals by algae. (1) Biosorption to cell surface, (2) Chelation with phytochelatins (PCs), (3) Assimilation into vacuoles.

the environment from agricultural, domestic, and industrial sources. Among them, wastewater generated from domestic and industrial sources is subjected to treatment. Micropollutants that are frequently detected in domestic wastewater are pharmaceutically active compounds (PhACs), active ingredients of personal care products (PCPs), and surfactants (Tran et al., 2018). Industrial micropollutants that pose risk to the environment and human health include corrosion inhibitors, plasticizers, fire retardants, disinfectants/biocides, antifreezing agents, and dyes (Dsikowitzky and Schwarzbauer, 2013).

Removal of micropollutants from wastewater by different bacterial and algal treatment systems has been reported in the literature (Table 2.1). Bacterial removal of many domestic and industrial micropollutants has been investigated in detail including removal mechanisms, biotransformation pathways, and kinetics. However, there is very limited number of studies on algal removal of micropollutants, since most of the algal studies are on the treatment of domestic wastewater. As a result, PhACs and PCPs are the common micropollutants monitored in both bacterial and algal treatment systems (Table 2.1) (Nguyen et al., 2020). The data consolidated from studies performed on HRAPs and PBRs (Norvill et al., 2016; Matamoros et al., 2016; Zhou et al., 2014), and conventional bacterial wastewater treatment systems (Gavrilescu et al., 2015; Margot et al., 2015) indicate that overall removal of antibiotics ranges between 20% and 100% in algal systems, and 35% and 95% in bacterial systems. Although biotransformation plays a role in the removal of antibiotics from wastewater in both systems, the majority of the antibiotics are removed by sorption to biomass. Given the proven sorption capacity of algal biomass, removal efficiency of high rate algal systems is higher than what is achieved in bacterial systems.

Table 2.1 Comparison of micropollutant removal efficiencies in algal and bacterial treatment systems (Biot.: Biotransformation; Sorp.: Sorption to biomass and Vol.: Volatilization).

Micropollutant	Removal efficiency (%)		Contribution of. mechanisms (%)		
	Algal	Bacterial	Biot.	Sorp.	Vol.
PhACs					
Antibiotics					
Ciprofloxacin	74–79	69	20	80	0
Clarithromycin	53–93	33	70	30	0
Erythromycin	37–79	45	98	2	0
Norfloxacin	41–53	69	20	80	0
Ofloxacin	43–52	58	20	80	0
Roxithromycin	87–94	40	—	—	—
Sulfapyridine	31–100	83.5	—	—	—
Sulfamethoxazole	19–48	44	100	0	0
Tetracycline	69	95	—	—	—
Trimethoprim	46–93	35	87	13	0
Analgesics and antiinflammatory drugs					
Acetaminophen	88–94	100	100	0	0
Clofibric acid	−1 to 30	39	—	—	—
Diclofenac	50–82	20	80	20	0
Ibuprofen	99	80	97	3	0
Naproxen	85	40	100	0	0
Salicylic acid	92–99	99	100	0	0
Tramadol	49–76	33	100	0	0
Others					
Atenolol	43–95	41	98	2	0
Bezafibrate	24–79	41	95	5	0
Caffeine	82	95	100	0	0
Carbamazepine	0–25	16	0	100	0
Codeine	69–92	32.5	—	—	—
Metoprolol	76–92	25	100	0	0
Oxazepam	24–34	13	0	100	0
Venlafaxine	55–81	40	100	0	0
Hormones					
17a-ethinylestradiol (EE2)	85	60	83	17	0
17b-estradiol (E2)	90	90	89	11	0
Estrone (E1)	84–88	76	87	13	0
Progesterone	83–87	97	52	48	0
Testosterone	100	99	99	1	0
PCPs					
Biocides					
Carbendazim	14–30	30	95	5	0
DEET	−59 to −34	62	100	0	0

Table 2.1 Comparison of micropollutant removal efficiencies in algal and bacterial treatment systems (Biot.: Biotransformation; Sorp.: Sorption to biomass and Vol.: Volatilization).—cont'd

Micropollutant	Removal efficiency (%)		Contribution of. mechanisms (%)		
	Algal	Bacterial	Biot.	Sorp.	Vol.
Fluconazole	0—28	15	100	0	0
Triclocarban	81—99	90	10	90	0
Triclosan	31—58	90	35	65	0
BACs	54—100	95	90	10	0
Preservatives/Additives					
Bisphenol A	99	80	95	5	0
Ethyl paraben	100	95	95	5	0
Methylparaben	100	95	95	5	0

High removal efficiencies reaching 100% have been reported for active substances of analgesic and antiinflammatory drugs in both systems. PhACs that undergo biotransformation have the highest removal efficiencies, whereas substances like carbamazepine and oxazepam, which are removed primarily by sorption, have the lowest removal efficiencies (Table 2.1).

Removal efficiency of hormones and preservatives from wastewater is almost always over 80% in both systems. The main removal mechanism is biotransformation. On the contrary, removal efficiency of biocides is poor in algal systems. Physiologically, bacteria are more tolerant to biocides than algae, and bacterial treatment systems perform higher biocide removal compared to algal systems. Some of the micropollutants are (re)generated or desorbed from the solids of wastewater while getting treated in algal ponds; therefore, removal efficiencies of chemical substances, e.g., clofibric acid and DEET, are reported in negative values implying regeneration of those substances during the treatment (Table 2.1).

Overall, algae dominant wastewater treatment systems remove domestic origin micropollutants from wastewater more efficiently than bacteria abundant systems. Given that algal treatment systems are not always bacteria free, therefore, the contribution of bacteria in the removal of micropollutants in algal treatment systems should not be underestimated. On the other hand, the studies on evaluating the details of micropollutant removal mechanisms in combined bacteria—algae treatment systems are currently not available. In these systems, micropollutants having octanol—water partition coefficients ($logK_{ow}$) higher than 4, and Henry's law constants bigger than $11—12$ Pa/m^3.mol are removed via sorption to biomass and volatilization, respectively (Fig. 2.3) (Nguyen et al., 2020). Other dominant nonbiological removal mechanism in open ponds and PBRs is photodegradation. Some micropollutants are sensitive to light. Given that light is essential for the growth of algae in treatment systems, the light-sensitive chemicals would go photodegradation via natural light in shallow HRAPs and artificial light in

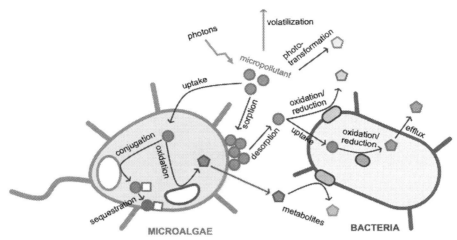

Figure 2.3 Mechanisms of micropollutant removal from wastewater by an algae—bacteria consortium.

PBRs. For instance, pseudofirst order photodegradation rate constants (per h) of several micropollutants are as follows: Ciprofloxacin (0.02—1.4), sulfamethoxazole (0.4—0.68), trimethoprim (0.03—0.18), tetracycline (0.32—1.74), and hormones (0.02—2.7). Photodegradation rates of some chemicals are comparative to biotransformation rates. For instance, biotransformation rate of sulfamethoxazole is in the range of 0.04—1.58 per h (assuming 5 g MLSS/L biomass in activated sludge) (Pomies et al., 2013), which suggests that biotransformation and photodegradation would compete in a photoreactor containing bacteria. Biotransformation of micropollutants can also be achieved by algae. Main mechanism of biotransformation includes conjugation via transferases such as glucosyltransferase and glutathione S-transferase followed by sequestration of the pollutant in the algal cell, and oxidation by mitochondrial cytochrome P450 oxygenases (Fig. 2.3) (Sutherland and Ralph, 2019). Hydroxylation, carboxylation, hydrogenation, (de)methylation, ring cleavage, decarboxylation, dehydroxylation, and bromination are other algal micropollutant transformation mechanisms reported in the literature (Xiong et al., 2018). Oxidoreductases such as mono-, dioxygenases, and hydrogenases, and hydrolases are major groups of enzymes that contribute biotransformation of micropollutants (Garcia-Becerra and Ortiz, 2018; Wackett and Hershberger, 2001). In addition, algae and bacteria may share their metabolites. For instance, glucosylated metabolite of a micropollutant which was formed by algae can serve as a substrate to bacteria that cannot degrade the micropollutant otherwise (Sutherland and Ralph, 2019).

The concentration of micropollutants in the wastewater is so low that they cannot serve as carbon and energy source for either bacteria or algae. Therefore, major fraction of micropollutant transformation is achieved by cometabolism in the bacteria—algae systems treating wastewater. Especially, autotrophic ammonia-oxidizing bacteria have a significant role in cometabolic oxidation of many micropollutants in wastewater while oxidizing reduced N-species by ammonia monooxygenase enzyme (Su et al., 2021). Moreover, sorption of micropollutants onto algae surface may help to overcome kinetic limitations of biotransformation at low substrate conditions by helping the bacteria to access high concentration of micropollutants at algae surface. As a result, bacteria can utilize micropollutants adsorbed on to algae as a carbon and energy source.

Taxonomic composition of the algae—bacteria consortium is crucial for efficient removal of micropollutants from wastewater in a combined biological treatment system. Especially, species of α, β, and γ-Proteobacteria (Wang and Wang, 2016), and *Chlamydomonas, Chlorella,* and *Scenedesmus* (Sutherland and Ralph, 2019; Xiong et al., 2018) are prominent micropollutant degrading bacteria and algae, respectively (Table 2.2). Maintaining a community with selected algae and bacteria species which cope well together and degrade multiple micropollutants may enhance the removal of micropollutants in a wastewater treatment system. For instance, sulfamethoxazole is an antibiotic frequently detected in wastewater. The removal of this moderately photosensitive, hydrophilic antibiotic from wastewater by algae and bacteria is low (Table 2.1). *Achromobacter denitrificans* PR1 and *Pseudomonas psychrophila* HA-4 can consume it as a source of carbon and nitrogen (Jiang et al., 2014; Reis et al., 2014). The biotransformation of sulfamethoxazole starts with hydrolysis and it is converted to 3-amino-5-methylisoxazole. On the other hand, a green algae *Chlamydomonas reinhardtii* converts sulfamethoxazole to pterin-sulfamethoxazole via glucuronide conjugation which accumulates in the cell. Clearly, bacteria and algae have different pathways of biotransformation which may facilitate complete removal of micropollutants from wastewater. Although micropollutant removal by bacteria—algae collaboration is promising, it needs systematic research on optimizing bacteria—algae interactions in taxonomic and molecular level for enhanced removal of micropollutants from wastewater.

3. Mechanism of symbiosis between algae and bacteria

Bacteria and algae coexist and function in natural systems. They often benefit from each other through interspecies material transfer, cell-to-cell communication, horizontal gene transfer, or adjustment of environmental conditions (Zhang et al., 2020). In a symbiotic system, microalgae are mainly responsible for removal of nutrients, organic hazardous contaminants, and heavy metals, whereas bacteria consume organic carbon including hazardous contaminants. Oxidation of organic carbon produces inorganic carbon, i.e., carbon dioxide, which is utilized by microalgae as carbon source. Balanced carbon cycling by bacteria and microalgae controls the alkalinity in the wastewater treatment systems having a natural symbiotic algae—bacteria consortium. Photosynthetic oxygen produced by microalgae during inorganic carbon sequestration is consumed by bacteria

Table 2.2 Taxonomic distribution of bacteria and algae that degrade micropollutants.

	Group
BACTERIA	**High G + C Gram positive** *Arthrobacter, Brevibacterium, Clavibacter, Corynebacterium, Dehalobacter, Nocardia, Rhodococcus, Streptomyces, Terrabacter* **Low G + C Gram positive** *Bacillus, Clostridium, Desulfitobacterium, Eubacterium, Staphylococcus* **α-Proteobacteria** *Agrobacterium, Ancylobacter, Brevundimonas, Chelatobacter, Hyphomicrobium, Methylobacterium, Paracoccus, Rhodobacter, Sphingomonas* **β-Proteobacteria** *Achromobacter, Alcaligenes, Azoarcus, Burkholderia, Comamonas, Hydrogenophaga, Ralstonia, Thauera, Thiobacillus* **γ-Proteobacteria** *Acinetobacter, Aeromonas, Azotobacter, Enterobacter, Escherichia, Klebsiella, Methylobacter, Methylococcus, Moraxella, Pseudomonas* **δ/ε-Proteobacteria** *Desulfovibrio* **Cytophagales–green sulfur** *Flavobacterium* **Green nonsulfur bacteria** *Dehalococcoides* **Cyanobacteria** *Microcystis, Spirulina*
ALGAE	**Bacillariophyta** *Chaetoceros, Nitzschia, Halamphora, Thalassiosira* **Euglenophyta** *Euglena* **Chlorophyta** *Chlamydomonas, Chlorella, Nannochloris, Stigeoclonium, Klebsormidium, Micractinium, Selenastrum, Scenedesmus, Ankistrodesmus, Cosmarium, Desmodesmus, Neochloris, Tetraselmis*

while oxidation of organic carbon, which reduces external aeration and limits the stripping of volatile organic contaminants from the wastewater. Bacteria and algae share vitamins and cofactors and support each other's growth. Selective metabolization of toxic contaminants by one member of the consortium may facilitate the survival and proliferation of the other member which otherwise would be inhibited by that contaminant (Munoz and Guieysse, 2006; Ramanan et al., 2016). Understanding these interactions would help to engineer microalgae—bacteria communities for efficient and sustainable wastewater treatment systems with high microalgal biomass productivity which can serve as biofuel and biochemical feedstock.

3.1 Carbon—oxygen recycle

Molecular oxygen (O_2) is the main electron acceptor of aerobic heterotrophic bacteria which removes major fraction of the organic carbon in the wastewater treatment systems. Therefore, supplying oxygen through aeration of the reactors is crucial for the treatment. On the other hand, aeration is the most energy demanding process which accounts for the 50% of the overall operational cost of a WWTP (Munoz and Guieysse, 2006). Microalgae use energy of photons in the light and convert inorganic carbon, i.e., CO_2 and nutrients to biomass. Therefore, inorganic carbon, nitrogen and phosphorus containing nutrients are essential for algal growth. Algal photosynthesis yields O_2 and organic biomass. Treatment of wastewater via microalgae—bacteria consortium is a very promising and sustainable alternative to conventional systems since bacteria would benefit from O_2 generated by microalgae, whereas CO_2 generated by bacteria after oxidation of the organic carbon would serve as the carbon source for the microalgae (Fig. 2.4). Overall, both aeration, thus the energy demand, and CO_2 emissions of the plant employing algae—bacteria treatment system decrease significantly compared to conventional treatment plants. Studies revealed that complete removal of easily biodegradable organics can be achieved without an external O_2 supply in an algae—bacteria reactor treating wastewater (Zhang et al., 2020). Moreover, bacterial consumption of O_2 generated by microalgae especially in closed PBRs avoids the inhibition of photosynthesis by O_2; thus, both enhance the growth of algae and improve the removal of nutrients and organics (Munoz and Guieysse, 2006). The highest treatment efficiency is obtained

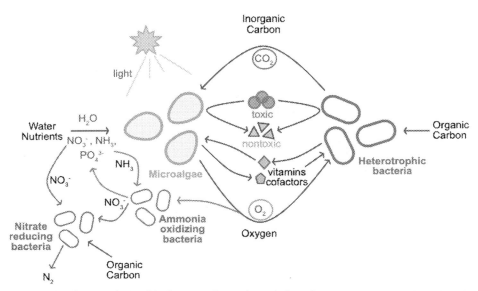

Figure 2.4 Symbiotic relationship between bacteria and algae in a wastewater treatment system.

when optimum carbon and oxygen cycling is achieved within algae and bacteria in the reactors. Key operational parameters for sustaining this optimum mutualistic relationship include light intensity, light-dark cycle periods, microbial composition of the consortium, and hydraulic/solid retention time (Flores-Salgado et al., 2021; Sanchez Zurano et al., 2021; Lee et al., 2016).

There are also competing processes within a diverse microalgae—bacteria consortium for nutrients that are essential for microalgae and O_2 which is the electron acceptor for aerobic heterotrophic bacteria (Fig. 2.4). For instance, nitrifying bacteria which oxidize ammonia (NH_3) to nitrate using O_2 as the electron acceptor are major competitors of microalgae and heterotrophic bacteria on NH_3 and O_2, respectively. Given that nitrifying bacteria play a crucial role in cometabolic oxidation of micropollutants in wastewater, their presence should be sustained in a consortium aiming efficient nutrient and micropollutant removal (Su et al., 2021; Sánchez-Zurano et al., 2020). In addition, mixotrophic algae consume O_2 during dark respiration which is a major path of oxygen utilization in closed PBRs and aerated HRAPs when light intensity is not sufficient enough to drive efficient photosynthesis (Flores-Salgado et al., 2021). Because of oxygen depletion in low-light conditions, anoxic conditions would prevail. Anoxic condition facilitates denitrification which results in conversion of nitrate (NO_3^-) to nitrogen gas (N_2). As a result, denitrifying bacteria compete with microalgae for limited nutrient under anoxic conditions. N_2 is an alternative source of nutrients for algae only if nitrogen fixing bacteria are present in the consortium. Nitrogen-fixing bacteria such as the species of *Rhizobium* and *Azotobacter* facilitate growth of algae by supplying inorganic nitrogen produced from N_2 (Villa et al., 2014). Similarly, mineralization of complex nitrogenous organic chemicals such as amino acids and peptides by bacteria releases inorganic nitrogen that promotes the growth of algae. An example of such mutualism has been reported between *Chlamydomonas* and methylobacteria on proline, hydroxyproline, and peptides (Calatrava et al., 2018). In summary, high performance wastewater treatment with the algae—bacteria consortia needs fine tuning of critical conditions to balance interspecies substrate transfer and microbial community structure.

3.2 Growth stimulation

Other than substrates, algae and bacteria share essential molecules promoting their growth and subsistence in the environment (Fig. 2.4). Algal blooms in natural waters represent a good example on bacteria and algae cooperation for growth stimulation (Buchan et al., 2014). Systematic research on algal blooms depicted that although algal diversity changes, very stable heterotrophic bacterial community, which is mainly composed of species of *Bacteroidetes,* e.g., *Flavobacterium,* α-, and γ-*Proteobacteria,* is responsible for algal proliferation (Buchan et al., 2014; Han et al., 2016; Krohn-Molt et al., 2013). In this interaction, bacteria supply fixed nitrogen, vitamins, phytohormones, and siderophores to algae, whereas algae provide easily biodegradable dissolved organics to bacteria. Sharing micronutrients, vitamins, and cofactors produced by members of the well-competing consortia promotes collateral mixotrophic community with high growth rates and tolerance to fluctuating environmental conditions (Zhang et al., 2020).

Many algal species cannot synthesize vitamin B-12 therefore acquire it from auxiliary sources. Similarly, about 25% of the existing microalgae require vitamin B1, and approximately 8% of them obtain vitamin B7 from other sources. Vitamin B-12 is obligatory for the functioning of methionine synthase enzyme of microalgae. This enzyme produces an essential amino acid called methionine (Cooper et al., 2019). Bacteria is the main source of vitamin B-12 for algae that cannot produce its vitamins. A good example on B-12 driven mutualistic relationship is between a marine Roseobacter *Ruegeria pomeroyi* and a diatom *Thalassiosira pseudonana*. In that consortium, *R. pomeroyi* supplies B-12 to *T. pseudonana*. The algae produce 2,3 dihydroxypropane-1-sulphonate which is used as a substrate by the bacteria (Durham et al., 2015). In a recent study, Higgins et al. reported that cofactors such as thiamine derivatives and transformation products produced by *E. coli* improved the substrate utilization efficiency of a green alga *Auxenochlorella protothecoides* and stimulated its growth (Higgins et al., 2016).

In addition to vitamins and cofactors, some chemical molecules synthesized by bacteria induce certain metabolic pathways in algae and promote production of substrates for bacteria or enhance bacterial adhesion to algal phycosphere (Dao et al., 2018; Seymour et al., 2017). One of algal growth stimulant molecule that is synthesized by bacteria is indole-3-acetic acid (IAA). Dao et al. stated that IAA produced by a number of different γ-*Proteobacteria* isolated from domestic wastewater increases the growth rate and biomass yield of a green alga *Scenedesmus* sp. (Dao et al., 2018). The study also implied that the algae also promote the synthesis of IAA by the bacteria. Acyl-homoserine lactone is another signaling molecule produced by bacteria which is involved in biofilm formation between bacteria and algae cells. Biofilms facilitate better transfer of substrates between bacteria and algae and promote growth (Fuentes et al., 2016).

Bacteria-stimulated algal growth yields biomass with higher lipid content. Lipids and fatty acids produced by the algae support the growth of symbiotic bacteria (Krohn–Molt et al., 2013). Those easily degradable energetic substrates promote bacterial growth and facilitate the cometabolic processes of bacteria. Cometabolism is especially important in biodegradation of toxic micropollutants present in domestic and industrial wastewater (Xiong et al., 2018).

3.3 Toxicity reduction

Bacteria and algae can detoxify many hazardous chemicals in wastewater. As discussed in Section 2.3, detoxification mechanisms may involve adsorption and biotransformation. In a bacteria—algae consortium treating wastewater, algal species are generally more sensitive to wastewater toxicants. While bacteria can be acclimated to hazardous chemicals easily due to their fast replication rates and elastic genome open for mutations and horizontal gene transfer, adaptation of algae is slow. Therefore, bacteria play a crucial role in protecting algae from inhibition by hazardous chemicals. For instance, disinfectants that have algicidal properties such as quaternary ammonium compounds can be actively degraded by γ-*Proteobacteria,* e.g., *Pseudomonas* spp., which are good mutualists of algae. Similarly, algae protect bacteria from antibiotics in the wastewater. Given that antibiotics are mainly designed to kill or suppress bacteria, their mode of actions does not target eukaryotic metabolism. As a result, algae are more resistant to antibiotics than bacteria. Algae can also degrade a whole spectrum of antibiotics present in the wastewater via oxidation or conjugation (Leng et al., 2020; Xiong et al., 2016, 2017, 2019; Stravs et al., 2017). Biotransformation of different toxic chemicals by at least one member of the consortia would protect the other members which are susceptible to those chemicals. As a result, the consortia could sustain and function normally (Fig. 2.4).

Wastewater is a heterogeneous medium which contains a wide spectrum of chemicals. Treatment of both domestic and industrial wastewater by algae—bacteria consortia is promising if a strong symbiotic relationship between those microorganisms prevails.

4. Examples of wastewater treatment by algae—bacteria consortia

4.1 Domestic wastewater

Algae—bacteria consortium can be superior for eliminating nutrients and chemical oxygen demand (COD) over CAS systems when suitable conditions are maintained. Tang et al. showed lower aeration demand in algae—bacteria consortia reactor when removal rates of nutrients are similar to CAS system (Tang et al., 2016). He et al. observed better treatment of nutrients and organic matter when algae were accompanied with coexisting bacteria compared to algae-only treatment (He et al., 2013). In a study on reclamation of sewage contaminated lake, Verma et al. found that maximum removal of nutrients and organic matter was achieved by the symbiotic coculture of algae and bacteria than standalone cultures of microalgae and activated sludge (Verma et al., 2020). Ji et al. observed effect of the type of algae in the consortia and found better treatment efficiency with *C. vulgaris* than *M. aeruginosa* in consortium with *Bacillus licheniformis* (Ji et al., 2018). Ji et al. reported that *C. vulgaris—B. licheniformis* consortium had the ability to degrade protein like substances, which are recalcitrant in natural environments (Ji et al., 2019). Algae—bacteria inoculum ratio is also important for achieving synergistic relationship and this ratio may change from species to species. Russel et al. reported an optimal inoculum ratio of 2:1 for *Scenedesmus obliquus*: *Acinetobacter pittii* consortium in wastewater nutrient and COD treatment (Russel et al., 2020). The optimum inoculum ratio was found to be 3:1 between a *Chlorella* sp. and activated sludge for both treatment and biomass generation (Nguyen et al., 2020). Without addition of algae inoculum,

Arango et al. observed development of algae within 12 days when illuminated (Arango et al., 2016). In a study of pilot scale HRAP for urban wastewater treatment, Robles et al. did not supply neither algae nor activated sludge bacteria, and observed development of a stable microalgae—bacteria consortium after 1 month (Robles et al., 2020). Bacterial composition in the consortium system may also affect the treatment quality and biomass generation. Paddock et al. observed unstable community development in algae-activated sludge systems which led to varying treatment efficiency and biomass production (Paddock et al., 2020). Authors also found a correlation between microalgae *C. sorokiniana* biomass generation rate and relative abundance of a bacterial genus, *Pusillimonas*.

Algae—bacteria treatment systems can be affected by abiotic factors which in turn shift the treatment efficiency. In a stagnant pond, for instance, temperature and light intensity, alkalinity, turbidity, initial ratio of biomasses, algal cell size, and pond depth all found to significantly affect the removal of substrates and microalgal biomass productivity (Cervantes-Gaxiola et al., 2020). Influent cyanobacteria content might affect the consortium diversity, and temperature might affect biomass density (Cho et al., 2015). Increasing pH from acidic to neutral increased ammonia removal efficiency from 78% to 86% (Liang et al., 2015). It was indicated that lower light intensity achieves better removal efficiencies for both lab-scale and outdoor consortia systems (Arcila and Buitrón, 2017; Pastore et al., 2018). The effect of irradiance is on activated sludge and nitrifying bacteria in consortium with algae are different. In their lab-scale experiment with fluorescent lamp, González-Camejo et al. found that higher illuminescence ($85-125\ \mu E/m^2 s$) was necessary for algae to predominate against denitrifying bacteria for removal of ammonia nitrogen (Gonzalez-Camejo et al., 2018). At $40\ \mu E/m^2 s$ light intensity, nitrifying bacteria predominated, and nitrate accumulation occurred. Gonzales-Camejo et al. also found that the nutrient removal efficiency was better at $22°C$ than $28°C$, where in the former temperature algae predominated.

In continuous systems, hydraulic retention time (HRT) is an important factor. Arcila et al. observed efficient removal of COD and nutrients at 6 and 10 d HRT with algae—bacteria consortium, but a poor treatment appeared at 2 d (Arcila and Buitrón, 2016). Pastore et al. observed a washout of microalgae and lower N, P removal at HRTs lower than 2.5 days (Pastore et al., 2018). At stable HRT, feeding regime may also affect the removal efficiency of the system. Serejo et al. reported better biomass yield for microalgae and better removal efficiencies when feed was shifted from continuous to semicontinuous (12 h/d or 1 h/d) (Serejo et al., 2020).

Various immobilization methods might be used to decrease the HRT and to improve the treatment efficiency. Early studies have tested nutrient removal from wastewater by coimmobilizing algae growth promoter bacteria with the algae in Ca-alginate beads (de-Bashan et al., 2004; de-Bashan et al., 2002). Those beads are permeable to organic and inorganic substrates for microorganisms but do not permeate microorganisms in or out protecting the immobilized cultures (Covarrubias et al., 2012). When immobilized algae was used for photosynthetic aeration of the system, suspended *P. putida* could degrade 100% of the substrate, 500 mg/L glucose, in an HRT of 3.3 h without external aeration (Praveen and Loh, 2015). This immobilized algae-suspended bacteria system is found to be also superior for N, P, and COD removal (Mujtaba et al., 2017). Nonetheless, alginate is not durable in real primary wastewater effluent and deteriorates through time (Cruz et al., 2013).

Alternative immobilization methods were also studied recently. Biofloc formation from microalgae and activated sludge can be one alternative to suspended systems. High harvesting ratio along with high nutrient and COD removal might be achieved through biofloc formation (Nguyen et al., 2019). Van Den Hende combined algae—bacteria biofloc formation with flue gas influx, yielding efficient nutrient removal along with 48%, 87%, and 99% reduction of CO_2, NO_x, and SO_2 from the flue gas, respectively (Van Den Hende et al., 2011). MBRs are utilized for continuous wastewater treatment operations. Sun et al. introduced microalgae to MBRs yielding 4.6%, 6.7%, 10.1%, and 8.2% increase in removal efficiencies of COD, NH_4^+-N, total N, and PO_4^{3-}, respectively, along with 25% faster growth of both algae and bacteria (Sun et al., 2018). Algae—bacteria consortia was also utilized in sequencing batch biofilm bioreactors (Tang and et al., 2018a). The underlying principle is readily harvesting the biomass when aeration is terminated. Tang et al. modified this system with a biofilm carrier lighter than water. Assimilation of nutrients into microalgae biomass highly escalated resulting in higher nutrient elimination (Tang and et al., 2018b). Recently, a novel oxygenic photogranules (OPGs) system was introduced where outer layer consists of mostly cyanobacteria, and inner layer consists of noncyanobacterial microorganisms (Abouhend et al., 2020). Initially small, these granules develop into 4—5 cm diameter beads with the help of extracellular polymeric substances, while filamentous cyanobacteria become enriched and other phototrophic microbes diminish (Abouhend et al., 2020). This study has shown that OPGs can be used to treat wastewater without aeration (Abouhend et al., 2020). In a similar study, Zhang et al. demonstrated that algal—bacterial granules can be used to treat synthetic domestic wastewater with over 95% removal of COD, where ammonia and phosphorus removal can be achieved between 20 and 100 days (Zhang et al., 2018). Alas, algal—bacterial granules collapsed between 100 and 120 days (Zhang et al., 2018).

In another study, Zhao et al. generated high-density microalgae—bacteria consortia for the treatment of landfill leachate (Zhao et al., 2014). It was found that up to 90% of the TN (out of a concentration of 221.6 mg/L) present in landfill leachate can be removed in culture with 10% leachate spike ratio onto municipal wastewater (Zhao et al., 2014). This ratio was also optimal for biomass growth, carbon fixation, and lipid productivity of *Chlorella pyrenoidosa* (FACHB-9), utilized as the microalgae species in the consortia (Zhao et al., 2014). In a study by Tighiri and Erkurt, above 90% removal efficiencies of nitrate, COD, and phenol were reported for undiluted leachate, treated with a community of microalgae in consortium with sewage activated sludge bacteria maintained at 3:1 ratio (Tighiri and Erkurt, 2019). Stable removal efficiency of >99% for nutrients and phenol, and >90% for COD was accomplished when the initial biomass was increased from 0.34 g/L to 1.34 g/L for algae and from 0.1 g/L to 0.4 g/L for bacteria, respectively (Tighiri and Erkurt, 2019).

4.2 Industrial wastewater

In contrast to urban wastewater, industrial wastewater streams bear toxic organic chemicals and/or heavy metals that cannot be treated with conventional wastewater treatment systems. Microalgae—bacteria—based synergistic treatment systems have the potential to treat various industrial wastewater systems. Algae can be utilized in such systems as oxygen supplier for the bacteria which has biodegradation potential of the organic contaminant in such wastewater systems. For instance, Muñoz et al. utilized algae to avoid aeration for acetonitrile degrading bacteria and yielded complete degradation of 2 g/L acetonitrile in continuous operation with an HRT of 1.6 days (Muñoz et al., 2005). Borde et al. conducted a similar study with phenol and phenol-like compounds and observed 89% removal using algal—bacterial microcosms established in 50 mL sealed flasks in 2 days (Borde et al., 2003). Inoculated with 5% v/v freshly grown algae—bacteria consortia, 100% and 85% removals for salicylate and phenanthrene were observed in 93 h, respectively. Another study showed a consortia immobilized onto various solid carriers (capron fibers for algae; ceramics, capron, and wood for bacteria) can be used for biotreatment of industrial wastewater in a pilot installation (Safonova et al., 2004). The consortia consisted of *Chlorella* sp., *Scenedesmus obliquus*, *Stichococcus* spp., and *Phormidium* sp. as algal species and *Rhodococcus* sp., *Kibdelosporangium aridum*, as well as two unidentified bacterial strains (Safonova et al., 2004). The study reported removal of phenols up to 85%, anionic surface active substances up to 73%, oil spills up to 96%, copper up to 62%, nickel up to 62%, zinc up to 90%, manganese up to 70%, and iron up to 64% (Safonova et al., 2004). The reductions of the BOD and COD were 97% and 51%, respectively (Safonova et al., 2004).

Recently, algae—bacteria consortia systems have been applied to a variety of industrial wastewater streams. Tang et al. investigated crude oil degradation, and found out that a stable consortium can be achieved by constructing it with axenic *Scenedesmus obliquus* GH2 and bacteria (*Sphingomonas*, *Burkholderia cepacia*, *Pseudomonas,* and *Pandoraea pnomenusa*) (Tang et al., 2010). The unialgal (one species of algae contaminated with bacteria) culture of *S. obliquus* GH2 was not suitable for consortium as it did not cooperate well with *B. cepacia,* and the trial with *Pseudomonas* and *P. pnomenusa.* However, the study has successfully demonstrated that utilizing axenic *S. obliquus* GH2 combined with the four bacteria elevated efficiency in degrading both aliphatic and aromatic hydrocarbons of crude oil can be achieved (Tang et al., 2010). Katam et al. stated that algae—bacteria consortia are more effective in degradation of linear alkyl sulfonates than bacterial consortia, whereas there was not much difference in caffeine removal (Katam et al., 2020). Cocultures may be more effective than pure cultures. Qi et al. found an overall 22.7% increase in removal efficiency of ethanol, butanol, acetic acid, and butyric acid from fermentation wastewater through addition of *C. sorokiniana* to the pure bacterial culture (Qi et al., 2018).

Maza-Márquez et al. applied algae—bacteria consortium for treatment of olive wash water (OWW), wastewater from washing olives before entering the olive mills, in a 14.5-L enclosed tubular PBR. Over 90% removal of phenolic compounds and over 80% removal for COD and turbidity were achieved in 3 days of HRT (Maza-Márquez et al., 2017). Same group also studied the treatment of OWW in full scale reactor system consisting of 80 tubular PBRs each with 98.17 L volume (Maza-Marquez et al., 2017). The system was able to treat 1570—3926 L/day of OWW under different HRTs (5—2 days) and reported 95% removal of phenolics, 86% removal of COD, and 96% removal of turbidity via algae—bacteria consortia treatment without external aeration (Maza-Marquez et al., 2017). Ryu et al. tested the feasibility of using an algal—bacterial process for removal of phenol and NH_4^+-N from differently diluted coke wastewater, and observed 100% removal of phenol by algae—bacteria consortia from diluted toxic coke water, while sole microalgal culture degraded a maximum of 27.3% of phenol with an initial concentration of 24.0 mg/L) (Ryu et al., 2017).

Mahdavi et al. studied naphthenic acids in situ biodegradation in oil-sands tailing pond (Mahdavi et al., 2015). Typically, this wastewater contains sand, clay, dissolved metals, unrecovered hydrocarbons, and acid extractable organics called naphthenic acids. Testing bacteria alone, with indigenous algae, or both with a diatom (*Navicula pelliculosa*) in consortia, the highest biodegradation rate was observed with bacteria growth only with a half-life of 203 days (Mahdavi et al., 2015). The algae—bacteria consortium enhanced the detoxification process; however, bacterial biomass played the main role in toxicity reduction (Mahdavi et al., 2015). Paulssen and Gieg investigated biodegradation of 1-adamantanecarboxylic acid, a model naphthenic acid among tricyclic naphthenic acids which were known with their toxicity and recalcitrance. Authors found up to 80% removal of 1-adamantanecarboxylic acid by algae—bacteria consortia in 90 days using 125 mL Erlenmeyer flasks operated in batch mode (Paulssen and Gieg, 2019).

Algae—bacteria treatment systems may also be a good candidate for metal bearing wastewaters. One of such wastewaters is mining wastewater. This matrix is also called as acid mine drainage (AMD) which is highly acidic wastewater containing substantial amounts of various metals. A biofilm dominated by *Chlorella* like microalgae was effectively utilized for the treatment of nickel refinery tailings where 24.8%, 10.5%, 24.8%, and 26.4% reduction in Ni, Co, Mn, and Sr were achieved, respectively (Palma et al., 2017). When experimental media was enriched with phosphorus, the lipid content of the biomass was similar to that grown in control nutrient media, which indicates potential use of biomass in biofuel production after wastewater treatment (Palma et al., 2017). In another lab-scale study, algal—microbial consortium was immobilized within a laboratory-scale photo-rotating biological contactor (PRBC) and used to investigate the potential for heavy metal removal from AMD (Orandi et al., 2012). The PRBC reactor was operated continuously with a 24 h HRT over a 10-week period and demonstrated that 20%—50% removal of metals, 20%—50% of the various metals in the order of $Cu > Ni > Mn > Zn > Sb > Se > Co > Al$ can be achieved (Orandi et al., 2012). Metals removal can be accomplished by sulfate reducing bacteria (SRB), potentially in consortium with acidophilic or acclimated algae (Abinandan, 2018). Li et al. fed SRB (*Desulfovibrio* sp.) with microalgae (*Chlorella vulgaris, Scenedesmus obliquus, Selenastrum capricornutum*, and *Anabaena spiroides*) by immobilizing all microorganisms together in a continuous reactor for the treatment of wastewater containing 60 mg/L Cu(II) and 600 mg/L sulfate in batch experiments. Authors observed 72.4%—74.4% of sulfate and over 91.7% of Cu(II) were removed (Li et al., 2018).

Textile wastewater, high in BOD, TN, and TP levels and containing various metals, was treated with microalgae (Oyebamiji et al., 2019). Along with color removal, partial removal of Al, V, Cu, and complete removal of Pb and Se was observed. An algae—bacteria consortia isolated from a WWTP efficiently removed Pb (97.4%) along with COD and nutrients from the matrix containing up to 50 mg/L Pb in 250 mL batch cultures containing 60% wastewater mixed onto Bolds Basal Medium (Hernandez-Melchor et al., 2018).

Another high salinity, high COD, metal bearing wastewater is wastewater produced in hydrocracking. Cracking the shales to retrieve shale oil or shale gas was done via pumping water into the shales horizontally. The return water is contaminated with various elements in the lithosphere. Akyon et al. utilized microbial mats, biofilms generated on solid mat surface, to remove COD from produced water and achieved 1.45 mg COD/g wet weight · day at 91,351 mg/L total dissolved solids level (Akyon et al., 2015). In a study on the treatment of fracturing wastewater, Li et al. observed that algae growth and COD removal increased when *Chlorella* was in consortia with *Bacillus* instead of *Chlorella* alone (Li et al., 2019). Authors also tested for treatment conditions by means of dilutions of fracturing wastewater, initial bacteria dose, and pH of the system, and found dilution ratio of two, bacterial dose of 72.13 mg/L, and pH of 6.5 as optimal for algae growth and treatment.

Algae—bacteria consortia can also be employed for treatment of agro-industrial wastewater streams with low C to N ratio. Compared to nitrification/denitrification mechanisms of conventional wastewater treatment, higher N ratio can be obtained once successfully assimilated into algae biomass. This mechanism generates possibility for the posttreatment algal biomass to be utilized as fertilizer or animal feed. Hernandez et al. treated two agroindustrial wastewater types with microalgae—bacteria consortia: 1)

Potato processing industry (PP) and 2) Treated liquid fraction of pig manure (TE) (Hernandez et al., 2013). Authors found total COD removal of 62.3% and 84.8%, ammonia removal of 82.7% and >95%, and soluble phosphorus removal of 58% and 80.7% for TE and PP, respectively (Hernandez et al., 2013). Nitrification/denitrification was main mechanism for ammonia removal from TE, whereas authors reported that no nitrification/denitrification occurred in PP (Hernandez et al., 2013). In a study by González-Fernández et al., four 3-L open ponds were inoculated with microalgae—bacteria consortium treating different swine slurries (fresh and anaerobically digested) (Gonzalez-Fernandez et al., 2011). The authors compared nitrogen transformation under optimal and real conditions of temperature and illumination and observed higher ammonia removal efficiencies (>90%) in 10 days of HRT in the reactors operated under more optimal temperature (31.5°C) and lighting (24 h illumination). When the temperature was around 16°C and illumination was 9 h mimicking the real conditions, the removals decreased to 58.2%—84.5%. For fresh slurry, main mechanism of removal was found to be denitrification and for anaerobically digested slurry, it was nitrification in real conditions (Gonzalez-Fernandez et al., 2011).

5. Circular economy, process design, and modeling aspects of algae—bacteria based wastewater treatment

5.1 Circular economy examples of algae—bacteria consortia in wastewater treatment

Algae generate a significant potential for circular economy applications where carbon dioxide from flue gases and nutrients from several waste streams can be removed. Several studies have proposed and demonstrated algae cultivation during wastewater treatment can be coupled with value-added product generation such as biofuels. In wastewater treatment systems, carbon dioxide requirements can be met from bacterial heterotrophic systems as mentioned earlier in Sections 3.1 and 3.2. The only caveat for such cases is the ratio of dissolved carbon to other required nutrients. In a notable study by Goswami et al., a consortium comprised of two algae—two bacteria species was cocultivated in four different wastewater streams including paper, textile, leather industry, as well as domestic wastewater, and production of hydrothermal liquefaction (HTL)—derived biocrude was tested (Goswami et al., 2019). Total biomass concentration and nutrient removal efficiencies were improved when the consortium of two microalgae (*Chlorella sorokiniana* and *Chlorella* sp.) and two bacteria (*Klebsiella pneumoniae* and *Acinetobacter calcoaceticus*) was grown together compared to individual cultivations. In a 7.5-L bioreactor operating in batch and fed-batch mode, 100% phosphate removal was achieved in all wastewater streams, while COD removal ranged from 52.8% to 94.2%, and TN removal from 55.7% to 89.3% (Goswami et al., 2019). Improved removal of nickel and chromium was also observed for mixed algae—bacteria consortia compared to individual cultivation.

In a similar study by Selvaratnam et al., *Galdieria sulphuraria*, an extremophilic algae capable of growing at low pH and high temperature conditions, was fed mixotrophically with primary effluent of domestic wastewater (Selvaratnam et al., 2015). Resulting algal biomass was used for HTL for bioenergy production as biocrude and biochar for soil amendment. It was also proposed an integrated process called Photosynthetically Oxygenated Waste to Energy Recovery (POWER) where nutrient-rich aqueous product of HTL is recycled back to PBRs to improve both biomass and energy recovery (Selvaratnam et al., 2015). The results were very promising with ammonia—nitrogen removal of more than 11 mg/L-d and phosphate removal more than 2.7 mg/L-d with *G. sulphuraria* biomass productivity of more than 0.2 g/L-d (Selvaratnam et al., 2015).

In a very recent study, Biswas et al. attempted to integrate phycoremediation and lipid production using a microalgae—bacterial consortium grown in dairy wastewater-fed aquaculture pond (Biswas et al., 2021). Among three different contaminant levels, the consortium was able to reduce 93% of nitrate (from 139 to 9.5 mg/L), 97% of phosphate (from 66.2 to 1.9 mg/L), 80% of ammonium (from 16 to 3.2 mg/L), and 89% of COD (from 1867 to 207 mg/L) from the highest contaminant rich dairy wastewater in 48 h (Biswas et al., 2021). Dry cell weight biomass was also enhanced by 67%, while 42% lipid, 55% carbohydrate, and 18.6% protein content was achieved upon treatment with ammonia-rich dairy wastewater indicating a solid potential for biofuel generation (Biswas et al., 2021).

In a bench-scale study by Bélanger-Lépine et al., wastewater samples (comprised of 45% pharmaceutical, 41% dairy, 10% chemical cleaners, and 4% leachate, all v/v) from an industrial park were tested with a native microalgae—bacteria consortium isolated from a wastewater stabilization pond. Although individual nutrient or contaminant removals were not reported in this study, the highest lipid content was achieved as $28 \pm 4.3\%$, offering a sustainable solution to the companies at the industrial park in terms of reduced treatment costs and environmental footprint through value-added product generation (Bélanger-Lépine et al., 2018).

5.2 Dynamic algae—bacteria models for wastewater treatment

Dynamic activated sludge models simulating biological removal of organic carbon and nutrients from the wastewater have been used since 1986. Since then, AS models have evolved by implementing different microbial processes in the wastewater treatment including nitrogen and phosphorus removal. As the most recent model, "Activated Sludge Model No.3 (ASM) includes 12 bacterial transformation processes involving hydrolysis, heterotrophic and autotrophic microbial growth by either aerobic or anoxic respiration. AS models mainly focus on bacterial growth and substrate utilization processes, and effect of operational factors such as temperature, electron acceptor availability on the process efficiency. On the contrary, chemical and photo-chemo and –autotrophic processes are excluded (Henze et al., 2006).

There are also models describing microalgae growth by coupling photosynthesis and nutrient uptake. Those models also consider the influence of light intensity, pH, temperature, and nutrient availability on the microalgae growth. In general, Monod growth equation is used in those models. Influence of limiting factors such as light intensity, nitrogen, phosphorus, and carbon dioxide is integrated into the Monod growth equation as "switching functions" in the dynamic microalgae models. Using "switching functions" is common in AS models in case the electron acceptors, such as oxygen and nitrate, are limited (Solimeno and Garcia, 2017).

$$\rho_{net,ALG} = \mu_{m,ALG} X_{ALG} \underbrace{\frac{S_{CO_2,aq}}{S_{CO_2,aq} + K_{CO_2,aq}}}_{\text{Dissolved carbon dioxide}} \underbrace{\frac{S_N}{S_N + K_N}}_{\substack{\text{Nitrogen}}} \underbrace{\frac{S_P}{S_P + K_P}}_{\text{Phosphorus}} \underbrace{\frac{I_{av}}{I_{av} + K_I}}_{\substack{\text{Light intensity}}} - \underbrace{b_{ALG} X_{ALG}}_{\text{Decay of biomass}} \tag{2.4}$$

Similar to the growth equations for bacteria in the AS models, $\rho_{net,ALG}$ is the net microalgae growth rate ($ML^{-3}T^{-1}$); $\mu_{m,ALG}$ is the maximum specific growth rate constant (T^{-1}) which changes with temperature and pH; X_{ALG} is the microalgae biomass concentration (ML^{-3}); b_{ALG} is the microalgae decay rate constant (T^{-1}); and the rest of terms are the switching functions (Eq. 2.4). The switching functions reflect the influence of dissolved carbon dioxide (S_{CO_2}), nitrogen (S_N), phosphorus (S_P) concentrations (ML^{-1}), and average light intensity (I_{av}, MT^{-3}) on the growth rate. K represents half-saturation coefficient for each growth limiting constituent in the equation. Expression of the switching functions may show variability depending on the approach in handling the limiting factor. As Eq. (2.4) implies, the growth of microalgae will slow down as the concentration of the constituents decreases during time, and when the growth limiting constituent diminishes in the wastewater, microalgae would not growth. In the dynamic models, concentration of every growth limiting constituent changes in time; therefore, they are presented with separate ordinary differential equations coupled to the growth equation. In addition, in case of simulating microalgae growth in continuous PBRs, expressions describing the hydrodynamics in the reactor system, such as advection and diffusion, are also included to the rate equations.

Relatively recently, integrated models combining both bacterial and algal biotransformation processes are emerging due to the great potential of this dual-kingdom biological systems for sustainable wastewater treatment as well as production of value-added products and energy. A module (ASM-A) containing microalgae transformation processes has been developed as an extension to the ASM-2d (Wagner et al., 2016). This module is a stand-alone module but easily integrated into the processes of ASM-2d to simulate wastewater treatment by microalgae—bacteria consortium. ASM-A is composed of six microalgae biotransformation processes including uptake and storage of nitrogen from ammonium and nitrate, uptake and storage of phosphorus from phosphate, autotrophic growth, heterotrophic growth, and decay. Five particulate and six soluble components are generated and consumed during these processes which are shared with bacterial processes defined in ASM-2d. ASM-A, either stand-alone or integrated with ASM-2d, can simulate microalgae processes in WSPs, HRAPs, and closed PBRs fed with wastewater in case hydrodynamics of the reactors are considered.

Another integrated microalgae—bacteria model, that is BIO_ALGAE, has been developed based on "River Water Quality Model No.1 (*RWQM-1*)" and ASM-3 after certain simplifications (Solimeno et al., 2017). BIO_ALGAE has been validated with the experimental data obtained from duplicate HRAPs (Solimeno and Garcia, 2019) and PBRs (Andreotti et al., 2019; Solimeno et al., 2019). BIO_ALGAE model is composed of 17 biological processes including growth, endogenous respiration and decay of microalgae, heterotrophic and autotrophic bacteria, and 8 physicochemical processes including hydrolysis, water—air chemical transfer and chemical equilibrium, some of which are missing in the ASM-A. BIO_ALGAE model successfully predicts the temporal profiles of soluble and particulate components in the bioreactors treating wastewater. In addition, the integrated model accurately estimates the microalgae proportion in the total biomass during both light and dark periods of bioreactor operation (Andreotti et al., 2019).

Current integrated benchmark models are useful to simulate wastewater treatment with microalgae—bacteria consortium, estimate organic, and nutrient removal efficiency and predict biomass yields from the reactor system. These models would help operators to control process parameters to achieve better treatment. However, the biomass generated during the treatment is a waste and needs handling. On the other hand, algal biomass is nutritious and high calorific; therefore, it is a valuable feedstock. Current models can be improved with additional process models that consider the utilization of the biomass generated in microalgae—bacteria bioreactor system. For instance, combining "Anaerobic Digestion Model No.1 (ADM1)" with benchmark-activated sludge and microalgae models would favor design and operation of sustainable wastewater treatment, aiming resource recovery and sustainable energy production (Fig. 2.5) (Batstone et al., 2002). Use of ADM1 in simulation of biogas production from microalgae has been successfully demonstrated (Mairet et al., 2011). ADM1 can easily be integrated to ASM-A or BIO_-ALGAE since they share the same structure and model ontology.

AlgaeSim is an example of such an integrated model that has been developed to explore the potential outcomes of implementing algal bioreactors for wastewater treatment. The model contains a module for advanced wastewater treatment composed of units for BOD removal, nitrification, and denitrification. The second module contains a dynamic model for microalgae in PBRs receiving only the secondary clarifier supernatant from the first module. Microalgae biomass generated in the PBRs was further processed for biodiesel, biogas, and fertilizer. The model showed that implementation

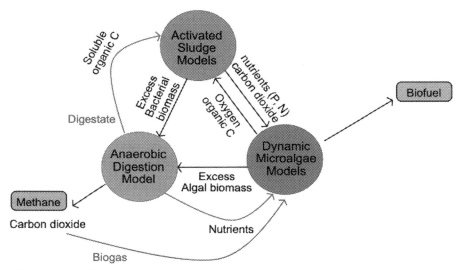

Figure 2.5 Integrated activated sludge, microalgae, and anaerobic digestion models to simulate sustainable wastewater treatment. Arrows show material flow between the models.

of algal PBRs into the current wastewater resource recovery facility (WRRF) in Tampa, Florida, significantly reduces the plant operation cost spent especially for nitrification and denitrification processes. Moreover, plant may make significant annual profit if excess algae is utilized for biodiesel, biogas, and fertilizer production (Drexler et al., 2014).

Microalgae—bacteria wastewater treatment systems are attractive alternatives to current conventional systems due to potential contribution of algae to nutrient removal and its commercial value as a feedstock for biofuels. Therefore, integrated models are useful to evaluate the treatment process performance and economic benefits of such hybrid systems. Better and more reliable models are on the edge as the circular economy goals are implemented on the wastewater management in the world.

6. Conclusions

Natural water bodies suffer from discharges of untreated or inadequately treated domestic/industrial wastewaters. In engineered systems, the potential to recover valuable resources such as carbon dioxide, nitrogen, and phosphorus is not harvested at desirable levels yet. The other outstanding issue is to decrease the costs of wastewater treatment systems by proving better aeration and removal efficiencies. In this chapter, recent literature was presented in detail on how algae—bacteria consortia contemplating on their metabolic synergies could be benefitted in domestic and industrial wastewater treatment systems to remove contaminants at different levels and characteristics. As such, many examples on the proven performance of algae—bacteria consortia for the removal of various nutrients, heavy metals, emerging contaminants, and similar are demonstrated compared to systems where algae or bacteria alone was utilized. Due to heterogeneity in wastewater parameters and wide diversity of organisms thriving in wastewater, there still exist many challenges to address with regards to performance optimization and durability of algae—bacteria consortia. Nevertheless, it is no doubt that well-contemplated algae—bacteria consortia will be one of the key toolsets for modern wastewater treatment systems.

References

Abinandan, S., et al., 2018. Microalgae-bacteria biofilms: a sustainable synergistic approach in remediation of acid mine drainage. Appl. Microbiol. Biotechnol. 102 (3), 1131—1144.

Abouhend, A.S., et al., 2020. Growth progression of oxygenic photogranules and its impact on bioactivity for aeration-free wastewater treatment. Environ. Sci. Technol. 54 (1), 486—496.

Ahner, B.A., Kong, S., Morel, F.M.M., 1995. Phytochelatin production in marine algae. 1. An interspecies comparison. Limnol. Oceanogr. 40 (4), 649—657.

Ahuja, S., et al., 2014. Comprehensive Water Quality and Purification. Elsevier, Amsterdam.

Akyon, B., et al., 2015. Microbial mats as a biological treatment approach for saline wastewaters: the case of produced water from hydraulic fracturing. Environ. Sci. Technol. 49 (10), 6172—6180.

Al Hamed, S., 2014. Dairy wastewater treatment using microalgae in Karbala city—Iraq. Int. J. Environ. Ecol. Fam. Urban Stud. 4, 13, 2014. 22.

Almomani, F., et al., 2019. Intergraded wastewater treatment and carbon bio-fixation from flue gases using Spirulina platensis and mixed algal culture. Process Saf. Environ. Protect. 124, 240—250.

Andreotti, V., et al., 2019. Growth of Tetraselmis suecica and Dunaliella tertiolecta in aquaculture wastewater: numerical simulation with the BIO_ALGAE model. Water Air Soil Pollut. 230 (3).

Arango, L., et al., 2016. Effect of microalgae inoculation on the start-up of microalgae-bacteria systems treating municipal, piggery and digestate wastewaters. Water Sci. Technol. 73 (3), 687—696.

Arcila, J.S., Buitrón, G., 2016. Microalgae-bacteria aggregates: effect of the hydraulic retention time on the municipal wastewater treatment, biomass settleability and methane potential. J. Chem. Technol. Biotechnol. 91 (11), 2862—2870.

Arcila, J.S., Buitrón, G., 2017. Influence of solar irradiance levels on the formation of microalgae-bacteria aggregates for municipal wastewater treatment. Algal Res. 27, 190—197.

Banu, J.R., et al., 2009. Effect of alum on nitrification during simultaneous phosphorous removal in anoxic/oxic reactor. Biotechnol. Bioproc. Eng. 14 (4), 543—548.

Banu, J.R., Yeom, I.-T., 2009. Nutrient removal in an A2O-MBR reactor with sludge reduction. Bioresour. Technol. 100 (16), 3820—3824.

Batstone, D.J., et al., 2002. The IWA anaerobic digestion model No 1 (ADM1). Water Sci. Technol. 45 (10), 65—73.

Bélanger-Lépine, F., et al., 2018. Cultivation of an algae-bacteria consortium in wastewater from an industrial park: effect of environmental stress and nutrient deficiency on lipid production. Bioresour. Technol. 267, 657—665.

Biswas, T., et al., 2021. An eco-friendly strategy for dairy wastewater remediation with high lipid microalgae-bacterial biomass production. J. Environ. Manag. 286, 112196.

Borde, X., et al., 2003. Synergistic relationships in algal—bacterial microcosms for the treatment of aromatic pollutants. Bioresour. Technol. 86 (3), 293—300.

Buchan, A., et al., 2014. Master recyclers: features and functions of bacteria associated with phytoplankton blooms. Nat. Rev. Microbiol. 12 (10), 686—698.

Calatrava, V., et al., 2018. OK, thanks! A new mutualism between Chlamydomonas and methylobacteria facilitates growth on amino acids and peptides. FEMS Microbiol. Lett. 365 (7).

Cervantes-Gaxiola, M.E., et al., 2020. In silico study of the microalgae—bacteria symbiotic system in a stagnant pond. Comput. Chem. Eng. 135, 106740.

Çetinkaya Dönmez, G., et al., 1999. A comparative study on heavy metal biosorption characteristics of some algae. Process Biochem. 34 (9), 885—892.

Chai, W.S., et al., 2021. Multifaceted roles of microalgae in the application of wastewater biotreatment: a review. Environ. Pollut. 269, 116236.

Cho, D.H., et al., 2015. Organic carbon, influent microbial diversity and temperature strongly influence algal diversity and biomass in raceway ponds treating raw municipal wastewater. Bioresour. Technol. 191, 481—487.

Christenson, L., Sims, R., 2011. Production and harvesting of microalgae for wastewater treatment, biofuels, and bioproducts. Biotechnol. Adv. 29 (6), 686—702.

Cooper, M.B., et al., 2019. Cross-exchange of B-vitamins underpins a mutualistic interaction between Ostreococcus tauri and Dinoroseobacter shibae. ISME J. 13 (2), 334—345.

Covarrubias, S.A., et al., 2012. Alginate beads provide a beneficial physical barrier against native microorganisms in wastewater treated with immobilized bacteria and microalgae. Appl. Microbiol. Biotechnol. 93 (6), 2669—2680.

Cruz, I., et al., 2013. Biological deterioration of alginate beads containing immobilized microalgae and bacteria during tertiary wastewater treatment. Appl. Microbiol. Biotechnol. 97 (22), 9847—9858.

Curtis, T.P., Mara, D.D., Silva, S.A., 1992. Influence of pH, oxygen, and humic substances on ability of sunlight to damage fecal coliforms in waste stabilization pond water. Appl. Environ. Microbiol. 58 (4), 1335—1343.

Dao, G.-H., et al., 2018. Enhanced microalgae growth through stimulated secretion of indole acetic acid by symbiotic bacteria. Algal Res. 33, 345—351.

de-Bashan, L.E., et al., 2002. Removal of ammonium and phosphorus ions from synthetic wastewater by the microalgae Chlorella vulgaris coimmobilized in alginate beads with the microalgae growth-promoting bacterium Azospirillum brasilense. Water Res. 36 (12), 2941—2948.

de-Bashan, L.E., et al., 2004. Microalgae growth-promoting bacteria as "helpers" for microalgae: a novel approach for removing ammonium and phosphorus from municipal wastewater. Water Res. 38 (2), 466—474.

Ding, G.T., et al., 2016. Biomass production and nutrients removal by a newly-isolated microalgal strain Chlamydomonas sp in palm oil mill effluent (POME). Int. J. Hydrogen Energy 41 (8), 4888—4895.

Drexler, I.L., et al., 2014. *AlgaeSim*: a model for integrated algal biofuel production and wastewater treatment. Water Environ. Res. 86 (2), 163—176.

Dsikowitzky, L., Schwarzbauer, J., 2013. Organic contaminants from industrial wastewaters: identification, toxicity and fate in the environment. In: Lichtfouse, E., Schwarzbauer, J., Robert, D. (Eds.), Pollutant Diseases, Remediation and Recycling. Springer International Publishing, Cham, pp. 45—101.

Durham, B.P., et al., 2015. Cryptic carbon and sulfur cycling between surface ocean plankton. Proc. Natl. Acad. Sci. U. S. A. 112 (2), 453—457.

Eckenfelder, W.W., Cleary, J.G., 2013. Activated Sludge Technologies for Treating Industrial Wastewaters. DEStech Publications, Inc.

Edwards, C.D., et al., 2013. Aerobic transformation of cadmium through metal sulfide biosynthesis in photosynthetic microorganisms. BMC Microbiol. 13 (1), 161.

Eerkes-Medrano, D., Leslie, H.A., Quinn, B., 2019. Microplastics in drinking water: a review and assessment. Curr. Opin. Environ. Sci. Health 7, 69—75.

EUROSTAT, 2021. Chemicals Production and Consumption Statistics. Eurostat Statistics Explained [cited 2021 29.03.2021]; Available from: https://ec.europa.eu/eurostat/statistics-explained/index.php?title=-Chemicals_production_and_consumption_statistics&oldid=493630.

Filippino, K.C., Mulholland, M.R., Bott, C.B., 2015. Phycoremediation strategies for rapid tertiary nutrient removal in a waste stream. Algal Res. 11, 125—133.

Flores-Salgado, G., et al., 2021. Kinetic characterization of microalgal-bacterial systems: contributions of microalgae and heterotrophic bacteria to the oxygen balance in wastewater treatment. Biochem. Eng. J. 165, 107819.

Fuentes, J.L., et al., 2016. Impact of microalgae-bacteria interactions on the production of algal biomass and associated compounds. Mar. Drugs 14 (5).

Ganeshkumar, V., et al., 2018. Use of mixed wastewaters from piggery and winery for nutrient removal and lipid production by Chlorella sp. MM3. Bioresour. Technol. 256, 254–258.

Gao, F., et al., 2014. Concentrated microalgae cultivation in treated sewage by membrane photobioreactor operated in batch flow mode. Bioresour. Technol. 167, 441–446.

Garcia-Becerra, F.Y., Ortiz, I., 2018. Biodegradation of emerging organic micropollutants in nonconventional biological wastewater treatment: a critical review. Environ. Eng. Sci. 35 (10), 1012–1036.

Gavrilescu, M., et al., 2015. Emerging pollutants in the environment: present and future challenges in biomonitoring, ecological risks and bioremediation. N. Biotech. 32 (1), 147–156.

Gekeler, W., et al., 1988. Algae sequester heavy metals via synthesis of phytochelatin complexes. Arch. Microbiol. 150 (2), 197–202.

Gonçalves, A.L., Pires, J.C., Simões, M., 2017. A review on the use of microalgal consortia for wastewater treatment. Algal Res. 24, 403–415.

Gonzalez-Camejo, J., et al., 2018. Wastewater nutrient removal in a mixed microalgae-bacteria culture: effect of light and temperature on the microalgae-bacteria competition. Environ. Technol. 39 (4), 503–515.

Gonzalez-Fernandez, C., Molinuevo-Salces, B., Garcia-Gonzalez, M.C., 2011. Nitrogen transformations under different conditions in open ponds by means of microalgae-bacteria consortium treating pig slurry. Bioresour. Technol. 102 (2), 960–966.

Goswami, G., Makut, B.B., Das, D., 2019. Sustainable production of bio-crude oil via hydrothermal liquefaction of symbiotically grown biomass of microalgae-bacteria coupled with effective wastewater treatment. Sci. Rep. 9 (1), 1–12.

Grandclément, C., et al., 2017. From the conventional biological wastewater treatment to hybrid processes, the evaluation of organic micropollutant removal: a review. Water Res. 111, 297–317.

Gross, M., Wen, Z., 2014. Yearlong evaluation of performance and durability of a pilot-scale revolving algal biofilm (RAB) cultivation system. Bioresour. Technol. 171, 50–58.

Gupton-Campolongo, T., et al., 2013. Characterization of a high affinity phytochelatin synthase from the Cd-utilizing marine diatom Thalassiosira pseudonana. J. Phycol. 49 (1), 32–40.

Han, J., et al., 2016. Co-culturing bacteria and microalgae in organic carbon containing medium. J. Biol. Res. 23, 8.

He, P.J., et al., 2013. The combined effect of bacteria and Chlorella vulgaris on the treatment of municipal wastewaters. Bioresour. Technol. 146, 562–568.

Henze, M., et al., 2006. Activated Sludge Models ASM1, ASM2, ASM2d and ASM3. IWA Publishing.

Hernandez, D., et al., 2013. Treatment of agro-industrial wastewater using microalgae-bacteria consortium combined with anaerobic digestion of the produced biomass. Bioresour. Technol. 135, 598–603.

Hernandez-Melchor, D.J., et al., 2018. Experimental and kinetic study for lead removal via photosynthetic consortia using genetic algorithms to parameter estimation. Environ. Sci. Pollut. Res. Int. 25 (22), 21286–21295.

Higgins, B.T., et al., 2016. Cofactor symbiosis for enhanced algal growth, biofuel production, and wastewater treatment. Algal Res. 17, 308–315.

Huy, M., et al., 2018. Photoautotrophic cultivation of mixed microalgae consortia using various organic waste streams towards remediation and resource recovery. Bioresour. Technol. 247, 576–581.

Irving, T.E., Allen, D.G., 2011. Species and material considerations in the formation and development of microalgal biofilms. Appl. Microbiol. Biotechnol. 92 (2), 283–294.

Ji, M.-K., et al., 2013. Removal of nitrogen and phosphorus from piggery wastewater effluent using the green microalga Scenedesmus obliquus. J. Environ. Eng. 139 (9), 1198–1205.

Ji, X., et al., 2018. The interactions of algae-bacteria symbiotic system and its effects on nutrients removal from synthetic wastewater. Bioresour. Technol. 247, 44–50.

Ji, X., et al., 2019. The collaborative effect of Chlorella vulgaris-Bacillus licheniformis consortia on the treatment of municipal water. J. Hazard Mater. 365, 483–493.

Jiang, B., et al., 2014. Biodegradation and metabolic pathway of sulfamethoxazole by Pseudomonas psychrophila HA-4, a newly isolated cold-adapted sulfamethoxazole-degrading bacterium. Appl. Microbiol. Biotechnol. 98 (10), 4671—4681.

Jimenez-Perez, M., et al., 2004. Growth and nutrient removal in free and immobilized planktonic green algae isolated from pig manure. Enzym. Microb. Technol. 34 (5), 392—398.

Kamyab, H., et al., 2017. Evaluation of Lemna minor and Chlamydomonas to treat palm oil mill effluent and fertilizer production. J. Water Process Eng. 17, 229—236.

Karna, D., Visvanathan, C., 2019. From conventional activated sludge process to membrane-aerated biofilm reactors: scope, applications, and challenges. In: Water and Wastewater Treatment Technologies. Springer, pp. 237—263.

Katam, K., et al., 2020. Performance evaluation of two trickling filters removing LAS and caffeine from wastewater: light reactor (algal-bacterial consortium) vs dark reactor (bacterial consortium). Sci. Total Environ. 707, 135987.

Komolafe, O., et al., 2014. Biodiesel production from indigenous microalgae grown in wastewater. Bioresour. Technol. 154, 297—304.

Krohn-Molt, I., et al., 2013. Metagenome survey of a multispecies and alga-associated biofilm revealed key elements of bacterial-algal interactions in photobioreactors. Appl. Environ. Microbiol. 79 (20), 6196.

Kumar, M.S., Miao, Z.H., Wyatt, S.K., 2010. Influence of nutrient loads, feeding frequency and inoculum source on growth of Chlorella vulgaris in digested piggery effluent culture medium. Bioresour. Technol. 101 (15), 6012—6018.

Lee, C.S., et al., 2016. Two-phase photoperiodic cultivation of algal-bacterial consortia for high biomass production and efficient nutrient removal from municipal wastewater. Bioresour. Technol. 200, 867—875.

Leng, L., et al., 2020. Use of microalgae based technology for the removal of antibiotics from wastewater: a review. Chemosphere 238, 124680.

Leonardo, T., et al., 2016. Silver accumulation in the green microalga Coccomyxa actinabiotis: toxicity, in situ speciation, and localization investigated using Synchrotron XAS, XRD, and TEM. Environ. Sci. Technol. 50 (1), 359—367.

Levy, J.L., et al., 2005. Toxicity, biotransformation, and mode of action of arsenic in two freshwater microalgae (Chlorella sp. and Monoraphidium arcuatum). Environ. Toxicol. Chem. 24 (10), 2630—2639.

Li, K., et al., 2019. Microalgae-based wastewater treatment for nutrients recovery: a review. Bioresour. Technol. 291, 121934.

Li, R., et al., 2019. Treatment of fracturing wastewater using microalgae-bacteria consortium. Can. J. Chem. Eng. 98 (2), 484—490.

Li, Y., Yang, X., Geng, B., 2018. Preparation of immobilized sulfate-reducing bacteria-microalgae beads for effective bioremediation of copper-containing wastewater. Water Air Soil Pollut. 229 (3), 1—13.

Liang, Z., et al., 2015. A pH-dependent enhancement effect of co-cultured Bacillus licheniformis on nutrient removal by Chlorella vulgaris. Ecol. Eng. 75, 258—263.

Ma, B., et al., 2016. Biological nitrogen removal from sewage via anammox: recent advances. Bioresour. Technol. 200, 981—990.

Mahdavi, H., et al., 2015. In situ biodegradation of naphthenic acids in oil sands tailings pond water using indigenous algae—bacteria consortium. Bioresour. Technol. 187, 97—105.

Mairet, F., et al., 2011. Modeling anaerobic digestion of microalgae using ADM1. Bioresour. Technol. 102 (13), 6823—6829.

Mara, D., 2003. Low-cost treatment systems. In: Horan, DM.a.N. (Ed.), The Handbook of Water and Wastewater Microbiology. Elsevier, London, pp. 441—448.

Marbelia, L., et al., 2014. Membrane photobioreactors for integrated microalgae cultivation and nutrient remediation of membrane bioreactors effluent. Bioresour. Technol. 163, 228—235.

Marella, T.K., Parine, N.R., Tiwari, A., 2018. Potential of diatom consortium developed by nutrient enrichment for biodiesel production and simultaneous nutrient removal from waste water. Saudi J. Biol. Sci. 25 (4), 704—709.

Margot, J., et al., 2015. A review of the fate of micropollutants in wastewater treatment plants. WIREs Water 2 (5), 457—487.

Matamoros, V., et al., 2016. Assessment of the mechanisms involved in the removal of emerging contaminants by microalgae from wastewater: a laboratory scale study. J. Hazard Mater. 301, 197–205.

Maza-Márquez, P., et al., 2017. Biotreatment of industrial olive washing water by synergetic association of microalgal-bacterial consortia in a photobioreactor. Environ. Sci. Pollut. Control Ser. 24 (1), 527–538.

Maza-Marquez, P., et al., 2017. Full-scale photobioreactor for biotreatment of olive washing water: structure and diversity of the microalgae-bacteria consortium. Bioresour. Technol. 238, 389–398.

Mera, R., Torres, E., Abalde, J., 2014. Sulphate, more than a nutrient, protects the microalga Chlamydomonas moewusii from cadmium toxicity. Aquat. Toxicol. 148, 92–103.

Mohsenpour, S.F., et al., 2020. Integrating micro-algae into wastewater treatment: a review. Sci. Total Environ. 142168.

Monteiro, C.M., Castro, P.M., Malcata, F.X., 2012. Metal uptake by microalgae: underlying mechanisms and practical applications. Biotechnol. Prog. 28 (2), 299–311.

Mujtaba, G., Rizwan, M., Lee, K., 2017. Removal of nutrients and COD from wastewater using symbiotic co-culture of bacterium Pseudomonas putida and immobilized microalga Chlorella vulgaris. J. Ind. Eng. Chem. 49, 145–151.

Muñoz, R., et al., 2005. Combined carbon and nitrogen removal from acetonitrile using algal–bacterial bioreactors. Appl. Microbiol. Biotechnol. 67 (5), 699–707.

Muñoz, I., et al., 2009. Chemical evaluation of contaminants in wastewater effluents and the environmental risk of reusing effluents in agriculture. Trac. Trends Anal. Chem. 28 (6), 676–694.

Munoz, R., Guieysse, B., 2006. Algal-bacterial processes for the treatment of hazardous contaminants: a review. Water Res. 40 (15), 2799–2815.

Nawaz, M.S., Ahsan, M., 2014. Comparison of physico-chemical, advanced oxidation and biological techniques for the textile wastewater treatment. Alex. Eng. J. 53 (3), 717–722.

Nguyen, T.D.P., et al., 2019. Bioflocculation formation of microalgae-bacteria in enhancing microalgae harvesting and nutrient removal from wastewater effluent. Bioresour. Technol. 272, 34–39.

Nguyen, H.T., et al., 2020. The application of microalgae in removing organic micropollutants in wastewater. Crit. Rev. Environ. Sci. Technol. 1–34.

Nguyen, T.-T.-D., et al., 2020. Co-culture of microalgae-activated sludge for wastewater treatment and biomass production: exploring their role under different inoculation ratios. Bioresour. Technol. 314, 123754.

Norvill, Z.N., Shilton, A., Guieysse, B., 2016. Emerging contaminant degradation and removal in algal wastewater treatment ponds: identifying the research gaps. J. Hazard Mater. 313, 291–309.

Orandi, S., Lewis, D.M., Moheimani, N.R., 2012. Biofilm establishment and heavy metal removal capacity of an indigenous mining algal-microbial consortium in a photo-rotating biological contactor. J. Ind. Microbiol. Biotechnol. 39 (9), 1321–1331.

Oyebamiji, O.O., et al., 2019. Green microalgae cultured in textile wastewater for biomass generation and biodetoxification of heavy metals and chromogenic substances. Bioresour. Technol. Rep. 7, 100247.

Paddock, M.B., Fernandez-Bayo, J.D., VanderGheynst, J.S., 2020. The effect of the microalgae-bacteria microbiome on wastewater treatment and biomass production. Appl. Microbiol. Biotechnol. 104 (2), 893–905.

Palma, H., et al., 2017. Assessment of microalga biofilms for simultaneous remediation and biofuel generation in mine tailings water. Bioresour. Technol. 234, 327–335.

Paskuliakova, A., et al., 2018. Phycoremediation of landfill leachate with the chlorophyte Chlamydomonas sp. SW15aRL and evaluation of toxicity pre and post treatment. Ecotoxicol. Environ. Saf. 147, 622–630.

Pastore, M., et al., 2018. Light intensity affects the mixotrophic carbon exploitation in Chlorella protothecoides: consequences on microalgae-bacteria based wastewater treatment. Water Sci. Technol. 78 (8), 1762–1771.

Paulssen, J.M., Gieg, L.M., 2019. Biodegradation of 1-adamantanecarboxylic acid by algal-bacterial microbial communities derived from oil sands tailings ponds. Algal Res. 41, 101528.

Pearson, H., 2003. Microbial Interactions in Facultative and Maturation Ponds. Handbook of Water and Wastewater Microbiology, pp. 449–458.

Peavy, H.S., Tchobanoglous, G., 1985. Environmental Engineering.

Perales-Vela, H.V., Pena-Castro, J.M., Canizares-Villanueva, R.O., 2006. Heavy metal detoxification in eukaryotic microalgae. Chemosphere 64 (1), 1–10.

Pomies, M., et al., 2013. Modelling of micropollutant removal in biological wastewater treatments: a review. Sci. Total Environ. 443, 733–748.

Praveen, P., Loh, K.C., 2015. Photosynthetic aeration in biological wastewater treatment using immobilized microalgae-bacteria symbiosis. Appl. Microbiol. Biotechnol. 99 (23), 10345–10354.

Qi, W., et al., 2018. Enhancing fermentation wastewater treatment by co-culture of microalgae with volatile fatty acid- and alcohol-degrading bacteria. Algal Res. 31, 31–39.

Rajkumara, R., Takriffab, M., 2015. Nutrient removal from anaerobically treated palm oil mill effluent by Spirulina platensis and Scenedesmus dimorphus. Der Pharm. Lett. 7 (7), 416–421.

Ramanan, R., et al., 2016. Algae-bacteria interactions: evolution, ecology and emerging applications. Biotechnol. Adv. 34 (1), 14–29.

Ranjit, P., Jhansi, V., Reddy, K.V., 2021. Conventional wastewater treatment processes. In: Advances in the Domain of Environmental Biotechnology. Springer, pp. 455–479.

Reis, P.J., et al., 2014. Biodegradation of sulfamethoxazole and other sulfonamides by Achromobacter denitrificans PR1. J. Hazard Mater. 280, 741–749.

Robles, A., et al., 2020. Microalgae-bacteria consortia in high-rate ponds for treating urban wastewater: elucidating the key state indicators under dynamic conditions. J. Environ. Manag. 261, 110244.

Rugnini, L., et al., 2018. Phosphorus and metal removal combined with lipid production by the green microalga Desmodesmus sp.: an integrated approach. Plant Physiol. Biochem. 125, 45–51.

Ruiz-Marin, A., Mendoza-Espinosa, L.G., Stephenson, T., 2010. Growth and nutrient removal in free and immobilized green algae in batch and semi-continuous cultures treating real wastewater. Bioresour. Technol. 101 (1), 58–64.

Russel, M., et al., 2020. Investigating the potentiality of Scenedesmus obliquus and Acinetobacter pittii partnership system and their effects on nutrients removal from synthetic domestic wastewater. Bioresour. Technol. 299, 122571.

Ryu, B.-G., et al., 2017. Feasibility of using a microalgal-bacterial consortium for treatment of toxic coke wastewater with concomitant production of microbial lipids. Bioresour. Technol. 225, 58–66.

Saavedra, R., et al., 2018. Comparative uptake study of arsenic, boron, copper, manganese and zinc from water by different green microalgae. Bioresour. Technol. 263, 49–57.

Safonova, E., et al., 2004. Biotreatment of industrial wastewater by selected algal-bacterial consortia. Eng. Life Sci. 4 (4), 347–353.

Sanchez Zurano, A., et al., 2021. Modeling of photosynthesis and respiration rate for microalgae-bacteria consortia. Biotechnol. Bioeng. 118 (2), 952–962.

Sánchez-Zurano, A., et al., 2020. A novel photo-respirometry method to characterize consortia in microalgae-related wastewater treatment processes. Algal Res. 47, 101858.

Satyanarayana, S.D., et al., 2018. In silico structural homology modeling of nif A protein of rhizobial strains in selective legume plants. J. Genet. Eng. Biotechnol. 16 (2), 731–737.

Selvaratnam, T., et al., 2015. Algal biofuels from urban wastewaters: maximizing biomass yield using nutrients recycled from hydrothermal processing of biomass. Bioresour. Technol. 182, 232–238.

Serejo, M.L., et al., 2020. Surfactant removal and biomass production in a microalgal-bacterial process: effect of feeding regime. Water Sci. Technol. 82 (6), 1176–1183.

Seymour, J.R., et al., 2017. Zooming in on the phycosphere: the ecological interface for phytoplankton-bacteria relationships. Nat Microbiol 2, 17065.

Sheng, A., et al., 2017. Sequencing batch membrane photobioreactor for real secondary effluent polishing using native microalgae: process performance and full-scale projection. J. Clean. Prod. 168, 708–715.

Solimeno, A., et al., 2017. Integral microalgae-bacteria model (BIO_ALGAE): application to wastewater high rate algal ponds. Sci. Total Environ. 601–602, 646–657.

Solimeno, A., Garcia, J., 2017. Microalgae-bacteria models evolution: from microalgae steady-state to integrated microalgae-bacteria wastewater treatment models - a comparative review. Sci. Total Environ. 607–608, 1136–1150.

Solimeno, A., Garcia, J., 2019. Microalgae and bacteria dynamics in high rate algal ponds based on modelling results: long-term application of BIO_ALGAE model. Sci. Total Environ. 650 (Pt 2), 1818–1831.

Solimeno, A., Gomez-Serrano, C., Acien, F.G., 2019. BIO_ALGAE 2: improved model of microalgae and bacteria consortia for wastewater treatment. Environ. Sci. Pollut. Res. Int. 26 (25), 25855–25868.

Sousa, J.C., et al., 2018. A review on environmental monitoring of water organic pollutants identified by EU guidelines. J. Hazard Mater. 344, 146–162.

Sridhara Chary, N., Kamala, C.T., Samuel Suman Raj, D., 2008. Assessing risk of heavy metals from consuming food grown on sewage irrigated soils and food chain transfer. Ecotoxicol. Environ. Saf. 69 (3), 513–524.

Stravs, M.A., Pomati, F., Hollender, J., 2017. Exploring micropollutant biotransformation in three freshwater phytoplankton species. Environ. Sci. Process Impacts 19 (6), 822–832.

Su, Q., et al., 2021. Role of ammonia oxidation in organic micropollutant transformation during wastewater treatment: insights from molecular, cellular, and community level observations. Environ. Sci. Technol. 55 (4), 2173–2188.

Su, Y., Mennerich, A., Urban, B., 2016. The long-term effects of wall attached microalgal biofilm on algae-based wastewater treatment. Bioresour. Technol. 218, 1249–1252.

Sumithrabhai, K., et al., 2016. Expedient study on treatment of dairy effulent in fluidized bed reactor using immobilized microalgae. Int. J. Adv. Eng. Technol. 7 (2), 231–235.

Sun, L., et al., 2018. A novel symbiotic system combining algae and sludge membrane bioreactor technology for wastewater treatment and membrane fouling mitigation: performance and mechanism. Chem. Eng. J. 344, 246–253.

Sutherland, D.L., Ralph, P.J., 2019. Microalgal bioremediation of emerging contaminants - opportunities and challenges. Water Res. 164, 114921.

Tang, X., et al., 2010. Construction of an artificial microalgal-bacterial consortium that efficiently degrades crude oil. J. Hazard Mater. 181 (1), 1158–1162.

Tang, C.C., et al., 2016. Effect of aeration rate on performance and stability of algal-bacterial symbiosis system to treat domestic wastewater in sequencing batch reactors. Bioresour. Technol. 222, 156–164.

Tang, C.C., et al., 2018a. Enhanced nitrogen and phosphorus removal from domestic wastewater via algae-assisted sequencing batch biofilm reactor. Bioresour. Technol. 250, 185–190.

Tang, C.C., et al., 2018b. Performance and mechanism of a novel algal-bacterial symbiosis system based on sequencing batch suspended biofilm reactor treating domestic wastewater. Bioresour. Technol. 265, 422–431.

Tao, Q., et al., 2017. Enhanced biomass/biofuel production and nutrient removal in an algal biofilm airlift photobioreactor. Algal Res. 21, 9–15.

Tighiri, H.O., Erkurt, E.A., 2019. Biotreatment of landfill leachate by microalgae-bacteria consortium in sequencing batch mode and product utilization. Bioresour. Technol. 286, 121396.

Ting, H., et al., 2017. Progress in microalgae cultivation photobioreactors and applications in wastewater treatment: a review. Int. J. Agric. Biol. Eng. 10 (1), 1–29.

Tran, N.H., Reinhard, M., Gin, K.Y., 2018. Occurrence and fate of emerging contaminants in municipal wastewater treatment plants from different geographical regions-a review. Water Res. 133, 182–207.

Tsuji, N., et al., 2003. Regulation of phytochelatin synthesis by zinc and cadmium in marine green alga, Dunaliella tertiolecta. Phytochemistry 62 (3), 453–459.

Van Den Hende, S., et al., 2011. Bioflocculation of microalgae and bacteria combined with flue gas to improve sewage treatment. N. Biotech. 29 (1), 23–31.

Verma, K., et al., 2020. Phycoremediation of sewage-contaminated lake water using microalgae–bacteria Co-culture. Water Air Soil Pollut. 231 (6).

Villa, J.A., Ray, E.E., Barney, B.M., 2014. Azotobacter vinelandii siderophore can provide nitrogen to support the culture of the green algae Neochloris oleoabundans and Scenedesmus sp. BA032. FEMS Microbiol. Lett. 351 (1), 70–77.

Vo, H.N.P., et al., 2019. A critical review on designs and applications of microalgae-based photobioreactors for pollutants treatment. Sci. Total Environ. 651, 1549–1568.

Wackett, L.P., Hershberger, C.D., 2001. Biocatalysis and Biodegration. American Society of Microbiology.

Wagner, D.S., et al., 2016. Towards a consensus-based biokinetic model for green microalgae - the ASM-A. Water Res. 103, 485–499.

Wang, J., Wang, S., 2016. Removal of pharmaceuticals and personal care products (PPCPs) from wastewater: a review. J. Environ. Manag. 182, 620–640.

Whitton, R., et al., 2015. Microalgae for municipal wastewater nutrient remediation: mechanisms, reactors and outlook for tertiary treatment. Environ. Technol. Rev. 4 (1), 133–148.

Wollmann, F., et al., 2019. Microalgae wastewater treatment: biological and technological approaches. Eng. Life Sci. 19 (12), 860–871.

Xiong, J.Q., et al., 2016. Biodegradation of carbamazepine using freshwater microalgae Chlamydomonas mexicana and Scenedesmus obliquus and the determination of its metabolic fate. Bioresour. Technol. 205, 183–190.

Xiong, J.-Q., et al., 2017. Biodegradation and metabolic fate of levofloxacin via a freshwater green alga, Scenedesmus obliquus in synthetic saline wastewater. Algal Res. 25, 54–61.

Xiong, J.Q., et al., 2019. Combined effects of sulfamethazine and sulfamethoxazole on a freshwater microalga, Scenedesmus obliquus: toxicity, biodegradation, and metabolic fate. J. Hazard Mater. 370, 138–146.

Xiong, J.Q., Kurade, M.B., Jeon, B.H., 2018. Can microalgae remove pharmaceutical contaminants from water? Trends Biotechnol. 36 (1), 30–44.

Yu, K.L., et al., 2017. Microalgae from wastewater treatment to biochar—feedstock preparation and conversion technologies. Energy Convers. Manag. 150, 1–13.

Zeraatkar, A.K., et al., 2016. Potential use of algae for heavy metal bioremediation, a critical review. J. Environ. Manag. 181, 817–831.

Zhang, B., et al., 2018. Enhancement of aerobic granulation and nutrient removal by an algal—bacterial consortium in a lab-scale photobioreactor. Chem. Eng. J. 334, 2373–2382.

Zhang, B., et al., 2020. Microalgal-bacterial consortia: from interspecies interactions to biotechnological applications. Renew. Sustain. Energy Rev. 118, 109563.

Zhao, X., et al., 2014. Characterization of microalgae-bacteria consortium cultured in landfill leachate for carbon fixation and lipid production. Bioresour. Technol. 156, 322–328.

Zheng, H., et al., 2018. Balancing carbon/nitrogen ratio to improve nutrients removal and algal biomass production in piggery and brewery wastewaters. Bioresour. Technol. 249, 479–486.

Zhou, G.J., et al., 2014. Simultaneous removal of inorganic and organic compounds in wastewater by freshwater green microalgae. Environ. Sci. Process Impacts 16 (8), 2018–2027.

CHAPTER 3

Integrated algal-based sewage treatment and resource recovery system

N. Nirmalakhandan, I.S.A. Abeysiriwardana-Arachchige,
S.P. Munasinghe-Arachchige and H.M.K. Delanka-Pedige
Civil Engineering Department, New Mexico State University, Las Cruces, NM, United States

1. Introduction

The nation's 14,748 Publicly Owned Treatment Works (POTWs) serving over 240 million Americans are considered the most basic and critical infrastructure system for protecting public health and the environment while providing the foundation for economic growth (ASCE, 2017). POTWs are responsible for collecting and treating municipal sewage and are mandated to control their effluent discharges to receiving water bodies. To comply with this mandate, POTWs have been utilizing a series of processes, each designed to remove a specific pollutant in sewage such as suspended solids, biochemical oxygen demand (BOD), nutrients, and pathogenic bacteria. Although these processes have served well in meeting the respective discharge standards, they are energy- and resource intensive (McCarty et al., 2011). Most of these processes were developed in the 1900s specifically to meet discharge standards through destructive removal of the pollutants, with little regard to their energy–intensity, carbon-footprint, or lifecycle impacts (US EPA, 2014); their immediate direct benefits of improved sanitation and ecosystem services were valued to be much greater (GYOBU et al., 2015; Jenssen et al., 2007; Lazarova et al., 2012; Soda et al., 2013, 2010). Current realizations and concerns about greenhouse gas emissions, natural resource depletion, sustainability, and affordability of sewage treatment technologies and the potential for recovering resources inherent in sewage have been driving the development of improved or alternate technologies for sewage management (Batstone et al., 2015; McCarty et al., 2011; Puyol et al., 2017; Shoener et al., 2014).

1.1 Concerns about traditional POTWs

The most common sewage treatment technology deployed at POTWs in the United States and worldwide to meet secondary level treatment is the activated sludge (AS) process. Developed a century ago, its specific design goal is to reduce BOD in settled sewage to the discharge level. It utilizes heterotrophic bacteria that oxidize about 40%−50% of

---

Integrated Wastewater Management and Valorization using Algal Cultures
ISBN 978-0-323-85859-5, https://doi.org/10.1016/B978-0-323-85859-5.00008-7

© 2022 Elsevier Inc.
All rights reserved.

51

the dissolved BOD in the settled sewage to carbon dioxide (CO_2) and assimilate about 40%–50% to synthesize biomass. During biomass synthesis, about 15% of the ammoniacal nitrogen and about 2% of the phosphates in the sewage are also assimilated. While the AS process serves well in meeting the discharge standards for BOD, it suffers from two major drawbacks: (i) high energy demand for oxidizing BOD and (ii) inability to meet discharge standards for nutrients.

The first drawback stems from the low solubility of oxygen (of about 8 mg/L). As such, significant energy has to be expended for artificially aerating the cultures to maintain the desired dissolved oxygen level (of about 2–4 mg/L) for BOD oxidation. The theoretical oxygen demand for the AS process is 1.85 kg O_2 per kg C which translates to about 1.08 kJ/L of sewage (McCarty et al., 2011). Because of the poor gas-to-liquid oxygen transfer efficiency of aeration devices, the energy expended for aeration is 1.1 kWh per kg O_2, accounting for nearly 50% of the cost of operation at most POTWs. Aeration can also promote stripping of volatile organics in the influent, creating air pollution concerns.

The second drawback stems from the disparity in the stoichiometric organic carbon-to-nitrogen (C:N) ratio in the settled sewage and that in heterotrophic bacteria (Selvaratnam et al., 2014). Because of this disparity, AS process is limited by organic carbon, falling short in reducing nitrogen levels in its effluent. This has necessitated a follow-up nitrification–denitrification (ND) process to reduce the remaining nitrogen to the discharge standards. The ND process also requires dissolved oxygen (4.57 kg O_2 per kg NH_4 nitrified), doubling the total energy input to 2.16 kJ/L of sewage treated (Batstone et al., 2015). In addition, the ND process requires supplemental dissolved organic carbon (e.g., methanol) to complete the process (2.9 kg methanol per kg NO_3–N denitrified). A lifecycle analysis study reported that use of methanol for denitrification results in greenhouse emission of 1.4 CO_2-eq. per kg NO_3 denitrified (Theis and Hicks, 2012). While this two-step AS/ND process meets the discharge standards, about 80% of nitrogen in sewage is lost in the ND process as gaseous nitrogen (N_2) and nitrous oxide (N_2O) without any possibility of recovery. Emission of N_2O by the ND process is also a concern as it is a greenhouse gas (~ 300 times more potent than CO_2) and an ozone destructor.

As an example, in a 5.5 MGD plant, energy consumption averaged 1.48 kW-hr per kg BOD removal; 13.44 kW-hr per kg total N removal; and 6.44 kW-hr per kg total P removal (US EPA, 2008). Overall, energy consumption by POTWs in the United States averages 1200 kW-hr per million gallons (MG) of sewage treated, amounting to nearly 1.5% of the country's total energy consumption, which is generated from fossil fuels. In 2000, energy-related emissions associated with POTW operations in the United States (excluding organic sludge degradation) have been estimated as 15.5 Tg CO_2- eq., with an acidification potential of 145 Gg SO_2-equivalents and eutrophication potential of 4 Gg PO_4^{3-} eq (US EPA, 2018).

Considering the energy demand and the associated environmental emissions on one hand, and the depletion and/or dissipation of nonrenewable natural resources on the other, the current sewage treatment practice can be seen to be unsustainable. An EPA report has acknowledged that new techniques are needed for sewage treatment to reduce emissions and to recover their nutrient contents at substantially less cost and with reduced carbon footprint (US EPA, 2014). It is also recognized that rehabilitation of the nation's aging wastewater infrastructure is a critical need; and reinvented technologies and techniques are becoming available to upgrade existing POTWs (US EPA, 2014).

1.2 Reinvention of POTWs

A limited number of recent studies have reported on innovative technologies for sewage treatment; however, most of them have focused on a specific contaminant, process, or performance metric. In 2016, WE&RF conducted a survey of large sewage treatment facilities across the United States and other countries to assess the status of innovative technologies being tested in the industry (WERF/WEF LIFT Guidance Manual, 2017). Responses from over 100 respondents showed that innovation is currently limited to individual subsystems and processes rather than on sewage treatment system as a whole. For example, 24 utilities were investigating advanced blower technologies and 6 utilities were investigating membrane aerated bioreactors for saving energy in aerating the AS process; 37 utilities were investigating biological nutrient removal; 36 utilities were evaluating enhancements of chlorine disinfection (WERF/WEF LIFT Guidance Manual, 2017). As can be noted from the results of this survey, current innovation in sewage treatment is aiming at incremental changes and enhancements to the classical technologies rather than on transformative, system-level reinvention.

In a similar vein, municipal sewage is now being viewed not just as a waste stream that has to be treated for disposal, but as a locally available, renewable resource from which reusable water, greener fuels and fertilizers, and other valuable commodities could be recovered for beneficial use (Hernández-Sancho et al., 2015; Jenssen et al., 2007; Puyol et al., 2017; Van Der Hoek et al., 2016). Currently, the following three pathways have emerged as the feasible ones for recovering resources from sewage (Mo and Zhang, 2012): (i) anaerobic digestion of primary and secondary sludges for recovering energy content of sewage as heat and electrical energy; (ii) land application of the digested sludge to utilize their nutrient content as fertilizers; and (iii) reuse of the treated effluent in non-potable applications. Recovering energy from sewage can offset the energy consumed for its treatment, conserve fossil fuels, and reduce greenhouse gas emissions; utilizing digested sludge as fertilizers can reduce the energy required for producing virgin (commercial) fertilizers utilizing fossil fuels and conserve the limited natural mineral ores; and reusing reclaimed water for nonpotable applications can reduce the energy and chemicals

1.2.1 Potential for energy recovery from sewage

Urban sewage contains internal energy of 6.3—7.6 kJ/L (Batstone et al., 2015; Heidrich et al., 2011), which is roughly 3—4 times the energy that is now being expended to treat them to meet discharge standards (McCarty et al., 2011); this energy demand amounts to 1% of the current global energy consumption or, 4% of the electricity production (International Energy Agency, 2015). Although the energy content of urban sewage is of low grade, current technologies in place at POTWs are able to recover some of it; however, the energy input for its treatment [∼2.16 kJ/L (McCarty et al., 2011)] exceeds that could be recovered, rendering POTWs energy negative. GYOBU et al. (2015), for example, reported that the average calorific inflow at three POTWs was 2.93 kJ/L, while biogas from anaerobic digestion of the primary and secondary sludges at these plants amounted to 1.15 kJ/L, for a recovery of 39% of the calorific content of sewage. Assuming biogas-to-electricity conversion efficiency of 39%, this recovery amounts to 0.45 kJ/L; this recovery is only 21% of the demand for aeration [∼2.16 kJ/L (McCarty et al., 2011)]. Clearly, reinvented technologies capable of minimizing energy input for sewage treatment and maximizing the energy recovery can make POTWs energy positive.

1.2.2 Potential for nutrient recovery from sewage

It has been estimated that sewage accounts for 50%—100% of lost waste resources. For example, human excreta are laden with 20% of all manufactured nitrogen and phosphorous that are currently wastefully destroyed or dissipated (Batstone et al., 2015; Matassa et al., 2015); a good percentage of this could be recovered by taking advantage of urban concentrations and the existing sewage collection infrastructure. Technologies currently in place for removing nitrogen and phosphorous from sewage and the potential for recovering them are summarized in Table 3.1. As can be noted from Table 3.1, current technologies are designed for destructive or wasteful removal of nitrogen and phosphorous from sewage with very low potential for any recoveries.

In a case study by Mo and Zhang (2012) of a state-of-the-art sewage treatment plant (average flow of 52.4 MGD), it was concluded that carbon neutrality is not possible with current technologies (primary settling, AS, nitrification, and denitrification) even if best-case scenarios are assumed for current levels of resource recovery (i.e., heat and energy recovery, land application of digested sludge, and reuse of reclaimed water). This finding reiterates the notion that reinvented sewage treatment technologies are to be integrated with efficient resource recovery methods for greening POTWs and making them sustainable.

Integrated algal-based sewage treatment and resource recovery system

Table 3.1 Common processes for nutrient removal in sewage treatment and potential for recovering the nutrients.

Common processes for nutrient removal	Type[a]	Input form of nutrient	Other chemical inputs	Resulting form of nutrient	Potential recovery of nutrient
(a) Nitrogen removal					
Ammonification	H	Org.$-$N	—	NH_3, NH_4-N	None
Uptake by activated sludge	B	NH_4-N	—	Cells (Org.$-$N)	AD > compost
Nitrification	B	NH_4-N	4.33 g O_2/g ΔNH_4	NO_3-N + cells (Org.$-$N) 0.15 g cells/g ΔNH_4-N	AD > compost
Nitrification	B	NH_4-N	—	NO_2-N	None
Denitrification	B	NO_3-N	2.47 g CH_3OH/g NO_3	N_2 + cells (Org.$-$N) 0.45 g cells/g ΔNO_3-N	AD > compost
Anammox	B	NH_4-N; NO_2-N	1.33 g O_2/ g ΔNH_4	N_2	None
Volatilization @ pH > 8	P	NH_4-N	—	NH_3	None
(b) Phosphorous removal					
Iron precipitation	C	PO_4	5.24 g $FeCl_2$/g P		None
Alum precipitation	C	PO_4	0.87 g Al/g P		None
Magnesium precipitation	C	PO_4	1.35 g Mg/g Δ P	Struvite	Fertilizer
Calcium (lime) precipitation	C	PO_4	(1.5 × Alk.) lime		None
Uptake by activated sludge *P in VSS = 2−2.5% dry wt*	B	PO_4	–	Cells (Org.$-$P) 0.012 g P/g ΔBOD_5	AD > compost
Enhanced BPR	B	PO_4	–	Cells (Org.$-$P)	AD > compost

[a]*AD*, Anaerobic digestion; *B*, Biological; *C*, Chemical; *H*, Hydrolysis; *P*, Physical.

1.3 Emerging approaches for sewage treatment and resource recovery

During the past decade, a handful of new approaches have been proposed for greening the traditional sewage treatment practice. Most of them, though, were aimed primarily at energy recovery, with a very few at enhanced recovery of carbon, nitrogen, and phosphorous (Puyol et al., 2017). In a review of platforms for recovering energy and nutrients from domestic sewage, Batstone et al. (2015) identified two major approaches for doing so: one reported by Verstraete et al. (2009) another by McCarty et al. (2011). The approach proposed by Verstraete et al. (2009) involves separation of the solids from sewage (by gravity, microfiltration MF, and reverse osmosis RO) for treatment by anaerobic digestion and recovery of the nutrients from digestate. Batstone et al. (2015) argued that use of MF/RO might be energy prohibitive unless the value of the reclaimed water is taken into consideration.

The approach proposed by McCarty et al. (2011) entails anaerobic treatment of the primary effluent in anaerobic membrane bioreactor (AnMBR) or anaerobic fluidized membrane bioreactor (AFMBR) to remove biosolids and dissolved organics. While avoiding the aeration energy input for BOD oxidation, this approach can generate energy via methane production. A related work (Kim et al., 2011) has reported that the energy input for the AFMBR is about 0.058 kW-hr m^{-3}, which is about 10% of that for the AS process. One of the concerns with this approach is the dissolved methane in the effluent that has to be recovered to improve energy recovery and more importantly, to minimize atmospheric release because methane has global warming potential almost 25 times that of CO_2. Even though this system can potentially be energy positive [1 kW-hr/m^3 (McCarty et al., 2011)], it is deficient in N and P removal. The N- and P-rich effluent might be suitable for irrigation applications or processed further to meet nutrient discharge standards. The effluent offers opportunities to recover N and P as struvite for use as fertilizers.

Building upon the above major approaches, Batstone et al. (2015) and Puyol et al. (2017) have presented the following modifications for enhancing recovery of energy and nutrients. Batstone et al. (2015) have proposed to modify the approach by McCarty's to eliminate primary settling and utilize an AnMBR for carbon removal from the raw sewage; treat the effluent of the AnMBR process by anammox process to remove nitrogen; and follow up with adsorptive process for phosphate recovery. Anammox is an emerging process that completely destroys all residual dissolved N to gaseous N with an energy requirement of 1.2 kW-hr per kg N removed. Again, dissolved methane in the effluent has to be recovered to enhance energy recovery and avoid methane emissions. While AnMBR can be energy intensive, it can serve as a pathogen barrier and yield high quality effluent with minimal disinfectant demand.

Puyol et al. (2017) and Batstone et al. (2015) have presented a partition—release—recover approach for sewage treatment and resources recovery. Following on the premise of Verstraete et al. (2009), Puyol et al. (2017) have suggested a partition step involving biological assimilation and accumulation to partition the dissolved constituents to a solid phase (biomass) in a membrane bioreactor (MBR), followed by a release step involving anaerobic digestion to release the energy content of the biomass as biogas (methane) and the recovery step where the nutrient-rich digestate is processed to recover phosphorous as struvite. Recovery of N is challenging because of the lack of suitable precipitates. Based on stoichiometry of the wastewater, bacteria, and algae, Batstone et al. (2015) have concluded that algal biomass offers the best option for the portioning step.

Conceptually, the above approaches are expected to be significantly more energy efficient providing options to recover much of the resources in the sewage rather than wasteful dissipation as in the current AS/ND processes. As cautioned by Batstone et al. (2015), these approaches have not yet been implemented in full scale or demonstrated at pilot scale, but individual subsystems have been demonstrated to be viable. Even though most of the analyses by McCarty et al. (2011), Batstone et al. (2015), and Puyol et al. (2017) are hypothetical and not verified beyond laboratory scales, these approaches are deemed to be practically viable and to have strong potential for greening the sewage management practice.

Recently, an integrated algal-based approach has been proposed for STaRR. Developed from laboratory treatability studies, this STaRR system has been validated at pilot scale at a POTW. Several previous reports have detailed the development of the STaRR system and its validation under field conditions. A compilation of pilot-scale performance of the STaRR system over 5 years is presented here.

1.4 Algal-based sewage treatment

Algal-based sewage treatment systems were introduced in the 1950s to minimize the energy intensity of the AS process. A common algal-based sewage treatment design is the high-rate algal pond (HRAP), where photoautotrophic algae and heterotrophic bacteria symbiotically remove the BOD and nutrients in a single step. In HRAPs, photoautotrophic algae produce oxygen by photosynthesis which is utilized by the heterotrophic bacteria to oxidize BOD to produce carbon dioxide which, in turn, serves as the inorganic carbon source for the photoautotrophic algae. This symbiosis overcomes the drawbacks of the AS system pointed out earlier, averting all external energy input for aeration and taking advantage of the high N and P uptake by algae to afford single-step removal of BOD and nutrients to meet discharge standards. Since the pH in photoautotrophic metabolism is typically alkaline (pH > 8), HRAPs also enable high degree of inactivation of pathogenic bacteria and viruses. Additional pathogen inactivation by sunlight can result from the long detention time and shallow depths in HRAPs.

Typically, HRAPs are implemented as open raceways, where the cultures are kept in circulation by paddlewheels. Since the photoautotrophic algae depend on sunlight,

HRAPs are designed with shallow culture depths (of ~ 20 cm) and operated at low biomass densities to ensure adequate sunlight penetration. Shallow culture depths translate to large footprint and hence high capital costs. On the other hand, low biomass densities lead to high biomass harvesting costs. Another shortcoming of HRAPs is the loss of ammoniacal-N by volatilization at the typical operating alkaline conditions (pH > 8). HRAPs suffer also from odor emissions, contamination, invasion, and evaporative water loss. In the case of HRAPs that are sparged with gaseous CO_2, the gas—liquid transfer is poor because of the short bubble detention time at the shallow depth, resulting in atmospheric loss of CO_2.

1.4.1 Mixotrophic sewage treatment and resource recovery system

In stark contrast to the current phototrophic algal sewage treatment systems, the recently introduced STaRR system utilizes a mixotrophic algal strain, *Galdieria sulphuraria*. The STaRR system consists of the following core subsystems: (i) a mixotrophic cultivation system of *G. sulphuraria* for single-step removal of BOD, N, P, and pathogens from primary effluent; (ii) hydrothermal liquefaction (HTL) of the resulting algal biomass; and (iii) processing of the products of HTL to harvest energy as liquid fuel and nutrients as fertilizers. In essence, the STaRR system can be categorized under the partition—release—recover approach outlined earlier. The partitioning is achieved mixotrophically meeting the discharge standards for N, P, and BOD in a single step concentrating them into algal biomass with minimal energy input and without any losses; the release is achieved through HTL (instead of anaerobic digestion) to yield products that can self-separate under gravity; and recovery of nutrients via precipitation and separation technologies. Fig. 3.1 depicts a schematic of the above process integrated in the STaRR system.

1.4.2 STaRR system versus high-rate algal ponds

The algal cultivation/sewage treatment step in the STaRR system is engineered to circumvent many of the shortcomings of the HRAP system. The two systems differ from each other in several aspects; fundamentally, the HRAPs rely on the symbiosis between heterotrophic bacterial and photoautotrophic algal communities, while the STaRR system relies primarily on mixotrophic algal cultures. The phototrophic HRAP system is solely dependent on sunlight as a source of energy for photosynthesis; the mixotrophic STaRR system, on the other hand, is able to utilize organics and/or sunlight as energy sources. As such, the STaRR system does not suffer from diurnal fluctuations of growth and pollutant removal rates.

Several studies have reported that mixotrophic cultures are capable of higher growth rates than the phototrophs for faster removal of the pollutants (Marquez et al., 1993; Ogawa and Aiba, 1981). Volumetric removal rates (VRRs) of organic carbon, ammoniacal nitrogen, and phosphates by the STaRR system have been shown to be comparable

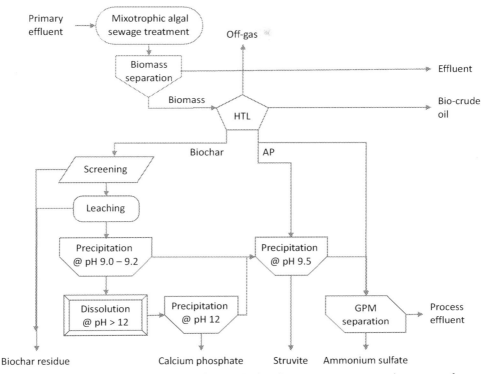

Figure 3.1 Schematic of the STaRR system for tertiary-level sewage treatment and recovery of energy as biocrude and nutrients as fertilizers. *GPM*, gas-permeable membrane; *HTL*, hydrothermal liquefaction.

to or better than those reported in the literature for HRAPs (Nirmalakhandan et al., 2019). It has also been reported that mixotrophs accumulate more lipids than the phototrophs (Cheirsilp and Torpee, 2012; Li et al., 2014; Liang et al., 2009). Higher lipid content can be beneficial in higher energy recovery, reducing the carbon footprint of POTWs. The pH in HRAPs is typically alkaline (pH > 8), while that in the STaRR system is acidic (pH = 2.5–4.0). The acidic environment is beneficial in resisting invasion and contamination by competing organisms. More importantly, unlike in the alkaline phototrophic systems where more than 75% of the ammonia content of the sewage is lost by volatilization (Martinez et al., 2000), the acidic STaRR system retains ammonia in the dissolved form for higher biomass production and maximize downstream recovery. The low pH can also contribute to inactivation of native organisms of concern in the sewage minimizing postdisinfectant demand. Salient features and performance characteristics of the two systems are summarized in Table 3.2.

Table 3.2 Comparison of salient features and performance characteristics of High-rate algal ponds (HRAPs) and the sewage treatment and resource recovery (STaRR) system.

Parameter	HRAP	STaRR	Refs.
Metabolism	Microbial heterotrophy + algal phototrophy	Algal mixotrophy	Nirmalakhandan et al. (2019)
Reactor design	Open raceway - Odor emissions - Evaporative losses - Potential for invasion	Enclosed raceway - No odor emissions - No evaporative losses - Minimal invasion	Shen et al., (2009), Richmond (2004)
Environment	Alkaline, pH > 8 - Potential for disinfection - Volatilization loss of NH_3-N - Precipitation of PO_4	Acidic, pH < 4 - Potential for disinfection - No volatilization loss of NH_3-N - No precipitation of PO_4	García et al. (2000), Nirmalakhandan et al. (2019), Young et al. (2016)
VRR of N	4.85 ± 3.92 mg/L-d[a]	6.02 ± 0.90 mg/L-d	Nirmalakhandan et al. (2019), Richmond (2004)
VRR of P	0.50 ± 0.64 mg/L-d[b]	1.38 ± 0.57 mg/L-d	
VRR of BOD	17.33 ± 15.85 mg/L-d	16.51 ± 3.23 mg/L-d	
HHV of resulting biomass	$18-24$ MJ/kg	24.2 ± 0.3 MJ/kg	Ceron Garcia et al. (2006), Shen et al. (2009)

HHV, higher heating vale; *VRR*, volumetric removal rate.
[a] Includes loss by volatilization.
[b] Includes loss by precipitation.

2. Details of the STaRR system

2.1 Mixotrophic sewage treatment by *Galdieria sulphuraria*

Feasibility of mixotrophic growth of *G. sulphuraria* fed with urban sewage was demonstrated for the first time by Selvaratnam et al. (2014) and Henkanatte-Gedera et al. (2015). Based on laboratory treatability studies, a pilot scale version of the STaRR system has been developed and deployed at the local sewage treatment plant (Las Cruces, NM, USA). Cultures expanded from axenic culture grown under controlled conditions were preadopted and cultivated in three 700 L bioreactors of the STaRR system, fed with primary effluent. The influent concentrations of BOD, NH_3-N, and $PO_4^{3-}-P$ to the reactor varied seasonally from 38 to 70 mg/L, $25-38$ mg/L, and $6-11$ mg/L, respectively (Henkanatte-Gedera et al., 2015, 2017). Upon reduction of BOD, N, and P to their respective local discharge levels (of 30 mg/L, 10 mg/L, and 1 mg/L), the culture medium was gravity settled, and the resulting algal biomass was centrifuged in 6 L batches at 10,000 rpm (18,000 RCF) for 5–10 min to a solid content of 15% by weight to generate the feed for the subsequent HTL process.

Samples from the cultivation system were analyzed to assess the reductions of N, P, BOD, bacteria, and viruses following the analytical procedures detailed in previous reports (Delanka-Pedige et al., 2019, 2020b; Tchinda et al., 2019). In the case of N, ammoniacal-nitrogen was the major constituent in the primary effluent at the test facility, with negligible levels of nitrates (<1 mg/L). Ammoniacal-nitrogen and phosphates were analyzed using an HACH DR 6000 spectrophotometer with salicylate method 8155 and phosver3 method 8048, respectively. Quantification of dissolved BOD_5 followed Standard Method 5210 B. Details of the procedures for assessing the performance of the STaRR system in sewage treatment have been reported elsewhere (Tchinda et al., 2019).

Fecal and total coliforms were assayed by membrane filtration technique using m FC and m Endo agar media (Difco, Becton Dickinson, Cockeysville, MD), respectively. Somatic and F-specific coliphages were enumerated using double agar layer method (US EPA method 1602). Incidences of total bacteria, several pathogenic bacteria, and enteric viruses of concern in the effluent were assessed by real-time PCR (qPCR) technique. Target genes of the bacteria and viruses of concern extracted from effluent samples were amplified and quantified in a Bio-Rad CFX real-time PCR system. In the case of virus detection, viruses in effluent samples were concentrated prior to genes extraction according to a previous method by Zhang et al. (2013). Detailed procedures of the microbial assessments have been documented elsewhere (Delanka-Pedige et al., 2019, 2020b).

2.2 Hydrothermal liquefaction of algal biomass

The STaRR system includes HTL to recover energy and nutrients from the biomass resulting from the cultivation/sewage treatment step. HTL was selected for the STaRR system because of its advantages over other biomass conversion technologies such as anaerobic digestion, combustion, and pyrolysis: faster reaction time; ability to process wet biomass to convert its lipid-, protein-, and carbohydrate contents to biocrude; ability of the byproducts to self-separate; and ability to recover the nutrients as high-quality fertilizers (Shanmugam et al., 2017). Studies by US Department of Energy (US DOE, 2017) and Maddi et al. (2017) have identified HTL as an appropriate technology for processing algal biomass as an alternative to the classical anaerobic digestion process.

The STaRR system included a 1.8 L HTL reactor (Parr Instrument Company, Moline, IL) to process the *G. sulphuraria* algal biomass cultivated in the bioreactor. Typically, the HTL was operated in batch mode, fed with 500 g biomass slurry, and pressurized to approximately 1.4 MPa with gaseous nitrogen before heating. Performance of the HTL step has been evaluated as a function of temperature; holding time; and algal solids content in the feed. Upon completion of the reaction, 200 mL hexane was added to the reactor, the contents emptied into a beaker, and three more 50 mL hexane reactor rinses added to the mixture. The char was removed from the liquid phases by filtration (11 μm pore size) and rinsed with dichloromethane to remove residual biocrude oil prior to drying overnight at room temperature. The aqueous phase was separated from the organic phase with a separatory funnel. The procedure adopted for the HTL process has been detailed elsewhere (Cheng et al., 2018, 2019).

2.3 Characterization of HTL byproducts

Samples of the HTL aqueous phase (AP) were analyzed for total nitrogen (TN) and total phosphorus (TP). Inorganic ammoniacal nitrogen (NH_3-N) and phosphate ($PO_4^{3-}-P$) were measured in all samples. Analyses of TN, TP, NH_3-N, and $PO_4^{3-}-P$ followed the persulfate digestion method 10072, molybdovanadate method 10127 and ascorbic acid method 8190, salicylate method 10031, and acid method 8048, respectively (Hach Company, Loveland, CO). Aqueous phase samples were also analyzed using a Dionex ICS-2100 Ion Chromatographic System (ICS) with an AS-DV auto sampler to quantify other anion nitrogen forms.

Biochar samples were analyzed for bioavailable P following the Olsen method (Smith, 2009). Ammonium molybdate and antimony potassium tartrate react in an acid solution with orthophosphate to form an antimony—phosphomolybdate complex. This complex is reduced by ascorbic acid to a blue-colored complex that is proportional to the P concentration and is quantified using a spectrophotometer (Thermo Spectronic 20). Biochar samples were analyzed for total phosphorus (TP) using microwave digestion with HNO_3 and H_2O_2 in a CEM MARS 5 microwave reactor system and quantified by inductively coupled plasma optical emission spectrometry (ICP-OES) (Perkin Elmer Optima 4300 DV) (EPA Method 200.7). Mass balance on P was completed by measuring the total P content of the HTL biocrude oil samples by ICP. A more general mass balances was completed by analyzing the char and biocrude oil samples for elemental CHNS using a Series II 2400 elemental analyzer (PerkinElmer, Waltham, MA). Details of tests used for characterizing HTL byproducts have been reported elsewhere (Abeysiriwardana-Arachchige et al., 2020; Karbakhshravari et al., 2020).

2.4 Leaching of phosphates from biochar

As most of the P in the algal biomass partitions during the HTL process into the biochar, a leaching process was developed to solubilize the P first for subsequent recovery by precipitation (Karbakhshravari et al., 2020). Acidic and alkaline leaching conditions were evaluated, with different concentrations of hydrochloric acid (HCl) and sodium hydroxide (NaOH) as a function of extraction temperature and extraction time. Polypropylene centrifuge tubes (15 mL) containing 0.01 g char were mixed with 10 mL of the extractant for a solid/liquid ratio of 1 g/L. These suspensions were equilibrated by end-over-end rotation at 10 rpm for 24 and 72 h. Upon equilibration, the eluate was separated by centrifugation at 6000 rpm (5500 RCF) for 5 min and analyzed for PO_4^{3-} concentrations recovered in the eluate.

2.5 Recovery of phosphate from eluate of biochar

A three-step procedure has been developed to recover high-purity P-fertilizers from the P-rich eluate of biochar (Abeysiriwardana-Arachchige et al., 2021). The first step in this procedure is to precipitate impurities along with some P in the eluate. The second step is to precipitate high-quality struvite from the supernatant of the first step. The third step is to dissolve the precipitate from the first step and reprecipitate its P-content as high-quality calcium phosphate. Struvite and calcium phosphate recovered by this three-step procedure have been evaluated by the following methods to ascertain their chemical purity and compliance with US EPA guidelines for land application: scanning electron microscope (SEM) imaging and energy dispersive X-ray (EDX) analysis of the precipitates using a model S—3400II SEM equipped with an EDX microanalysis system (Noran System Six 300, Thermo-Electron Corp., Madison, WI). Crystalline structures of the three precipitates have been analyzed using X-ray powder diffraction on a PANalytical Empyrean diffractometer which excited samples with Cu Kα radiation ($\lambda = 1.5406$ Å) at a 45 kV anode voltage and 40 mA beam current. The patterns obtained from XRD analysis have been compared to the reference data in the International Center for Diffraction Data (ICDD) database and literature.

2.6 Recovery of nitrogen from aqueous product of HTL

A tubular gas-permeable membrane reactor (GPMR) has been developed for inclusion in the STaRR system to recover N from the N-rich aqueous HTL product (Munasinghe-Arachchige et al., 2020c, 2021). In essence, the GPMR is composed of a tubular hydrophobic gas-permeable membrane (GPM) immersed in the N-rich feed, while sulfuric acid is circulated through the tube. By adjusting the pH of the feed to a value greater than the pK_a value of 9.26 bulk of the N in the feed dissociates to the $NH_3(g)$ form, which preferentially diffuses through the gas-filled pores in the wall of the tubular GPM to its inside, where it is instantaneously converted to ammonium sulfate by the reaction with the sulfuric acid. The transmembrane concentration gradient maintained in this manner enables spontaneous recovery of $NH_3(g)$ from the feed under ambient pressures and temperatures (Munasinghe-Arachchige et al., 2020b).

3. Performance of the STaRR system

3.1 Sewage treatment

Results from long-term stable operation of the pilot scale version of the STaRR system in batch and fed-batch modes have been documented (Abeysiriwardana-Arachchige and

Nirmalakhandan, 2019; Delanka-Pedige et al., 2020a; Henkanatte-Gedera et al., 2017; Tchinda et al., 2019). Performance of this system in removing N, P, and BOD to near-tertiary levels in a single step to yield discharge-ready effluent is summarized as follows.

3.1.1 Nitrogen removal

Nitrogen in the primary effluent at the test facility was essentially in the ammoniacal form with negligible nitrate levels (<1 mg/L). The STaRR system achieved discharge standard for NH_3-N (10 mg/L) in 2–3 days of fed-batch processing, irrespective of the fluctuations of the influent NH_3-N concentration (initial concentration $= 23.7 \pm 3.5$ mg/L). This reduction could be approximated by a first order rate of 0.505 day^{-1}; considering the average influent concentration and the first order rate constant, the fed-batch process time required to meet the discharge standard of 10 mg/L could be estimated as 1.71 days (Abeysiriwardana-Arachchige and Nirmalakhandan, 2019).

This N-removal rate constant of *G. sulphuraria* is comparable to or better than those reported in the literature (0.07–0.90 day^{-1}) for other algal species, all of which were from laboratory studies (Abeysiriwardana-Arachchige and Nirmalakhandan, 2019). VRR of NH_3-N in STaRR (6.1 ± 0.9 mg/L-d) is higher than that reported in the literature (4.8 ± 3.9 mg/L-d) for HRAPs (Nirmalakhandan et al., 2019). However, the reductions of NH_3-N in the alkaline HRAPs reported in the literature include abiotic losses. It has been reported that loss of NH_3-N by volatilization in HRAPs can be as high as 75%, which significantly reduces the potential for recovery of N from the resulting algal biomass (Martınez et al., 2000). In contrast, since the *G. sulphuraria*- based STaRR system is acidic (pH $= 2.5–4.0$), volatilization of NH_3-N is minimal, and the reduction reported above is entirely biotic. As such, the STaRR system affords opportunities for higher levels of N-recovery.

3.1.2 Phosphate removal

The STaRR system achieved the discharge standard for phosphate (1 mg/L) in 2–3 days of fed-batch processing, irrespective of the fluctuations of the influent phosphate concentration (initial concentration $= 4.6 \pm 1.8$ mg/L). This reduction could be approximated by a first order rate of 0.663 day^{-1}; considering the average initial concentration and the first order rate constant, the fed-batch process time required to meet the discharge standard of 1 mg/L could be estimated as 1.66 days (Abeysiriwardana-Arachchige and Nirmalakhandan, 2019).

This P-removal rate constant of *G. sulphuraria* is comparable to or better than those reported in the literature (0.07–0.90 day^{-1}) for other algal species, all of which were from laboratory studies (Abeysiriwardana-Arachchige and Nirmalakhandan, 2019). VRR of phosphate in the STaRR system (6.1 ± 0.9 mg/L-d) is higher than that reported in the literature (4.8 ± 3.9 mg/L-d) for HRAPs (Nirmalakhandan et al., 2019). Compared to the current practice of chemical precipitation, the STaRR system offers better opportunities for recovering P from the algal biomass.

3.1.3 BOD removal
The STaRR system achieved the discharge standard of 30 mg/L in less than 2 days, irrespective of the fluctuations of the influent (initial concentration = 54.8 ± 8.7 mg/L). This reduction could be approximated by a first order rate of 0.311 day^{-1}. Considering the average initial concentration and the first order rate constant, the fed-batch process time required to meet the discharge standard of 30 mg/L could be estimated as 0.73 days.

3.1.4 Comparison of STaRR system with other technologies
Sewage treatment performance of the STaRR system is comparable of better than that of the more common HRAPs and tertiary treatment systems. While a rigorous comparison is not feasible because of wide variance in the literature reports (strains used, types of wastewaters assessed, the influent concentrations, modes of operation, the types of analysis followed, and the data reported), a holistic comparison has been made on the basis of VRRs of N, P, and BOD irrespective of mode of operation (batch, semicontinuous, continuous) and environmental conditions (temperature, season, location). Fig. 3.2 summarizes these comparisons based on data on STaRR and HRAP systems detailed in Nirmalakhandan et al. (2019) and on the tertiary treatment systems reported in (US EPA, 2008). Average VRRs of NH$_4$–N (6.09 ± 0.92 mg/L-d) and of BOD$_5$ (16.5 ± 3.6 mg/L-d) in the STaRR system were found to be not different from the

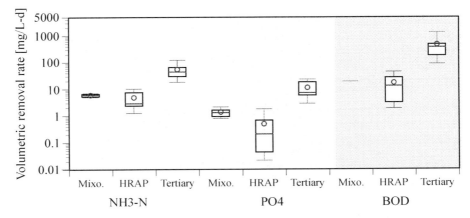

Figure 3.2 Comparison of volumetric removal rates of ammoniacal-N, phosphates, and BOD in *Galdieria sulphuraria*—based STaRR system (STaRR), high-rate algal ponds (HRAP), and tertiary treatment systems (Tertiary).

averages reported for the photoautotrophic HRAPs at a significance level of 0.05. Average VRR of PO_4 (1.40 ± 0.57 mg/L-d) was found to be greater than that in HRAPs at a significance level of 0.05).

Since VRR is calculated from the initial and final liquid phase concentrations of N and P, the above values include their removals by algal uptake as well as by other abiotic mechanisms. As mentioned earlier, volatilization is the most significant abiotic mechanism of N reduction in the liquid; this loss increases with pH and temperature. Since the pH in HRAPs is typically greater than 9, significant amounts of ammonia are lost to the atmosphere; In contrast, as the operating pH in the *G. sulphuraria*-based STaRR system is 4.0, most of the recorded N-reduction is attributed to algal uptake. Maximizing algal uptake can increase biomass production, a desirable outcome if downstream energy- and nutrient recovery are integrated with wastewater treatment (Park et al., 2011). Atmospheric pollution by ammonia emission is also minimized at this low pH. Similarly, the above VRR values for PO_4 also include removal by algal uptake as well as by other abiotic mechanisms. Abiotic reduction of P in the liquid phase could be due to chemical precipitation with polyvalent cations such as calcium or physical adsorption to biomass and calcium carbonate crystals, both of which increase with pH (Nurdogan and Oswald, 1995). Since the pH in HRAPs is typically greater than 9, removal of P by precipitation contributes significantly to VRR. As the operating pH in the *G. sulphuraria*-based STaRR system is < 4.0, most of the P-reduction reported here is associated with algal uptake. Maximizing algal uptake can increase biomass production, again a desirable outcome if downstream energy- and nutrient recovery are integrated with wastewater treatment.

3.2 Bacteria and virus removal

The inclement culture conditions in the STaRR system have been found favorable in contributing to inactivation of bacteria and virus simultaneous to reduction of N, P, and BOD. Comparisons of the inactivation performance of the STaRR system relative to literature data on traditional secondary treatment systems and phototrophic algal systems, and the emerging MBR systems have confirmed the superiority of the STaRR system.

3.2.1 Bacteria removal

The STaRR system was able to reduce bacterial indicators to nondetectable levels within a day of fed-batch processing. Complete removal of *Enterococcus faecalis* and *Escherichia coli* was recorded at the end of fed-batch processing in the STaRR system (Delanka–Pedige et al., 2019, 2020c). Log removal values of fecal coliform, total coliform, total bacteria, *Enterococcus faecalis*, and *Escherichia coli* in the STaRR system are compared in Fig. 3.3 with those in the parallel system consisting of a trickling filter followed by AS, both fed with the same primary effluent. As can be deduced from Fig. 3.3, the single-step STaRR

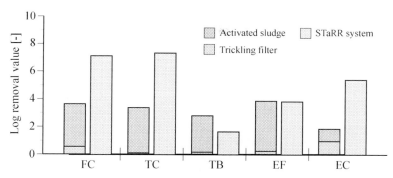

Figure 3.3 Comparison of log removal values of fecal coliform (FC); total coliform (TC); total bacteria (TB); *Enterococcus faecalis* (EF); and *Escherichia coli* (EC) in trickling filter followed by activated sludge system versus *G. sulphuraria*-based STaRR system, both fed with the same primary effluent.

system affords higher log removals than the traditional secondary system (with simultaneous BOD-, N-, and P-reductions). This benefit can translate to significant reduction of postdisinfection demand. This superior inactivation performance of the STaRR system is attributed to the inclement culture conditions in the algal system—low pH (2–4), high DO (6.26 ± 0.53 mg/L), above ambient temperature (27–46 °C), and exposure to sunlight (Munasinghe-Arachchige et al., 2019).

3.2.2 Virus removal

Somatic coliphages that infect host cells via the membrane were reduced by 3.13 logs in 3 days of fed-batch processing; at the same time, F-specific coliphages that infect host cells via pili were reduced by 1.23 logs. Among six different enteroviruses, only *Enterovirus* and *Norovirus GI* were detected in the primary effluent; the STaRR system was able to reduce them by 1.05 and 1.49 logs. Log removal values of somatic coliphages, F-specific coliphages, *Enterovirus* (EV), and *Norovirus GI* (NV) in the algal system are compared in Fig. 3.4 against those in secondary treatment (trickling filter followed by AS) with postchlorination. The notable performance shown in Fig. 3.4 in outranking the traditional practice in virus inactivation without the use of any disinfectants adds further credence to the STaRR system (Delanka-Pedige et al., 2020b).

3.2.3 Comparison of STaRR system with other technologies

Inactivation results recorded by the STaRR system are compared in Fig. 3.5 with data reported in the literature for typical secondary treatment systems (i.e., AS and trickling filter followed by AS), an emerging one-membrane bioreactor (MBR) system, and the conventional phototrophic algal sewage treatment system—the HRAP. When compared to the phototrophic HRAP systems, the STaRR system attained better or comparable removal of *E. coli* and fecal coliforms irrespective of the design configuration,

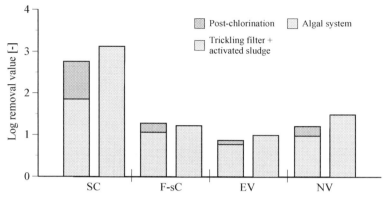

Figure 3.4 Comparison of log removal values of somatic coliphages (SC); F-specific coliphages (F-sC); *Enterovirus* (EV), and *Norovirus GI* (NV) in trickling filter followed by activated sludge system with post-chlorination versus *G. sulphuraria*-based STaRR system without any chlorination, both fed with the same primary effluent.

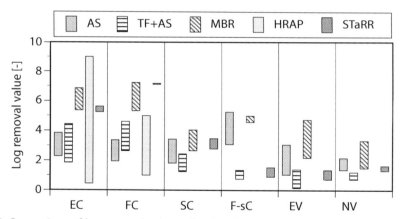

Figure 3.5 Comparison of log removal values of *Escherichia coli* (EC); fecal coliform (FC); somatic coliphages (SC); F-specific coliphages (F-sC); *Enterovirus* (EV) and *Norovirus GI* (NV) in activated sludge (AS), trickling filter followed by activated sludge (TF + AS), membrane bioreactors (MBR), high-rate algal pond (HRAP) versus *G sulphuraria*-based STaRR system.

hydraulic retention time, algal species, etc. The STaRR system was superior in removing *E. coli*, fecal coliforms, and somatic coliphages compared with the conventional AS-based sewage treatment systems, and the existing system in the City of Las Cruces, NM, US (Delanka-Pedige et al., 2020f).

Removal of bacterial indicators in the STaRR system was comparable with that reported for the emerging MBR systems. The reduction of F-specific coliphages, *Enterovirus*, and *Norovirus GI* in the algal systems was lower compared to MBR and traditional

AS-based systems, while higher/comparable with respect to existing trickling filter plus AS system. Even though rigorous comparisons among the alternate sewage treatment systems are not possible due to the wide variance in the metadata underlying the literature reports, this comparison presents a generic assessment of the STaRR system relative to the more familiar ones.

3.3 Energy recovery

3.3.1 Recovery of biocrude oil

Galdieria sulphuraria biomass cultivated in the STaRR system fed with primary effluent served as the feedstock to the HTL process that enabled recovery of energy and nutrients from its products. Higher heating value (HHV) of this feed biomass averaged 24.2 ± 0.3 MJ/kg; while its lipid-, protein-, and carbohydrate contents on % ash-free dry weight basis averaged 0.1 ± 0.0, 48.5 ± 1.0, and 7.9 ± 0.4, respectively (Fig. 3.6a). HTL product yields from the sewage-generated biomass are summarized and compared with typical literature results in Fig. 3.6b. HHVs of the feed biomass, biocrude oil, and biochar are summarized in Fig. 3.6c.

3.3.2 Comparison of STaRR system with other technologies

Fig. 3.6 also shows comparisons of the above results with literature data (Cheng et al., 2017, 2018, 2019). While the biochemical composition of algal biomass can vary depending on the species and the culturing medium and conditions and the HTL product characteristics can vary with the HTL operating conditions, this comparison is presented to assess the relative merit of the STaRR system. As can be noted from Fig. 3.6A, lipid content of *G. sulphuraria* biomass from the STaRR system is significantly less than the average literature values ($14.1 \pm 11.4\%$), while its biocrude yield is comparable to the average of the literature values ($35.98 \pm 8.57\%$). The HHV of the biocrude resulting from *G. sulphuraria* biomass (37.65 ± 1.80 MJ/kg) is about 10% more than the average of the literature values (34.53 ± 2.13 MJ/kg). ICP analysis of biocrude oil samples indicated negligible P levels in the biocrude oil; as such, the biocrude oil was ignored in the P material balance calculations presented later. Detailed characterization of the crude oil derived from sewage-cultivated *G. sulphuraria* can be found in (Cheng et al., 2017, 2018, 2019).

3.4 Nutrient recovery

HTL of the algal biomass cultivated in the STaRR system yields two nutrient-rich byproduct streams—a nitrogen-rich aqueous phase and a phosphate-rich solid phase, biochar; N and P concentrations in the aqueous phase and biochar indicated that most of the N and P in the primary-settled sewage assimilated by the algal biomass could be

70 | Integrated Wastewater Management and Valorization using Algal Cultures

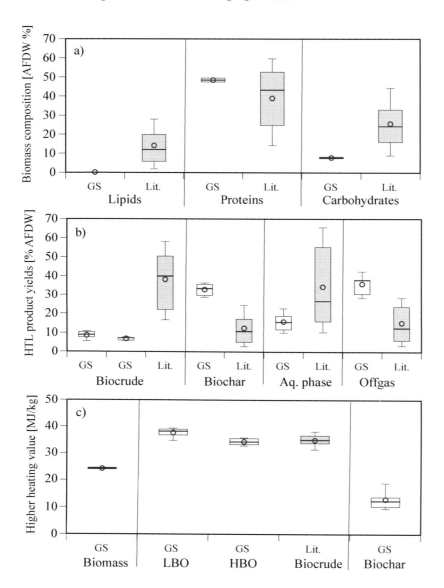

Figure 3.6 (A) Composition of *Galdieria sulphuraria* biomass cultivated in primary effluent; (B) product yields by hydrothermal liquefaction of *Galdieria sulphuraria* biomass cultivated in primary effluent; (C) higher heating value of *Galdieria sulphuraria* biomass cultivated in primary effluent and of products of hydrothermal liquefaction. *GS*, *G. sulphuraria* in STaRR system; *HBO*, heavy biocrude oil; *LBO*, light biocrude oil; *Lit.*, Literature data, details included as supplemental information.

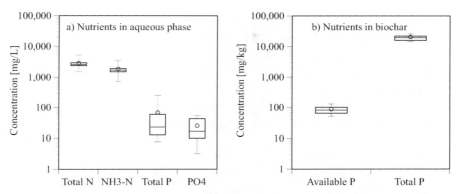

Figure 3.7 (A) Nutrient-content of aqueous phase of HTL of *G. sulphuraria* biomass; (B) Nutrient-content of biochar of HTL of *G. sulphuraria* biomass.

released through HTL, offering a promising pathway for recovery of these nutrients for beneficial reuse. Nitrogen levels in biochar were found to be insignificant. Nutrient contents of the aqueous phase and biochar are summarized in Fig. 3.7.

3.4.1 Aqueous phase characterization

Yield of aqueous phase resulting from HTL of *G. sulphuraria* biomass cultivated in the STaRR system fed with primary effluent averaged 82%. Total N was nearly the same in filtered and unfiltered samples of the aqueous phase (1400–5700 mg/L) indicating that the N in the aqueous phase is primarily in the dissolved form. Nearly 50% of the N in the aqueous phase is in the form of ammoniacal nitrogen (NH_3-N), ranging from 600 to 4100 mg/L with up to 20 mg/L of nitrate, and negligible amounts of nitrite. Speciation of N in the aqueous phase was found to depend on the solids content in the feed to the HTL process and the process temperature and hold time. Total phosphorus in the filtered samples of the aqueous phase ranged from 1.7 to 76 mg/L, consisting mainly of PO_4^{-3}.

3.4.2 Biochar characterization

Yield of biochar resulting from HTL of *G. sulphuraria* biomass cultivated in the STaRR system fed with primary effluent averaged 30% by weight. Bioavailable-P was less than 200 mg/kg of P for all the samples, whereas total P content determined by microwave digestion was as high as 42,730 mg/kg of P (4.3 wt.%). These results indicate that most of the P (in excess of 95%) is incorporated into the biochar. This necessitated a leaching process to solubilize the phosphates for recovery as struvite.

The leaching process for solubilizing P in biochar has been optimized as a function of leaching time, mixing during leaching, concentration of base eluent, liquid:solid (L:S) ratio in leaching, and temperature (Abeysiriwardana-Arachchige et al., 2021).

Considering the efficiency and cost of leaching, the following have been identified as optimal: leaching time of 72 h with mixing; base concentration of 0.5 M NaOH; L:S ratio of 20:1; and a leaching temperature of 60°C. Under these optimal conditions, >86% of P in the biochar could be extracted at an estimated cost of $4.54/kg P (Abey-siriwardana-Arachchige et al., 2021). The leaching time of 72 h was selected to synchronize the leaching process with the fed-batch cycle time required for the upstream sewage treatment step. This optimization procedure assumed that solar heating could be utilized to raise the temperature of the feed to the leaching temperature of 60°C. Under the above optimal conditions, average concentrations of NH_3-N and PO_4^{3-} in the eluate were 300.3 ± 18.7 mg/L and 4826.8 ± 34.3 mg/L, respectively; and its pH was 12.7 ± 0.1.

3.4.3 Recovery of P

The three-step procedure for P-recovery from the above eluate was developed to yield high-purity struvite and to maximize the recovery. The first step involves adjustment of pH to a range of 9.0—9.2 whereby most of the impurities as well as about 24% of the P in the eluate are precipitated. The supernatant from the first step is then mixed with a fraction of the aqueous phase and dosed with magnesium chloride to attain a P:N:Mg ratio of 1:1.5:1.5 and the pH of the mixtures is adjusted to 9.5 to precipitate struvite crystals. Based on mass balance, it has been estimated that >72% of P in the biochar eluate could be recovered as struvite in the second step. The third step entailed dissolution of the precipitate from the first step and reprecipitating its P-content as calcium phosphate by adding calcium chloride at pH > 12. This three-step process enabled 95.4% recovery of P from the biochar eluate ($72.7 \pm 1.0\%$ as struvite and $22.7 \pm 1.2\%$ as calcium phosphate). Considering the leaching efficiency of 86.1%, the above recovery efficiencies translate to overall P-recovery from algal biomass via HTL of 82.2%. Based on the price of $MgCl_2.6H_2O$ as $0.08/kg (Huang et al., 2014) and that of CaCl2 as $0.22/kg (Intratec, 2007), the total unit cost of recovering P by this three-step procedure has been estimated at $5.29/kg P, which is 21% lower than that estimated in Karbakhshravari et al. (2020) for recovering struvite alone. Given the market price of common P-fertilizers ($1.65—7.07/kg P), P-recovery by the STaRR could be economically feasible, when its technical and ecological benefits are taken into consideration.

Fig. 3.8 shows the concentrations of nine regulated heavy metals in struvite and calcium phosphate recovered by the STaRR system relative to their respective ceiling concentrations as per EPA/832/R-93/003 for sewage-derived biosolids (US Environmental Protection Agency, 1994). Heavy metal concentrations in both struvite and calcium phosphate were below the limits for all nine metals. Heavy metal contents of these precipitates were also 10—300 times lower compared to the regulatory limits set by Directive 86/278/EEC (Collivignarelli et al., 2019) for the European region. Detailed

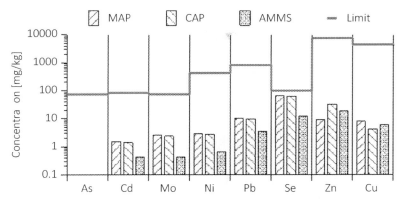

Figure 3.8 Heavy metals of concern in struvite and calcium phosphate samples. The limit corresponds to the ceiling for land application of sewage-derived biosolids in the United States.

characterization of struvite and calcium phosphate recovered from sewage-cultivated *G. sulphuraria* in the STaRR system can be found in (Abeysiriwardana-Arachchige et al., 2021).

3.4.4 Recovery of N

The GPMR developed for recovering N as ammonium sulfate from the N-rich process streams in the STaRR system (Fig. 3.1) has been optimized in terms of product-side pH, acid circulation rate, feed-side pH, and the degree of mixing on the feed side. Considering N-recovery, the following conditions have been identified as optimal: product side pH of 2.0; acid circulation rate of 25 mL/min; feed-side pH of no more than 10; and moderate mixing of the feed side.

Ammonia removals of 80%–100% have been recorded with 92%–100% of recovery in the product side, with minimal fugitive losses. Mass removal rates of 0.08–1.52 g N/d and transmembrane flux (N removal rate per square meter of the effective membrane area) of 29.2–300.1 g N/m^2-d have been recorded, yielding 2–86 g of ammonium sulfate from 1 L of the feed. Chemical and microscopic analysis of the recovered ammonium sulfate crystals confirmed their purity and compliance with the US EPA limits for land application as a fertilizer (Fig. 3.8). Microscopic images of the used membrane indicated minimal membrane fouling that could diminish the transmembrane flux. Detailed characterization of the ammonium sulfate recovered from sewage-cultivated *G. sulphuraria* in the STaRR system can be found in Munasinghe-Arachchige et al. (2021).

3.4.5 Comparison of STaRR system with other technologies

Performance of the STaRR system in recovering N and P from primary effluent as struvite has been compared with that of existing and emerging technologies reported in the literature. Nutrient removal per unit energy input in the STaRR system is estimated as 257.1 g N/kWh and 36.6 g P/kWh, while that in eight full-scale POTWs averaged 74.3 g N/kWh and 135.1 g P/kWh (Abeysiriwardana-Arachchige and Nirmalakhandan, 2019). Energy required to treat primary effluent in the STaRR system (531.5 kWh/MG) is estimated to be lower than the average in the eight POTWs (1037.9 ± 503.3 kWh/MG). While existing technologies had been originally designed for removal of nutrients rather than any recovery, a review of the literature revealed 12 emerging technologies for recovering nutrients. Performance of the STaRR system in nutrient recovery (5.9% N and 71.6% P) has been shown to be superior to that of those 12 emerging technologies (Abeysiriwardana-Arachchige et al., 2020). Recoveries recorded in the STaRR system translate to a yield of 1.62 kg struvite and 0.34 kg calcium phosphate.

Performance of the STaRR system in recovering N from primary effluent as ammonium sulfate via the GPMR has been evaluated against literature studies in terms of NH_3 transfer flux. Fluxes estimated from literature studies that had used ePTFE membranes ranged $2.0-89.0$ g N/m^2-d; in comparison, the highest value recorded in the STaRR system (300.1 g N/m^2-d in Case L) is more than three times the highest reported value. High flux attained in the GPMR is attributed to the reduced feed-side mass transfer resistance achieved by continuous mixing and to the initial pH adjustment of the feed to 10. Among the reported studies, Garcia-González and Vanotti (2015) mixed the feed using a magnetic stirrer as in the current study, but the transmembrane flux was considerably lower ($2.0-4.5$ g N/m^2-d), likely due to the low pH ($7.5-8.2$) maintained in their study ($30-44$ days). Most of the previous studies incorporated mild aeration to increase feed-side pH (Daguerre-Martini et al., 2018; Dube et al., 2016; García-González et al., 2015; Riaño et al., 2019; Vanotti et al., 2017); in the case of Dube et al. (2016) and García-González et al. (2015), nitrification inhibitors were used to avoid biological loss of NH_4^+. In contrast, in the STaRR system, the final pH values on the feed side were above 9.26 ensuring abundance of $NH_3(g)$ throughout the experiment, without the need for continuous pH adjustment or any inhibitors. Although NaOH was used in the STaRR system for pH adjustment, slaked lime ($Ca(OH)_2$) could be used as a cost-effective alternative. In any case, readjustment of the pH of the treated effluent might be required depending on the discharge standards or any downstream processing requirements.

4. Outlook

Technical performance of the STaRR system indicates that it can potentially be a resource-efficient alternative to the current sewage treatment practice and other emerging alternatives. Comparisons of the STaRR system with current sewage treatment alternatives based on multicriteria evaluations have affirmed that the STaRR system could be a more sustainable pathway. A recent study reported on a multicriteria comparison of the following sewage treatment systems: AS treatment followed by anaerobic digestion; photoautotrophic algal treatment followed by HTL or by anaerobic digestion; and the STaRR system including HTL or anaerobic digestion. This evaluation based on 15 criteria derived from the United Nation's Sustainable Development Goals (SDGs) ranked the STaRR system as the most preferred one (Munasinghe-Arachchige et al., 2020c).

Another multicriteria comparison of technologies for recovering nitrogen from anaerobically digested domestic sludge had ranked the GPM incorporated into the STaRR system above the available technologies such as air-stripping, ultrafiltration/ion exchange, struvite precipitation, and ultrafiltration/reverse osmosis (Munasinghe-Arachchige and Nirmalakhandan, 2020).

The manner in which the STaRR system could contribute toward meeting the four targets of Goal # 6 of the SDGs of improving water quality, wastewater treatment, and safe reuse; increasing water use efficiency and ensuring freshwater supplies; implementing water reuse management; and protecting and restoring water-related ecosystems has been reported (Delanka-Pedige et al., 2020e). Additionally, potential of the STaRR system in contributing toward most of the other 16 SDGs directly or indirectly has also been pointed out. In a similar vein, a multicriteria comparison of alternate sewage treatment infrastructure systems for meeting the UN SDG for sustainable cities (SDG #11) based on 36 sustainability-related process parameters ranked the STaRR system as the preferred one (Delanka-Pedige et al., 2020d).

Based on the documented technical and environmental performance of the pilot scale version, the STaRR system can be seen to hold promise as a strong candidate for further evaluations as an alternate to the current sewage treatment practice. These studies warrant an integrated model for optimizing and scaling up the system and a life cycle analysis for advancing the STaRR system toward industry adoption.

Acknowledgments

This study was supported in part by the National Science Foundation Engineering Research Center for Reinventing the Nation's Urban Water Infrastructure (ReNUWIt) (#EEC 1028968), the NSF New Mexico EPSCOR "Energize New Mexico" RIIA grant (#1301346), and the NMSU Ed & Harold Foreman Endowed Chair. Support provided by the City of Las Cruces Utilities Division in accommodating the algal testbed at the Jacob A. Hands Wastewater Treatment Facility is also acknowledged.

References

Abeysiriwardana-Arachchige, I.S.A., Nirmalakhandan, N., 2019. Predicting removal kinetics of biochemical oxygen demand (BOD) and nutrients in a pilot scale fed-batch algal wastewater treatment system. Algal Res 43, 101643. https://doi.org/10.1016/j.algal.2019.101643.

Abeysiriwardana-Arachchige, I.S.A., Chapman, G.W., Rosalez, R., Soliz, N., Cui, Z., Munasinghe-Arachchige, S.P., Delanka-Pedige, H.M.K., Brewer, C.E., Lammers, P.J., Nirmalakhandan, N., 2020. Mixotrophic algal system for centrate treatment and resource recovery. Algal Res 52. https://doi.org/10.1016/j.algal.2020.102087.

Abeysiriwardana-Arachchige, I.S.A., Munasinghe-Arachchige, S.P., Delanka-Pedige, H.M.K., Nirmalakhandan, N., 2020. Removal and recovery of nutrients from municipal sewage: algal vs. conventional approaches. Water Res. 175, 115709. https://doi.org/10.1016/j.watres.2020.115709.

Abeysiriwardana-Arachchige, I.S.A., Samarasinghe, N., Rosalez, R., Munasinghe-Arachchige, S.P., Delanka-Pedige, H.M.K., Zollner, S., Brewer, C.E., Nirmalakhandan, N., 2021. Maximizing phosphorus recovery from municipal sewage in an algal wastewater treatment system. Resour. Conserv. Recycl. 170, 105552.

ASCE, 2017. ASCE's 2017 Infrastructure Report Card | GPA: D+ [WWW Document]. URL. https://www.infrastructurereportcard.org/ (Accessed 12.5.18).

Batstone, D.J., Hülsen, T., Mehta, C.M., Keller, J., 2015. Platforms for energy and nutrient recovery from domestic wastewater: a review. Chemosphere 140, 2—11. https://doi.org/10.1016/J.CHEMOSPHERE.2014.10.021.

Ceron Garcia, M.C., Camacho, F.G., Sanchez Miron, A., Fernandez Sevilla, J.M., Chisti, Y., Molina Grima, E., 2006. Mixotrophic production of marine microalga phaeodactylum tricornutum on various carbon sources. J. Microbiol. Biotechnol. 16, 689—694.

Cheirsilp, B., Torpee, S., 2012. Enhanced growth and lipid production of microalgae under mixotrophic culture condition: effect of light intensity, glucose concentration and fed-batch cultivation. Bioresour. Technol. 110, 510—516. https://doi.org/10.1016/J.BIORTECH.2012.01.125.

Cheng, F., Cui, Z., Chen, L., Jarvis, J., Paz, N., Schaub, T., Nirmalakhandan, N., Brewer, C.E., 2017. Hydrothermal liquefaction of high- and low-lipid algae: bio-crude oil chemistry. Appl. Energy 206, 278—292. https://doi.org/10.1016/J.APENERGY.2017.08.105.

Cheng, F., Cui, Z., Mallick, K., Nirmalakhandan, N., Brewer, C.E., 2018. Hydrothermal liquefaction of high- and low-lipid algae: mass and energy balances. Bioresour. Technol. 258, 158—167. https://doi.org/10.1016/J.BIORTECH.2018.02.100.

Cheng, F., Mallick, K., Henkanatte Gedara, S.M., Jarvis, J.M., Schaub, T., Jena, U., Nirmalakhandan, N., Brewer, C.E., 2019. Hydrothermal liquefaction of Galdieria sulphuraria grown on municipal wastewater. Bioresour. Technol. 292, 121884. https://doi.org/10.1016/J.BIORTECH.2019.121884.

Collivignarelli, M., Abbà, A., Frattarola, A., Carnevale Miino, M., Padovani, S., Katsoyiannis, I., Torretta, V., 2019. Legislation for the reuse of biosolids on agricultural land in Europe: overview. Sustainability 11, 6015. https://doi.org/10.3390/su11216015.

Daguerre-Martini, S., Vanotti, M.B., Rodriguez-Pastor, M., Rosal, A., Moral, R., 2018. Nitrogen recovery from wastewater using gas-permeable membranes: impact of inorganic carbon content and natural organic matter. Water Res. 137, 201—210.

Delanka-Pedige, H.M.K., Munasinghe-Arachchige, S.P., Cornelius, J., Henkanatte-Gedera, S.M.S.M., Tchinda, D., Zhang, Y., Nirmalakhandan, N., 2019. Pathogen reduction in an algal-based wastewater treatment system employing Galdieria sulphuraria. Algal Res 39, 101423. https://doi.org/10.1016/j.algal.2019.101423.

Delanka-Pedige, H.M.K., Abeysiriwardana-Arachchige, I.S.A., Tchinda, D., Munasinghe-Arachchige, S.P., Henkanatte-Gedera, S.M., Nirmalakhandan, N., 2020a. Algal process for single-step removal of organic carbon and nutrients: modeling and optimization of fed-batch operation. J. Water Process Eng. 38, 101622. https://doi.org/10.1016/j.jwpe.2020.101622.

Delanka-Pedige, H.M.K., Cheng, X., Munasinghe-Arachchige, S.P., Abeysiriwardana-Arachchige, I.S.A., Xu, J., Nirmalakhandan, N., Zhang, Y., 2020b. Metagenomic insights into virus removal performance of an algal-based wastewater treatment system utilizing Galdieria sulphuraria. Algal Res 47, 101865. https://doi.org/10.1016/j.algal.2020.101865.

Delanka-Pedige, H.M.K., Cheng, X., Munasinghe-Arachchige, S.P., Bandara, G.L.C.L., Zhang, Y., Xu, P., Schaub, T., Nirmalakhandan, N., 2020c. Conventional vs. algal wastewater technologies: reclamation of microbially safe water for agricultural reuse. Algal Res 51, 102022. https://doi.org/10.1016/j.algal.2020.102022.

Delanka-Pedige, H.M.K., Munasinghe-Arachchige, S.P., Abeysiriwardana-Arachchige, I.S.A., Nirmalakhandan, N., 2020d. Wastewater infrastructure for sustainable cities: assessment based on UN sustainable development goals (SDGs). Int. J. Sustain. Dev. World Ecol. https://doi.org/10.1080/13504509.2020.1795006 (in press).

Delanka-Pedige, H.M.K., Munasinghe-Arachchige, S.P., Abeysiriwardana-Arachchige, I.S.A., Zhang, Y., Nirmalakhandan, N., 2020e. Algal pathway towards meeting United Nation's sustainable development goal 6. Int. J. Sustain. Dev. World Ecol. 1—9. https://doi.org/10.1080/13504509.2020.1756977.

Delanka-Pedige, H.M.K., Munasinghe-Arachchige, S.P., Zhang, Y., Nirmalakhandan, N., 2020f. Bacteria and virus reduction in secondary treatment: potential for minimizing post disinfectant demand. Water Res. 177, 115802. https://doi.org/10.1016/j.watres.2020.115802.

Dube, P.J., Vanotti, M.B., Szogi, A.A., García-González, M.C., 2016. Enhancing recovery of ammonia from swine manure anaerobic digester effluent using gas-permeable membrane technology. Waste Manag. 49, 372—377.

García, J., Mujeriego, R., Hernández-Mariné, M., 2000. High rate algal pond operating strategies for urban wastewater nitrogen removal. J. Appl. Phycol. 12, 331—339. https://doi.org/10.1023/A:1008146421368.

Garcia-González, M.C., Vanotti, M.B., 2015. Recovery of ammonia from swine manure using gas-permeable membranes: effect of waste strength and pH. Waste Manag. 38, 455—461. https://doi.org/10.1016/j.wasman.2015.01.021.

García-González, M.C., Vanotti, M.B., Szogi, A.A., 2015. Recovery of ammonia from swine manure using gas-permeable membranes: effect of aeration. J. Environ. Manag. 152, 19—26. https://doi.org/10.1016/j.jenvman.2015.01.013.

GYOBU, T., INOUE, M., SODA, S., IKE, M., 2015. Energy content of organics in municipal wastewater treatment streams at Tsumori wastewater treatment plant. J. Water Environ. Technol. 13, 89—97. https://doi.org/10.2965/jwet.2015.89.

Heidrich, E.S., Curtis, T.P., Dolfing, J., 2011. Determination of the internal chemical energy of wastewater. Environ. Sci. Technol. 45, 827—832. https://doi.org/10.1021/es103058w.

Henkanatte-Gedera, S.M., Selvaratnam, T., Caskan, N., Nirmalakhandan, N., Van Voorhies, W., Lammers, P.J., 2015. Algal-based, single-step treatment of urban wastewaters. Bioresour. Technol. 189, 273—278. https://doi.org/10.1016/J.BIORTECH.2015.03.120.

Henkanatte-Gedera, S.M., Selvaratnam, T., Karbakhshravari, M., Myint, M., Nirmalakhandan, N., Van Voorhies, W., Lammers, P.J., 2017. Removal of dissolved organic carbon and nutrients from urban wastewaters by Galdieria sulphuraria: laboratory to field scale demonstration. Algal Res. 24, 450—456. https://doi.org/10.1016/J.ALGAL.2016.08.001.

Hernández-Sancho, F., Lamizana-Diallo, B., Mateo-Sagasta, J., Qadir, M., 2015. Economic Valuation of Wastewater: The Cost of Action and the Cost of No Action..

Huang, H., Xiao, D., Zhang, Q., Ding, L., 2014. Removal of ammonia from landfill leachate by struvite precipitation with the use of low-cost phosphate and magnesium sources. J. Environ. Manag. 145, 191—198. https://doi.org/10.1016/J.JENVMAN.2014.06.021.

International Energy Agency, 2015. OECD/IEA Key World Energy Statistics.

Intratec, 2007. Calcium Chloride Prices: Historical & Current [WWW Document].

Jenssen, P.D., Vråle, L., Lindholm, O., 2007. Sustainable wastewater treatment. In: International Conference on Natural Resources and Environmental Management and Environmental Safety and Health. Kuching, Malaysia, pp. 27—29.

Karbakhshravari, M., Abeysiriwardana-Arachchige, I.S.A., Henkanatte-Gedera, S.M., Cheng, F., Papelis, C., Brewer, C.E., Nirmalakhandan, N., 2020. Recovery of struvite from hydrothermally processed algal biomass cultivated in urban wastewaters. Resour. Conserv. Recycl. 163, 105089. https://doi.org/10.1016/j.resconrec.2020.105089.

Kim, J., Kim, K., Ye, H., Lee, E., Shin, C., McCarty, P.L., Bae, J., 2011. Anaerobic fluidized bed membrane bioreactor for wastewater treatment. Environ. Sci. Technol. 45, 576—581. https://doi.org/10.1021/es1027103.

Lazarova, V., Peregrina, C., Dauthuille, P., 2012. Toward energy self-sufficiency of wastewater treatment. In: Water—energy Interactions in Water Reuse. IWA publishing, London, UK.

Li, T., Zheng, Y., Yu, L., Chen, S., 2014. Mixotrophic cultivation of a Chlorella sorokiniana strain for enhanced biomass and lipid production. Biomass Bioenerg. 66, 204—213. https://doi.org/10.1016/J.BIOMBIOE.2014.04.010.

Liang, Y., Sarkany, N., Cui, Y., 2009. Biomass and lipid productivities of Chlorella vulgaris under autotrophic, heterotrophic and mixotrophic growth conditions. Biotechnol. Lett. 31, 1043–1049. https://doi.org/10.1007/s10529-009-9975-7.

Maddi, B., Panisko, E., Wietsma, T., Lemmon, T., Swita, M., Albrecht, K., Howe, D., 2017. Quantitative characterization of aqueous byproducts from hydrothermal liquefaction of municipal wastes, food industry wastes, and biomass grown on waste. ACS Sustain. Chem. Eng. 5, 2205–2214. https://doi.org/10.1021/acssuschemeng.6b02367.

Marquez, F.J., Sasaki, K., Kakizono, T., Nishio, N., Nagai, S., 1993. Growth characteristics of Spirulina platensis in mixotrophic and heterotrophic conditions. J. Ferment. Bioeng. 76, 408–410. https://doi.org/10.1016/0922-338X(93)90034-6.

Martnez, M., Sánchez, S., Jiménez, J., El Yousfi, F., Muñoz, L., 2000. Nitrogen and phosphorus removal from urban wastewater by the microalga Scenedesmus obliquus. Bioresour. Technol. 73, 263–272. https://doi.org/10.1016/S0960-8524(99)00121-2.

Matassa, S., Batstone, D.J., Hülsen, T., Schnoor, J., Verstraete, W., 2015. Can direct conversion of used nitrogen to new feed and protein help feed the world? Environ. Sci. Technol. 49, 5247–5254. https://doi.org/10.1021/es505432w.

McCarty, P.L., Bae, J., Kim, J., 2011. Domestic wastewater treatment as a net energy producer—can this be achieved? Environ. Sci. Technol. 45, 7100–7106. https://doi.org/10.1021/es2014264.

Mo, W., Zhang, Q., 2012. Can municipal wastewater treatment systems be carbon neutral? J. Environ. Manag. 112, 360–367. https://doi.org/10.1016/j.jenvman.2012.08.014.

Munasinghe-Arachchige, S.P., Nirmalakhandan, N., 2020. Nitrogen-fertilizer recovery from the centrate of anaerobically digested sludge. Environ. Sci. Technol. Lett. 7, 450–459. https://doi.org/10.1021/acs.estlett.0c00355.

Munasinghe-Arachchige, S.P., Delanka-Pedige, H.M.K., Henkanatte-Gedera, S.M., Tchinda, D., Zhang, Y., Nirmalakhandan, N., 2019. Factors contributing to bacteria inactivation in the Galdieria sulphuraria-based wastewater treatment system. Algal Res. 38, 101392. https://doi.org/10.1016/j.algal.2018.101392.

Munasinghe-Arachchige, Srimali, P., Abeysiriwardana-Arachchige, I.S.A., Delanka-Pedige, H.M.K., Nirmalakhandan, N., 2020b. Sewage treatment process refinement and intensification using multi-criteria decision making approach: a case study. J. Water Process Eng. 37, 101485. https://doi.org/10.1016/j.jwpe.2020.101485.

Munasinghe-Arachchige, S.P., Cooke, P., Nirmalakhandan, N., 2020c. Recovery of nitrogen-fertilizer from centrate of anaerobically digested sewage sludge via gas-permeable membranes. J. Water Process Eng. 38, 101630. https://doi.org/10.1016/j.jwpe.2020.101630.

Munasinghe-Arachchige, S.P., Abeysiriwardana-Arachchige, I.S.A., Delanka-Pedige, H.M.K., Cooke, P., Nirmalakhandan, N., 2021. Nitrogen-fertilizer recovery from urban sewage via gas permeable membrane: process analysis, modeling, and intensification. Chem. Eng. J. 411, 128443. https://doi.org/10.1016/j.cej.2021.128443.

Nirmalakhandan, N., Selvaratnam, T., Henkanatte-Gedera, S.M.S.M., Tchinda, D., Abeysiriwardana-Arachchige, I.S.A.I.S.A., Delanka-Pedige, H.M.K., Munasinghe-Arachchige, S.P.S.P., Zhang, Y., Holguin, F.O., Lammers, P.J.P.J., 2019. Algal wastewater treatment: photoautotrophic vs. mixotrophic processes. Algal Res. 41. https://doi.org/10.1016/j.algal.2019.101569.

Nurdogan, Y., Oswald, W.J., 1995. Enhanced nutrient removal in high-rate ponds. Water Sci. Technol. 31, 33–43. https://doi.org/10.1016/0273-1223(95)00490-E.

Ogawa, T., Aiba, S., 1981. Bioenergetic analysis of mixotrophic growth in Chlorella vulgaris and Scenedesmus acutus. Biotechnol. Bioeng. 23, 1121–1132. https://doi.org/10.1002/bit.260230519.

Park, J.B.K., Craggs, R.J., Shilton, A.N., 2011. Wastewater treatment high rate algal ponds for biofuel production. Bioresour. Technol. 102, 35–42. https://doi.org/10.1016/J.BIORTECH.2010.06.158.

Puyol, D., Batstone, D.J., Hülsen, T., Astals, S., Peces, M., Krömer, J.O., 2017. Resource recovery from wastewater by biological technologies: opportunities, challenges, and prospects. Front. Microbiol. 7, 2106. https://doi.org/10.3389/fmicb.2016.02106.

Riaño, B., Molinuevo-Salces, B., Vanotti, M.B., García-González, M.C., 2019. Application of gas-permeable membranes for-semi-continuous ammonia recovery from swine manure. Environments 6, 32. https://doi.org/10.3390/environments6030032.

Richmond, A., 2004. Handbook of Microalgal Culture: Biotechnology and Applied Phycology. Oxford, UK.

Selvaratnam, T., Pegallapati, A.K.K., Montelya, F., Rodriguez, G., Nirmalakhandan, N., Van Voorhies, W., Lammers, P.J.J., 2014. Evaluation of a thermo-tolerant acidophilic alga, Galdieria sulphuraria, for nutrient removal from urban wastewaters. Bioresour. Technol. 156, 395—399.

Shanmugam, S.R., Adhikari, S., Shakya, R., 2017. Nutrient removal and energy production from aqueous phase of bio-oil generated via hydrothermal liquefaction of algae. Bioresour. Technol. 230, 43—48. https://doi.org/10.1016/j.biortech.2017.01.031.

Shen, Y., Yuan, Y.,W., Pei, W.,Z.J., Wu, Z.J.,Q., Mao, Q.,E., 2009. Microalgae mass production methods. Trans. ASABE 52, 1275—1287. https://doi.org/10.13031/2013.27771.

Shoener, B.D., Bradley, I.M., Cusick, R.D., Guest, J.S., 2014. Energy positive domestic wastewater treatment: the roles of anaerobic and phototrophic technologies. Environ. Sci. Process. Impacts 16, 1204—1222. https://doi.org/10.1039/c3em00711a.

Smith, S., 2009. Standard Operating Procedure for Bicarbonate Extractable Phosphorus, Swat Laboratory New Mexico State University. https://doi.org/10.1002/ejoc.201200111.

Soda, S., Iwai, Y., Sei, K., Shimod, Y., Ike, M., 2010. Model analysis of energy consumption and greenhouse gas emissions of sewage sludge treatment systems with different processes and scales. Water Sci. Technol. 61, 365—373. https://doi.org/10.2166/wst.2010.827.

Soda, S., Arai, T., Inoue, D., Ishigaki, T., Ike, M., Yamada, M., 2013. Statistical analysis of global warming potential, eutrophication potential, and sludge production of wastewater treatment plants in Japan. J. Sustain. Energy Environ. 4, 33—40.

Tchinda, D., Henkanatte-Gedera, S.M., Abeysiriwardana-Arachchige, I.S.A., Delanka-Pedige, H.M.K., Munasinghe-Arachchige, S.P., Zhang, Y., Nirmalakhandan, N., 2019. Single-step treatment of primary effluent by Galdieria sulphuraria: removal of biochemical oxygen demand, nutrients, and pathogens. Algal Res 42, 101578. https://doi.org/10.1016/j.algal.2019.101578.

Theis, T.L., Hicks, A., 2012. White Paper: Methanol Use in Wastewater Denitrification. Alexandria, USA.

US DOE, 2017. Biofuels and Bioproducts from Wet and Gaseous Waste Streams: Challenges and Opportunities. (DOE/EE 1472).

US Environmental Protection Agency, 1994. EPA A Plain English Guide to the EPA Part 503 Biosolids Rule Excellence in Compliance through. Epa-832/R-93/003.

US EPA, 2008. Municipal Nutrient Removal Technologies Reference Document. September 2008-EPA 832-R-08-006.

US EPA, 2014. Promoting Technology Innovations for Clean and Safe Water - Water Technology Innovation Blueprint–Version 2.

US EPA, 2018. Inventory of U.S. Greenhouse Gas Emissions and Sinks: 1990-2016.

Van Der Hoek, J.P., De Fooij, H., Struker, A., 2016. Wastewater as a resource: strategies to recover resources from Amsterdam's wastewater. Resour. Conserv. Recycl. 113, 53—64. https://doi.org/10.1016/j.resconrec.2016.05.012.

Vanotti, M.B., Dube, P.J., Szogi, A.A., García-González, M.C., 2017. Recovery of ammonia and phosphate minerals from swine wastewater using gas-permeable membranes. Water Res. 112, 137—146.

Verstraete, W., Van de Caveye, P., Diamantis, V., 2009. Maximum use of resources present in domestic "used water.". Bioresour. Technol. 100, 5537—5545.

WERF/WEF LIFT Guidance Manual [WWW Document], 2017. URL. http://www.werf.org/lift (Accessed 3.16.19).

Young, P., Buchanan, N., Fallowfield, H.J., 2016. Inactivation of indicator organisms in wastewater treated by a high rate algal pond system. J. Appl. Microbiol. 121, 577—586. https://doi.org/10.1111/jam.13180.

Zhang, Y., Riley, L.K., Lin, M., Purdy, G.A., Hu, Z., 2013. Development of a virus concentration method using lanthanum-based chemical flocculation coupled with modified membrane filtration procedures. J. Virol. Methods 190, 41—48. https://doi.org/10.1016/J.JVIROMET.2013.03.017.

Further reading

Munasinghe-Arachchige, S.P., Abeysiriwardana-Arachchige, I.S.A., Delanka-Pedige, H.M.K., Nirmalakhandan, N., 2020a. Algal pathway for nutrient recovery from urban sewage. Algal Res 51, 102023. https://doi.org/10.1016/j.algal.2020.102023.

CHAPTER 4

Microalgae-based technologies for circular wastewater treatment

Tânia V. Fernandes[1], Lukas M. Trebuch[1,2] and René H. Wijffels[2,3]

[1]Department of Aquatic Ecology, Netherlands Institute of Ecology (NIOO-KNAW), Wageningen, the Netherlands; [2]Bioprocess Engineering, AlgaePARC, Wageningen University, Wageningen, the Netherlands; [3]Faculty of Biosciences and Aquaculture, Nord University, Bodo, Norway

1. Introduction to circular wastewater treatment

In the coming decades we will see a shift in the way we approach production. As we enter the era of circular economy, where reuse and recycling are high priorities in the visions and missions of governments and private companies, our linear production systems will change into circular ones. As wastewater is increasingly recognized as a source of raw materials, our current wastewater treatment plants will shift from being linear pollutant removal to resource recovery facilities. This will allow us to produce reclaimed water for a world that is increasingly suffering from water stress and macronutrients scarcity, such as phosphate that is a finite element within the coming century and is essential for food production. It will also allow us to tackle microelements scarcity, such as zinc that is diminishing in soils and is decreasing the yield and quality of crops and develop all other new materials that can be used in circular economy.

1.1 Principles of circularity in wastewater

The principle of circularity is as old as the planet itself. Circularity is the only way nature works. The detritus of one is the resource of the other and no element is wasted.

In the last decade, the concept of circular economy has raised attention across all of societies' private and public sectors, with increasing peer-reviewed publications and reports that serve as the basis for economic growth strategies (Reike et al., 2018). The concept dates back to the mid-1700s when Quesnay' Tableau Économique introduced the notion of a circular flow of income (Murray et al., 2017). For material flows, this concept was only used a century after, when academics, such as Peter Lund Simmonds and August Wilhelm von Hofman, supported that manufacturing processes needed to utilize its waste, therefore closing the material loops (Murray et al., 2017; Reike et al., 2018).

For human wastewater, the concept of circularity dates back to the Ancient Greeks (around 300 BCE) when the brick-lined sewer directed the wastewater to agricultural fields, where it was used for irrigating and fertilizing the crops and orchards. In the

Integrated Wastewater Management and Valorization using Algal Cultures
ISBN 978-0-323-85859-5, https://doi.org/10.1016/B978-0-323-85859-5.00001-4

© 2022 Elsevier Inc.
All rights reserved.

most sophisticated dual water supply—sewer system of the Roman period, the (waste)water was reclaimed according to its quality (e.g., the wastewater of the spas was used to flush the latrines). However, after the fall of the Roman Empire, and for more than 1000 years thereafter, water was no longer identified as a source of health and wellness (Giusy and Brown, 2010). The lack of development of engineering techniques for appropriate wastewater management led to increased natural water pollution and outbreaks of water borne diseases. The very few initiatives that used wastewater as a source of irrigation and fertilization were *Chinampas*, which are artificially created plots built over wetlands and marshes used in central and South America, and *Sewage farms*, which are agricultural lands mostly used in Europe and the United States America (Angelakis et al., 2018). By mid-1800s, the environmental pollution was regarded with disgust by the western society, resulting in renewed progress within sanitary engineering (Aiello et al., 2008). At that time, the rapid industrialization and urbanization brought by the industrial revolution resulted in public health deterioration. In order to prevent massive disease outbreaks, sewer systems were built in European cities and the United States America (Giusy and Brown, 2010). In the beginning of the 1900s, sewage was commonly used for agricultural irrigation worldwide due to the high water demands. However, increasing concerns on public health and environmental risks started to raise objections (Jaramillo and Restrepo, 2017). The 20th century was marked by the implementation of wastewater treatment plants and the introduction of world standards for treated sewage and reuse of effluents (Giusy and Brown, 2010).

Only in the beginning of the 21st century, with rapidly increasing water and resources consumption of growing population, the world governments started to embrace the dual function of wastewater treatment plants, for pollutants removal and resource recovery.

As wastewater is still mostly perceived as a polluted stream, source of diseases, and ecological threat, scientists and practitioners developing technologies for resource recovery need to show that the dual focus of wastewater treatment plants will not only provide a future sustainable circular economy but also a safe one.

1.2 Recoverable resources from wastewater

To change the linear focus of current wastewater treatment into circular, we need to assess which elements can be recycled from which type of wastewater. We further need to define how to do this in the most sustainable manner without jeopardizing public health. Finally, we also need to identify how to recover the resources and how to market them in the circular economy.

Wastewater composition varies greatly according to the origin, environmental conditions, and collection system. It is usually divided into municipal, industrial, and agricultural and is treated by biological or physical—chemical treatment or a combination of both. It includes a complex matrix of particulate, colloidal, and soluble organic and inorganic chemical compounds, and microorganism that are present at different concentrations according to the type of wastewater. The soluble biodegradable fraction of the organic compounds and the inorganic nutrients are used as the carbon and energy sources for the microorganisms that are responsible for biological wastewater treatment, which is the most commonly used method to treat municipal wastewater.

From this array of compounds, there are several possible applications. The first and most explored one is water recovery. Undoubtedly, we will need to reuse wastewater to fulfill the water needs of the future societies, and this is already in practice in many regions with water scarcity. Using the treated wastewater for agricultural purposes will significantly decrease the pressure on freshwater sources (Angelakis et al., 2018; Jaramillo and Restrepo, 2017).

There are important aspects to address for using reclaimed water, which include accurate risk assessments, regulation, and guidelines. The biggest challenge however is public acceptability, as people feel the consequences of water stress and therefore support water reuse. Yet, they are still uncertain about the quality of this water and that results in lower acceptability. In addition, the more informed part of the population is aware of the presence of contaminants of emerging concern (CEC) in the wastewater and therefore will need convincing arguments from the governmental organizations to accept water reuse (Ricart et al., 2019; Rice et al., 2016; Villarín and Merel, 2020).

Recovering energy from wastewater, mostly in the form of biogas, has also been implemented worldwide due to the already established anaerobic digestion technology and the promise of reaching energy neutrality at the wastewater treatment plants (Campello et al., 2021; Capodaglio and Olsson, 2020; Kehrein et al., 2020). However, recent studies show that recovering the heat of the wastewater (thermal energy) is much more efficient than recovering its COD (chemical energy), meaning that wastewater treatment plants would no longer be aiming at energy neutrality, but would become energy suppliers (Hao et al., 2019; Kehrein et al., 2020).

Nutrients are the follow up candidates for recovery. With phosphate being the most wanted due to the worldwide depletion of this nonrenewable macronutrient and our dependency of it (Chrispim et al., 2019). Secondly, nitrogen due to the high energy production costs and the need to reduce anthropogenic effects on terrestrial nitrogen cycle (Winkler and Straka, 2019; Ye et al., 2018). Microelements are also important to recover as they are increasingly recognized as being missing in agriculture soils (de Haes et al., 2012). The recovery of these elements from wastewater and conversion into organic fertilizers will increase macronutrients (N, P, K) and the much-needed carbon and microelements (Mg, Fe, Co, Zn, etc.) that are currently missing in conventional inorganic fertilizers. These will enhance soil biodiversity and improve soil carbon, macronutrients and microelements cycling, soil structure, and overall soil health (Silva et al., 2019).

Even though the recovery of nutrients, and other elements, is gaining attention, it has been slower due to the complexity new technologies implementation entangles. This includes technology scale up, market development, and public acceptance. So far, phosphorus is the most explored resource for recovery with full-scale recovery technologies mainly delivering struvite and CaP as the commercial product (Amann et al., 2018; Chrispim et al., 2019). These P-rich products are then used as fertilizer, therefore replacing inorganic fertilizers. The challenge of using P-rich organic residues as fertilizer is that the chemical composition (N, P, K, C, etc.) does not always meet the soils and crops needs, therefore potentially leading to under- or overfertilization (Ylivainio et al., 2021). The other challenges are greenhouse gas emissions and contamination level, where some heavy metals, CEC, and human pathogens are above legal discharge levels. Therefore, this might lead to soil contamination and public health risks (Amann et al., 2018; Muys et al., 2021). Microplastics that have been shown to accumulate in agricultural soils fertilized with sewage sludge present another related concern (Corradini et al., 2019). These challenges are however not exclusive for P-rich products such as struvite, as all resulting wastewater treatment biomass types (sewage sludge, or microalgal biomass) have the potential to accumulate contaminants.

Other resources that can be recovered from wastewater are cellulose, metals, hydrogen, proteins, lipids, Polyhydroxyalkanoates (PHAs), extracellular polymeric substances (EPSs), and volatile fatty acids (Chrispim et al., 2019; Kehrein et al., 2020; Mannina et al., 2020; Puyol et al., 2017; van der Hoek et al., 2016; Wang and Ren, 2014; Winkler and Straka, 2019; Ye et al., 2018). As many technologies are being developed for the recovery of these resources, also the economic feasibility studies are starting to arise. They indicate that some resources like water are already favorable to recover, while others are in development (Kehrein et al., 2020; Ye et al., 2020). In addition, other studies indicate that energy is a resource easy to recover at large scale, while nutrients have the highest recovery potential at small scale, in source separated sanitation systems (Diaz-Elsayed et al., 2019).

It is clear that wastewater is a stream rich of an array of potential compounds for the circular economy; however, it is also a stream potentially polluted with recalcitrant contaminants (depending on wastewater origin). These contaminants characteristics and their potential risks will be discussed in the following sections.

2. Microalgae use for circularity

2.1 Advantageous physiology and biochemistry

Over 2 billion years ago, cyanobacteria developed the enzymatic machinery to generate metabolic energy from photonic energy by inorganic carbon (CO_2) fixation and the use of water as electron acceptor (Burris, 1977; Rasmussen et al., 2008). Compared to anoxygenic photosynthesis (e.g., by green and purple sulfur bacteria), the oxygenic photosynthesis generates oxygen as a side product. The biological formation of O_2 marked the beginning of a cascade of evolutionary steps that ultimately resulted in life conquering both water and land and enriching the earth's atmosphere with oxygen. The first eukaryotic algae are widely accepted to have emerged from the primary endosymbiosis event in which an early eukaryotic ancestor engulfed cyanobacteria (Raven and Allen, 2003; Yoon et al., 2004). The ability of these primary producers to harness the sun's energy via photosynthesis enabled the thriving of biological live on earth. Today, as for the past millions of years, algae are still the most essential primary producer and they are responsible for approximately half of all organic carbon and oxygen produced on our planet (Field, 1998).

Microalgae have evolved to use different energy and carbon sources. While most phototrophic autotrophs use light as energy source and inorganic carbon as carbon source (CO_2 fixation), some species can also take up organic carbon as carbon source (mixotrophs). Some microalgae can even grow in the dark, using organic carbon as energy and carbon source (chemoorganotrophs). This versatile metabolism, developed throughout billions of years of evolution, allows them to adapt to a large array of environmental conditions, ranging from extreme temperatures (below $10°C$ and above $50°C$), irradiation ($5000\ \mu mol\ m_2^{-1}\ s^{-1}$), salinity (up to 90‰), pH (below 5 and above 9), and toxic compounds (heavy metals, phenols, etc.) (Borowitzka et al., 2016; Kirst, 1990; Schönknecht et al., 2013; Singh and Singh, 2015; Wollmann et al., 2019). Microalgae are also found in the driest deserts, where they form biofilms on spider webs with coastal water evaporation being the sole source of nutrients and water (Azua-Bustos et al., 2012) or form a cyanobacterial layer in close association with sepiolite inclusions within gypsum deposits with an upper algal layer acted as buffer to heat and evaporation (Wierzchos et al., 2015). Like bacteria, microalgae produce EPSs to secure attachment, create a protective barrier, and enhance their local environment by creating an external digestion system and sink for excess of energy. The secretion of EPS enables the survival, metabolic efficiency, and adaptation of cells to all types of environmental conditions (Decho and Gutierrez, 2017).

Next to metabolic flexibility depending on the environmental conditions, microalgae also show various different morphological traits. Cells can be coccoid or filamentous and can occur as a single cell, as cell strands or in colonies. Filamentous microalgae can be single stranded or branched. In order to move through the water column, microalgae developed various strategies for motility. Coccoid algae can have flagella that allow them to move in any direction, while filamentous cyanobacteria can have gliding capabilities due to specialized glands along their sheath layer. Another strategy developed by cyanobacteria is gas vacuoles that allow to adjust their buoyancy (Borowitzka et al., 2016). Motility is an important trait as allows them to move in response to a chemical (chemotaxis), temperature (thermotaxis), or light (phototaxis) stimulus. For example, cyanobacteria can adjust their buoyancy to be close to the water surface during day where light is available and sink to the bottom of the water body during night to be close to high nutrient availability (as phosphorus released from the sediments) (Fang et al., 2014).

Oxygenic photosynthesis led to the biological production of oxygen and further to the accumulation of oxygen in the earth's atmosphere. However, a major challenge in the occurrence of elevated oxygen concentrations was to deal with this molecule in first place. Oxygen is very reactive and, more importantly, can form highly reactive derivatives called reactive oxygen species (ROS). Well-known ROS are superoxide, hydrogen peroxide, or hydroxyl radicals. This leads to oxidative stress for biological systems, which can damage components of a cell such as proteins, lipids, and DNA (Edreva, 2005). Therefore, cyanobacteria developed an enzymatic machinery of superoxide dismutase and catalases to counteract the damaging effect of oxygen and "neutralize" ROS (Sies, 1997). As oxygen level rose about 2 billion years ago, other organisms evolved to utilize oxygen and established the aerobic metabolism next to the widely spread anaerobic.

Many microalgae can form smaller colonies or even macroscopic structures that can be spherical or exhibited as mats. The colony forming *Volvox* forms spherical structures in which cells have a distinct function (motility, reproduction) depending on the location within the colony. In evolutionary biology, this differentiation of cell morphology and function could have triggered multicellularity (Grosberg and Strathmann, 2007). Filamentous cyanobacteria can form mats on solid surfaces (sediments, rocks), macroscopic aggregation on the surface of water (scum layers), or gelatinous macroscopic colonies (*Nostoc pruniforme*) (Sand-Jensen, 2014).

Microalgae often cooperate with other organisms by sharing metabolites or providing shelter from abiotic and biotic stressors. A common example for known mutualism between microalgae and bacteria is the supply of vitamin B12 from the bacteria in exchange for fixed carbon (Croft et al., 2005). In many cases, microalgae and bacteria form a confined physical structure to be in close proximity and to create favorable conditions for each other (shelter, pH, light, and nutrient availability). The formation of granular aggregates, such as the green berries from the Sippewissett salt marsh, enabled a unique ecosystem consisting of unicellular diazotrophic (nitrogen fixing) cyanobacteria, diatoms, and heterotrophic bacteria vividly exchanging O_2, CO_2, organic carbon, and fixed nitrogen (Wilbanks et al., 2017). Similar functioning systems can be found on glaciers in the form of cryoconites, which are stromatolite-like algal mats consisting of cyanobacteria and heterotrophic bacteria cooperating to withstand the harsh conditions in ice (Irvine-Fynn et al., 2010; Takeuchi et al., 2001). Microalgae not only form colonies or communities with themselves or bacteria but are also often found in symbiosis with other eukaryotic organisms. Examples are corals, sponges, sea anemones in aquatic systems or lichen on land (Barott et al., 2012; Lutzoni and Miadlikowska, 2009). They can also form photogranules, which are dense, highly diverse spherical agglomerates of phototrophic (cyanobacteria, eukaryotic green algae), and nonphototrophic (prokaryotic bacteria) microorganisms, and these can be cultivated in artificial ecosystems (photobioreactors) (Trebuch et al., 2020).

Microalgae, like most microorganisms, communicate by the exchange of chemicals that enables them to detect and respond to cell population density. They do this by regulating the expression of specific genes, known as quorum sensing. Quorum sensing has been shown to help with microalgal flocculation and controlling pathogenic bacteria in aquaculture systems (Ramanan et al., 2016). Even though it has been recognized to coordinate community behavior in natural and engineered ecosystems, it is not sufficiently investigated (Tan et al., 2014).

Exploring the syntrophic relation between phototrophs and chemotrophs for microalgae-based wastewater treatment will enable simultaneous oxygen (photosynthesis) and carbon dioxide (respiration) production and consumption. It will also enable the exchange of metabolites, shelter from biotic stressors, and protect each other from unfavorable environmental conditions. When growing them in granules, excellent settling properties will result in easy harvesting. The increased biodiversity of these microalgae—bacteria communities increased the array of functional traits within the wastewater technologies, making them more versatile and effective at resources recovery, and more robust to deal with the wastewater composition variation.

2.2 Ecosystem functioning approach

Microalgae are the foundation of life on earth and fuel nutrient cycling (mainly carbon) by the sun's energy. Therefore, it is only logical to embrace the ability of these photosynthetic organisms and incorporate them into our future wastewater treatment systems, making them energy efficient, carbon neutral, and resource factories.

As explained in the previous section, microalgae, together with bacteria, are highly efficient at assimilating macronutrients (N, P) and microelements (Co, Zn, Cu, Fe, Mo), incorporating these essential elements and that are beneficial for healthy soils and plant growth when used as organic fertilizers. In addition, they increase soil biodiversity, moisture, and stability. They can also accumulate intra- and extracellular compounds, such as PHAs, omega-3, carotenoids, polyunsaturated fatty acids, and pigments that can be used in industry (Shahid et al., 2020). Additionally, they can be cultivated in engineered novel ecosystems, where we can steer its functions based on its community structure and interactions. By applying the reactors operational stressors, such as washout and feast-famine conditions, we can stimulate traits that are favorable to process operation, such as granulation for easier biomass harvesting or specific compounds production for higher biomass value.

Having the fundamental understanding of the diverse intra- and interspecific community dynamics in these novel ecosystems is crucial for technology development. Understanding how hazard contaminants, such as CEC and heavy metals (HM), are biodegraded or accumulated by the microalgal—bacteria community will allow us to design mitigating strategies and/or complementing with other removal technologies. The same approach should be taken for human pathogen removal, where the biocide ability of microalgae should be better understood and explored for safe water ecotechnological solutions.

Incorporating nature-based solutions like microalgae—bacteria technology into the future cities, where more than two thirds of the population will be living by 2050 (United Nations et al., 2019), will enable us to create circular, robust, and sustainable urban ecosystem. In these novel ecosystems, humans are part of the natural systems again, and with the help of microorganisms, will couple the manufacturing and waste treatment cycles.

2.3 Biotechnology approach

Microalgae have been cultivated on a small scale for decades, mainly in Asia and North America, mostly for applications in (animal) nutrition. However, this concerns a small number of species. *Spirulina* and *Chlorella* are well-known examples of this. Growers produce these varieties especially for the food supplement industry. At the moment, only about 25,000 tons of dried algae are produced per year worldwide (Marketwatch, 2019).

One of the major bottlenecks in microalgal biomass use in the circular economy is its production costs, as up to now highest biomass yields have been reached in more expensive microalgal cultivation systems. There are four different microalgal cultivation systems in use worldwide. These are open ponds or raceways, single-layer or horizontal tube reactors, three-dimensional tubular reactors, and flat plate reactors (Norsker et al., 2011).

2.3.1 Open ponds or raceways

These are shallow annular channels in which mixing takes place by means of paddle wheels. These are the most widely used cultivation systems worldwide. An important advantage of this simple design is the still relatively low costs. On the other hand, such an open, large pond is less easy to control than a closed cultivation system. Water evaporates and this system is also susceptible to contamination, so that the choice of the algae species to be cultivated is limited to resistant, fast-growing species.

Algae ponds are said to be the most effective cultivation system, but there are counter arguments (Norsker et al., 2011). Based on calculations, photosynthesis can utilize a maximum of 10% of the energy in sunlight and convert it into biomass. The rest of the light energy is lost as heat. In practice, however, the percentage of solar energy used is much lower. For an algal pond, this is approximately 1.5%. This is partly because sunlight only penetrates a few centimeters into the pond where algae are present as suspended solids. As a result, only the algae cells on the surface receive sufficient of light.

Algae cultivation in sunny regions is potentially much more productive, provided that the use of solar energy becomes more efficient. This can be done, among other things, by having the algae absorb less light, for example, by reducing the antenna size of algae using genetic modification. Due to this reduced light absorption, light will also penetrate deeper into the pond, so that more algae receive sunlight for photosynthesis. A possible disadvantage of this method is that the genetically modified algae lose the competition with wild-type algae.

2.3.2 Single-layer or horizontal tube reactors

A closed culture system made up of a single layer of horizontal tubes. Such a tube reactor allows more control and is more productive per square meter. In addition, this system is easy to scale up: you simply extend the tube length. A major drawback of this design is that the light intensity falling on the algae-filled tube is very high. Algae cannot cope with this and then grow more slowly, which decreases productivity. A disadvantage of all tubular reactors is the high energy costs of pumping the algae around. The gas exchange is also not optimal in these systems; O_2 builds up and is toxic to algae at high concentrations. In addition, the construction costs may be higher.

2.3.3 Three-dimensional tubular reactors

This system is made up of several layers of tubes that are placed vertically on top of each other and thus form a kind of vertical panels, made up of tubes. This type of reactor has partly the same advantages and disadvantages as the single layer tube reactor. An important difference is that the problem of too high light intensity is much less important here. This is because the vertically stacked tubes are, as it were, in each other's shadow, which decreases the light intensity. The productivity is also higher than that of the single layer tube reactor, because more tubes fit on the same surface, which increases the yield per square meter.

2.3.4 Flat plate reactors

These are closed reactors made up of series of flat, parallel plates. In theory, those systems are the most productive. There is no accumulation of the toxic O_2 and the light intensity is not too high. The disadvantage of this system is that a relatively large amount of energy is required to mix nutrients and to keep the algae in suspension. In addition, it is more difficult to add extra CO_2 and to scale up Fig. 4.1.

The costs of algae biomass production would always be lower in open systems than in closed systems. However, the comparison is only partly valid because closed photobioreactors have only been used for the production of algae with very high market values (for

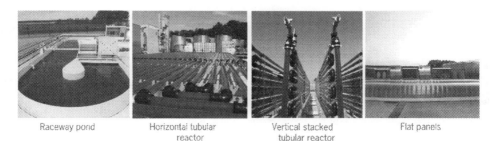

Figure 4.1 Microalgae cultivation systems operated at AlgaePARC, Wageningen, The Netherlands (Bosma, 2014).

niche markets). As a result, designs have never been adapted for large-scale production of bulk products. However, an extensive calculation of the production costs of open ponds and closed photobioreactors showed that the two systems did not differ much in that respect (Ruiz et al., 2016).

Considering only the biomass production costs, where generally open systems are cheaper than closed systems (Acién Fernández et al., 2019), might be insufficient when designing the future microalgae cultivation systems that will be incorporated in urban landscape (see Section 3.3: Futuristic perspective). With climate change (warming, rainfall patterns) and further urbanization (larger cities), humanity will need to adapt infrastructure and technology to mitigate the environmental effects, therefore both open and closed cultivation systems will have a purpose and need to be technologically improved.

2.4 Socio-economical approach

In the development of new technologies, the Technology Readiness Level (TRL) is the tool used for measuring feasibility, relevance, and closeness to market application. However, TRL alone is insufficient for a holistic and impact aware evaluation approach. The first steps of new technologies are commonly funded by national or international research funding organizations, such as the European Union (EU), that has an array of programs to stimulate application of scientific findings. However, in these funding schemes, multidisciplinary and holistic approaches that result in high societal impact are expected, and that is not included in a one-dimensional evaluation tool such as the TRL. Therefore, other tools need to be included, such as Societal Readiness Level (SRL), Organizational Readiness Level (ORL), and Legal Readiness Level (LRL). SRL assesses the level of societal acceptance of a certain technology, product, process, or intervention, therefore the readiness of the society to adopt the proposed solution. While SRL is already starting to be used in literature and practice, ORL and LRL are still on the discussion table. ORL assesses the impact of the new technology in the depending public and private organizations (e.g., human skills required). LRL is also relevant as often new technologies are hindered by governmental regulations that eventually need to be adapted to the future societal ambitions (Bruno et al., 2020).

On a world perspective, microalgae-based technologies offer solutions for many of the United Nations sustainable development goals (SDGs) in terms of closing carbon and nutrient cycles, resource recovery, sustainability, and nature-based solutions. Besides the evident SDGs 6 (Clean water and sanitation) and 13 (Climate action), microalgae-based technologies address SDGs 2 (Zero hunger) and 3 (Good health and well-being) as healthier soils, rich in microbes, carbon, nutrients, and essential elements, and without contaminants result in healthier crops. Additionally, they contribute to SDGs 8 (Decent work and economic growth), 9 (Industry, innovation, and infrastructure), and 11 (Sustainable cities and communities) as it enables innovative nature-based solutions for a sustainable agricultural practice based on long-term ecosystems services.

On a European perspective, microalgae-based technologies fit with the EU missions on circularity as they promote interventions for increased biodiversity and ecosystem services, resource efficiency, circular economy, and clean environment thereby, reversing soils impoverishment and promoting sustainable agriculture. The EU aims for a 50% reduction of nutrient loss leading to a 20% reduction in fertilizer use and 50% reduction of harmful pesticide use (EC, 2020).

It is time to realize that circular economy does not need to be a utopian dream, but that it can deliver a "win—win—win" situation in terms of macroeconomic, social, and environmental impacts. A metaanalysis study of 324 circular economy scenarios from 2020 to 2050 indicated that these scenarios are expected to increase employment and GDP while decreasing CO_2 emissions (Aguilar-Hernandez et al., 2021).

2.5 Uncertainties and challenges when using microalgae-based technology for circularity

2.5.1 Technology application challenges

The high market price for the small-scale production of high-quality algae products for niche markets does not make cost reductions a top priority. In this case, large-scale production makes little sense because that market is quite small and therefore quickly saturated. In the much larger market for wastewater treatment, cost reduction is much more important. After all, the technology must be able to compete economically with existing wastewater treatment technologies.

Apart for the cost of wastewater treatment, the microalgal biomass has a value. For that, it is important to use all components from the biomass in order to be able to make wastewater treatment with microalgae economically profitable. By dividing the algae biomass into different components, each with a specific value, a calculation of the total value of 1000 kg of algae can be made. As an example, we assume an algal biomass that consists of 40% oil, 50% protein, and 10% sugars. In this example, part of the oil is used to make biofuels and part is used as raw material for the chemical industry. Water-soluble proteins can be used in foodstuffs instead of soy protein and partly in animal feed. Sugars also have many uses. In addition, the O_2 produced by algae and the removal of nutrients from wastewater also generate income. In total, the algae biomass is worth 1.65 € kg^{-1}, at a cost of 0.50 €. This means that in principle it is worthwhile to produce algae for a combination of products in addition to the treatment of wastewater. However, the costs for biorefining have not yet been included in this, which are approximately 1 € kg^{-1} biomass for nonfood bulk products (Ruiz et al., 2016).

There are also risks when introducing a new crop like microalgae. Products intended for the food chain must be safe, and large-scale cultivation must not lead to an overgrowth of natural microalgae populations. Most microalgae do not contain toxic components and are cultivated under special conditions (high salt concentrations, high pH) and have no competitive advantage over natural populations. Of course, the development of the technology requires careful handling of these risks and mitigation of risks is part of research.

Algae cultivation is also attractive as a sustainable alternative because they can grow on CO_2 and excess nutrients in waste flows. As an example, the production of biofuel is taken, which can be an important bulk product produced with microalgae. To produce 1 L of biofuel using oil-containing crops requires about 5000 L of fresh water. According to calculations, algae cultivation requires only 1.5 L of fresh water per liter of oil produced.

If fuel based on algae oil is to replace all transport fuels in Europe (0.4 billion m^3 per year) and we assume a yield of approximately 40,000 L of oil ha^{-1} $year^{-1}$, an area of 9.25 million hectares is required. This is roughly the size of a country like Portugal. Although that is a large area, it fits well within Europe. The big advantage is that Europe will then be self-sufficient for fuel transport. As a by-product, 0.3 billion tons of protein is then produced. This is 40 times the amount that Europe currently imports as soy protein. In other words, algae can produce both fuels and food.

In addition to providing fuel for European transport, large-scale algae cultivation has positive consequences for both the CO_2 balance and the recovery of nutrients from wastewater and manure. For the growth of these large quantities of algae, 1.3 billion tons of CO_2 is needed and 25 million tons of nitrogen. In comparison, the total amount of CO_2 produced annually in Europe is 3.9 billion tons and 8 million tons of nitrogen is released via wastewater and manure annually.

The above figures show that the cultivation of algae does not have to be at the expense of biodiversity and food production and can make a substantial contribution to the sustainability of our economy (Ruiz et al., 2016).

2.5.2 Contamination of algal biomass

When cultivating microalgae industrially, microbial contamination by competing microalgae and grazers (ameba, ciliates, flagellates) can significantly reduce microalgal yield and harvested biomass quality. It can even lead to culture "crashes" that result in halting biomass. As in terrestrial crops, the mainstream strategy for preventing contamination is addition of biocides. However, like in organic agriculture, other alternatives are being tested, such as resistant strain selection, genetically modified organisms, and early detection signals followed by best management strategies to prevent infection (Day et al., 2017).

For wastewater treatment, grazers can also hinder microalgae cultivation, especially in open systems, however, not as detrimental as in axenic cultivation due to the already highly diverse community (higher competition) and the unfavorable composition of the cultivation media (inhibiting compounds) (Salama et al., 2019).

In addition, wastewater itself incorporates several contaminants depending on the type of wastewater used for cultivation. Domestic and municipal wastewater include CEC, HM, and human pathogens. Certain industrial wastewaters contain high concentrations of specific heavy metals or CEC.

2.5.2.1 Contaminants of emerging concern

CEC include pharmaceutical and personal care products (PPCPs) and are currently found in all ecosystems and living organisms of the globe, at concentrations ranging from $ng\,L^{-1}$ to $\mu g\,L^{-1}$. The major concern is that many tend to persist, bioaccumulating in the environment, and in living organisms, leading to toxic environments. In living organisms, they can lead to nervous system damage, disease of immune system, reproductive and development disorders, and other conditions. Some contaminants are also responsible for creating antibiotic resistance. They can be classified as follows:

- pharmaceuticals (compounds used for therapeutic purpose that include antiinflammatory drugs, antibiotics, β-blockers, stimulants, etc.),
- personal care products (compounds used in personal hygiene, protection, and beautification that include fragrances, disinfectants, UV filters, insect repellents, etc.),
- hormones (estrogens),
- surfactants (detergents, emulsifiers, foaming agents, etc.),
- industrial chemicals (flame retardants, plasticisers, conductors, etc.), and
- pesticides (inc. herbicides, fungicides, and insecticides).

Every day a large variety of new chemicals used in industry and agriculture enter the market without risk assessment and without being monitored in the environment and living organisms. From more than 100,000 chemicals on the market, only about 500 have been characterized for their hazard and exposure (EEA, 2019). This results in large knowledge gaps in the adverse effect of individual and combined CEC in ecosystems and human health. A gap that will not decrease in the coming decades as global chemical production is projected to triple between 2010 and 2050, mainly outside Europe.

An increasing number of studies show that CEC adversely affect aquatic ecosystems. It is clear that there is still much uncertainty on the hazardousness of individual versus mixtures of CEC in aquatic ecosystems and its consequences across aquatic food webs, so more studies are needed (Liu et al., 2020; Nilsen et al., 2019; Quadra et al., 2017; Schoenfuss et al., 2020). However, there is even a larger lack of data on the fate of CEC on soils, plants, herbivores, and humans (Bartrons and Peñuelas, 2017) and this is crucial for when we want to use the wastewater-derived biomass for resource recovery in circular systems.

Microalgae are efficient at removing a wide range of CEC from wastewater as is shown in the increasing number of studies (Hena et al., 2021; Maryjoseph and Ketheesan, 2020; Matamoros et al., 2015; Sutherland and Ralph, 2019; Tolboom et al., 2019). Some compounds, such as carbamazepine, are harder to remove, reaching 30% removal; however, many other compounds, such as metoprolol and ibuprofen, reach 60%—100% removal (de Wilt et al., 2016; Hena et al., 2021). The mechanisms of removal are biodegradation (enzymatic process), biosorption, bioaccumulation, photodegradation, and volatilization. The efficiency of CEC removal is dependent on abiotic factors such as temperature, light intensity, pH, hydraulic retention time, and biotic factors such as phylogenetic diversity of the microalgae and cocultured bacteria. Having a healthy microalgae culture will result in higher CEC removal (Hena et al., 2021).

2.5.2.2 Heavy metals

Some heavy metals, such as copper, selenium, nickel, boron, iron, molybdenum, and zinc, are essential for living organism metabolism at trace amounts; however, at high concentrations can be toxic (Vardhan et al., 2019). This hormesis phenomenon makes microalgal biomass valuable for metals to return to soils. Due to the persistent nature of heavy metals, they tend to bioaccumulate in living organisms, as they can be assimilated and stored faster than metabolized or excreted. Lead, cadmium, arsenic, and mercury are among the most concerning heavy metals found in wastewater as they lead to high environmental and public health risks.

The harmfulness of heavy metals is dependent on concentration, oxidation state, and its bioavailability. In addition, the pH and composition of the soil or water matrix will define the bioavailability and persistency of heavy metals in living organisms (Vardhan et al., 2019; Mohammad et al., 2015; Vardhan et al., 2019; Zeitoun and Mehana, 2014).

Heavy metals enter the food chain (including water and air) through bioaccumulation mechanisms. However, the contamination pathways are not fully understood (Kumar et al., 2019). The organism living in the aquatic and terrestrial ecosystems is exposed over long periods to the contaminated agricultural runoffs, but also municipal and industrial wastewaters that fail to retain them at the wastewater treatment plant. Sewage sludge that has been used directly on the fields or disposed in landfills also contributes to the heavy metal contamination of soils (Dregulo and Bobylev, 2020). In the EU, about 40% of the sludge generated from wastewater treatment plants is applied on agricultural land (Kehrein et al., 2020).

Microalgae are efficient at removing a wide range of HM from wastewater as is shown in the several review articles (Bulgariu and Bulgariu, 2020; Leong and Chang, 2020; Salama et al., 2019; Suresh Kumar et al., 2015). Compounds of worldwide concern such as arsenic, cadmium, chromium, lead, and mercury, due to their carcinogenic nature and toxicity at trace amounts, have been shown to bio adsorp and bioaccumulate on microalgal biomass suggesting microalgae-based technologies to be used for heavy metals bioremediation. However, further research is needed in terms of downstream purification of biomass for simultaneous bioremediation and product creation (e.g., fertilizer).

2.5.2.3 Human pathogens

Other contaminants of concern are human pathogens. These include bacteria (*E. coli* and *Salmonella*), virus (e.g., rotavirus and coronavirus), protozoans (e.g., *Giardia* spp.), helminths (e.g., *Ascaris lumbricoides*), hookworms (e.g., *Strongyloides stercoralis*), and trophozoites (e.g., *Trichomonas* and *Enterobius vermicularis*). The assessment of human pathogens contamination is done by monitoring indicator microorganism, such as total coliforms, fecal coliforms, *E. coli*, Enterococcus, enterovirus, and nematode eggs (helminth eggs and protozoan cysts) (WHO, 2006). Exposure of human pathogens by ingestion (water, food) or inhalation can result in a number of diseases, such as gastrointestinal and respiratory illnesses and death. Removing these human pathogens from wastewater remains a priority, as in the early days of wastewater treatment. Due to the recent Coronavirus disease outbreak in 2019, pathogens removal became an even higher priority, as we realize that these outbreaks might become more frequent due to extensive animal husbandry.

Microalgae-based technologies are efficient at human pathogens removal as microalgae compete for nutrients and they can release natural antibiotics and other metabolites that adhere to the cell surface of microalgae that aid in the reduction of pathogenic organisms (Kuramae et al., 2021). In addition, some microalgae culturing conditions, such as high pH, dissolved oxygen (e.g., *E.coli*, and *Salmonella* are facultative anaerobes), temperature, and UV, can lead to pathogenic organisms decay (Chai et al., 2021; Dar et al., 2019).

Unlike CEC and HM, pathogenic organisms do not bioaccumulate in the microalgal cells, therefore facilitating removal from the harvested biomass. However, the wet biomass should be hygienized before using as a fertilizer and/or during downstream processes. Thermal treatment, 1 h at $70°C$, is commonly used for removing pathogens from digestate and sludge (Kaur et al., 2020). Biodrying (using heat generated from anaerobic digestion plus aeration) and air–drying beds are also an alternative for regions with high temperature and solar irradiance, further increasing pathogenic organisms lysis, and even improving CEC removal (Lopes et al., 2020; Pilnáček et al., 2019).

The major concern for resource recovery is the risk of contamination of the reclaimed water as some pathogenic organisms remain in the water. Several technologies with varying efficiencies can be used for disinfection of the reclaimed water, including chlorination, ozonation, UV, filtration, pasteurization, solar, and storage. The downsides of some technologies are the formation of harmful metabolites or process residues (e.g., chlorine residues) and the enhancement of antimicrobial resistant which are harmful for aquatic organisms and public health (Al-Gheethi et al., 2018).

3. Microalgae-based technologies for wastewater treatment

Microalgae has been recognized by William J. Oswald and collaborators in early 20th century as a promising wastewater technology (Oswald et al., 1957). Already at that time Oswald identified the advantages of combining phototrophic with heterotrophic treatment (O_2 supply, nutrient uptake, and pollutants removal) and the challenges for large-scale application (productivity, harvesting) (Oswald, 2003). The, at that time, stabilization ponds were upgraded to high-rate open ponds with paddle wheel to improve gasses exchange (CO_2 and O_2) and productivity. Meantime, closed photobioreactors were developed further increasing productivity, yet also higher production costs. The harvesting challenge remained however unsolved. This challenge is currently addressed by innovative reactor designs and operation strategies, such biofilm cultivation systems and photogranules (Boelee et al., 2013; Brockmann et al., 2021; Miranda et al., 2017; Trebuch et al., 2020; Zhang et al., 2021).

Microalgae-based technologies can be used for all types of wastewater, from low strength municipal wastewater to high strength agro-based industries. Under sufficient light and moderate temperatures, microalgae-based technologies can recover a wide range of nutrients and trace elements, while remediating pollutants, such as CEC, heavy metals, and human pathogens. In the recent years, due to the increasing interest of using microalgae-based technologies for domestic, industrial, and agricultural wastewater treatment, quite a few reviews have been published on the opportunities and challenges (Amenorfenyo et al., 2019; Gupta et al., 2019; Gupta and Bux, 2019; Hena et al., 2021; Hussain et al., 2021; Kumar et al., 2021; Li et al., 2019; Maryjoseph and Ketheesan, 2020; Yong et al., 2021).

3.1 Currently technologies—large-scale application

In recent years, many reviews have been published on available and upcoming microalgae-based technologies for wastewater treatment, some focused on biodiesel production, others on nutrients recovery, biogas production, and high-value compounds (Li et al., 2019).

Currently, there are several research projects and companies around the world that test various algal treatment systems at demonstration stage. The treatment is performed with open raceway ponds, tubular photobioreactors, or biofilm systems, both with and without artificial illumination. The most commonly used system is the raceway ponds that is applied to a wide range of wastewater types ranging from municipal wastewater, piggery sewage, digestate, or tannery wastewater.

In Chiclana de la Frontera, Spain, the primary effluent of the local municipal wastewater treatment plant is partially treated by four 1 Ha raceway ponds each. The harvested algal biomass is subsequently codigested with conventional activated sludge to produce biogas as a sustainable car fuel. The algal treatment system was commissioned as part of the EU project All-gas in collaboration with Aqualia (https://www.all-gas.eu/en/). Raceway ponds are also used in the EU project SABANA (http://www.eu-sabana. eu/). The goal of this project is to ultimately run a 5 Ha raceway to produce algal biomass as biostimulants, biopesticides, biofertilizers, feed additives, and aquafeed. Other examples for raceway ponds used at demonstration scale are within the projects Water2Return (https://water2return.eu/) and Saltgae (http://saltgae.eu/).

The US company CLEARAS focused on closed three-dimensional tubular reactors specifically designed for effluent polishing of municipal wastewater treatment plants (https://www.clearaswater.com/). They have over 50 project sites in the United States and are aiming to bring down effluent levels especially of phosphorus to lower than 0.02 mg P L^{-1}. Their biggest demonstration facility can handle about 27 m^3 per day in West Bountiful, Utah. A similar goal is pursued by the British company Severn Trent (https://www.stwater.co.uk/) that wants to use microalgae as tertiary treatment for phosphorus removal, therefore meeting the upcoming stringent standards for discharge. Their aim is to lower phosphorus levels to below 0.1 mg P L^{-1} using closed nontranslucent vessels with submerged lamps for artificial illumination. At the moment, they have a demonstration site at Broadwindsor, UK, where they treat the effluent of the local municipal wastewater treatment plant. The installation is constructed together with the company I-PHYC (https://i-phyc.com/broadwindsor/).

Biofilm reactor types are used by Algaewheel, a US company that focuses on rotating biofilm systems for wastewater treatment (https://algaewheel.com/). These are similar to the algal cultivation systems developed under the EU project AlgaDisk (Blanken et al., 2017). The algal biomass grows on continuously rotating disks, which are partially submerged in wastewater.

Another system that is commercially applied in several locations in the United States is the algal turf scrubber (ATS) (https://hydromentia.com/technologies/algal-turf-scrubber/). This system consists of a large, inclined surface on which benthic algae attach and form a biofilm. The wastewater runs over this biofilm and is treated. The ATS requires large surfaces similar to raceway ponds, but has rather low investment costs and technological requirements (Adey et al., 2011).

3.2 Developing technologies—lab scale

Heterotrophic cultivation of microalgae is suggested by some researchers as an alternative to tackle the high light intensity needed and therefore high footprint of photoautotrophic cultivation. The cost saved by not having a light-dependent process is used for keeping the reactors aseptic and the cultures axenic, therefore avoiding the bacterial contamination and grazers (Di Caprio et al., 2021).

Biofilm systems show promising results for wastewater treatment. Algae grows on a support material (e.g., polycarbonate membrane, filters from different materials, woven or nonwoven fabrics, stainless steel mesh) and can easily be retained in the cultivation system. There are different types of biofilm systems based on the configuration: (1) Perfused biofilm systems, (2) constantly submerged biofilm systems, (3) Intermittently submerged biofilm systems with either rotating or rocker support material (Choudhary et al., 2017). Although there are several studies on lab scale that substantiate the use of biofilm system for wastewater treatment, it is still challenging to scale up the technology due to the harvesting of the algal biofilm, which often has to be done manually (Choudhary et al., 2017; Sukačová et al., 2015).

In liquid foam-bed photobioreactor, small gas bubbles are passed through a thin liquid layer which results in foam formation (Janoska et al., 2018). The liquid is either self-foaming or a foam stabilizing agent (surfactant) has to be added. The main advantages are high biomass productivity, high biomass concentrations, and an improved mass transfer from gas to liquid due to the increased surface area and contact time. This reactor has not yet been applied for wastewater treatment but was suggested as a valid option especially if the wastewater has self-foaming properties (Ngo et al., 2018).

Another culturing system is the thin layer cascade. It is comparable to a raceway pond with a very short light path (<10 mm) and an inclined surface. The liquid runs over this surface due to the gravitational force and is recirculated back to the top by a pumping system. Thin layer cascades combine the advantages of open systems (direct sun irradiance, evaporative self-cooling, simple cleaning, and low maintenance costs) with the advantages of closed systems (operation at high biomass densities and high biomass productivities) (Grivalský et al., 2019). The first thin layer cascades emerged in the 1940s in the United States and were further developed from the 1960s onward at the Centre Algatech (former Laboratory of Algal Research) in Trebon, Czech Republic (https://www.alga.cz/en/). These systems have been tested with different wastewaters including municipal wastewater or anaerobically treated piggery effluent (Raeisossadati et al., 2019; Sánchez Zurano et al., 2020). The system shows promising results for wastewater treatment, but high areal footprint and lack of control (pH, temperature, DO) remain as challenges that need to be tackled.

Another innovation for wastewater treatment is the microalgal—bacterial granules, known as photogranules, due to their functional versatility (N, P, COD removal) and excellent settling properties. In addition, no or little aeration is needed as in situ simultaneous oxygen (photosynthesis) and carbon dioxide (respiration) production and consumption take place. Photogranules are obtained by hydrostatic or hydrodynamic reactor operation. They have been tested under different condition at lab scale in bubble column reactors, with headspace recirculation, or in mechanically mixed reactors (Brockmann et al., 2021; Trebuch et al., 2020; Zhang et al., 2021).

3.3 Futuristic perspective—drawing board

Innovation in circularity comes with understanding and creativity, and one does not work without the other. When designing the future microalgae-based technologies for circular wastewater treatment, we need to understand the ecology and process technology of the systems we set to design.

This process can start from large to small scale or vice versa. For example, we can add, in a beaker, microalgae and bacteria species that are highly efficient in nutrient assimilation, provide then with the optimal conditions for growth, and design the next steps by extrapolating the results of the previous scale step. We can also start with the architectonic and engineering vision of the full-scale system, like a drawing of the future, and identify the system borders for application. The second option leaves the natural scientific disciplines of fundamental understanding to enter the creativity fields of architecture, art, and civil engineering. When considering the claim of Beijerinck (O'Malley, 2008) that "everything is everywhere but the environment selects," we should be able to design systems immediately in full scale. The challenges would then be to operate the technology in a manner that the targeted functions are fulfilled. This approach would however be unfeasible as the personnel, material cost, and the time needed would be unrealistic. However, we should strive for something in between, where both concepts come to play. When designing our fundamental experimental work in a lab setting, we should have the vision of the futuristic implementation to guide us through the needed step-wise testing.

Incorporating microalgae photobioreactors in the urban architecture is something we will see more in the coming century, in future sustainable cities. A few case studies have integrated photobioreactors in the facades of buildings for carbon sequestration (Cervera Sardá and Vicente, 2016; Oncel et al., 2020). Currently, there are still too many bottlenecks to make these systems sustainable (too much and expensive material use, high-energy costs, etc.). However, when the material and energy costs decrease, together with the change of resources and societal values, such architectural beauty, greener cities, and local resource generation will prevail. Then, these systems should be in an advanced development stage for fast implementation.

We should of course, concurrently, develop algae-based technology for the current reality and coming decennia, where space and light intensity are crucial for large-scale application. Therefore, creating reactors that enhance light distribution (Moreno-SanSegundo et al., 2020). Increasing biomass yield by improving the efficiency of photosynthetic carbon reactions (Weber and Bar-Even, 2019). Exploring microalgae—bacteria agglomerates to form photogranules for nutrients recovery, easy harvesting, and aeration-free cultivation systems (Brockmann et al., 2021; Trebuch et al., 2020; Zhang et al., 2021). This also improves the current open cultivation systems for higher nutrients removal and recovery and pollutants removal.

3.4 Advantages of using microalgae-based technologies

Microalgae are highly efficient at nutrients recovery, under sufficient light and moderate temperatures, easily reaching wastewater treatment effluent standards. Unlike the most commonly used technology for P recovery, struvite precipitation, microalgae can completely remove the P present in the wastewater even at low P concentrations and does not need pH correction of the wastewater. In addition, microalgae biomass cultivated on wastewater has a high potential for resource recovery as it assimilates an array of elements (C, N, P, K, Co, Mo, Cu, Zn, Fe, etc.) and molecules (PHAs, omega-3, carotenoids, polyunsaturated fatty acids and pigments, etc.) that can be used as fertilizer and in other application in our future circular economy.

Microalgae can be cultivated in climatically adverse conditions as desert-like areas, therefore not competing with food and feed production. It can also be cultivated on contaminated soils sites or roadsides, especially when cultivated in photobioreactors. It can be integrated in urban architecture, offering local treatment of wastewater, resource generation, and consumption while embellishing our cities.

Microalgae can mitigate contamination and produce a wide variety of products. In addition, microalgae and heterotrophic bacteria can synergistically produce oxygen (photosynthesis) and carbon dioxide (respiration), therefore eliminating the need of aeration in wastewater treatment plants. This relation also enables the exchange of metabolites, shelter from biotic stressors, and protects each other from unfavorable environmental conditions, resulting in more versatile resource recovery and robust wastewater treatment.

In microalgae-based wastewater treatment, the biogas can be recovered by anaerobically digesting the residual microalgae biomass after extraction of high-value compounds, such as pigments. By anaerobically digesting microalgae biomass instead of activated sludge, one can increase biogas production up to 10% (Arashiro et al., 2020). However, the results on biogas production vary greatly depending on the microalgae species used as harder and thicker cell walls result in lower anaerobic biodegradability. One option to improve digestibility of microalgae is thermal pretreatment of the biomass, as is nowadays also done in sewage sludge and animal manure (Mahdy et al., 2015; Solé-Bundó et al., 2020; Thorin et al., 2018).

Brewery has favorable nutrient concentrations for microalgae growth without having the downside of persistent contaminants such as heavy metals, CEC, and human pathogens (Amenorfenyo et al., 2019). When coupled with heterotrophic bacteria, both carbon (COD) and nutrients (N and P) can be removed and recovered into a contaminant-free biomass. This biomass can then be used as fertilizer or animal feed. Other examples of contaminant-free wastewaters that could use microalgae-based technologies for resource recovery include organic greenhouse wastewater and dairy wastewater (Gurunathan et al., 2018).

3.5 Risks involved in microalgae technology implementation

The use of resources recovered from wastewater entangles environmental and public health risks. These risks are associated with the contamination level of the wastewater to be treated and the efficiency of the microalgae-based technology, alone or in combination with other wastewater technologies, implemented. Some industrial wastewaters, such as the food processing industry, contain very low amount of hazardous contaminants; however, for other industrial wastewaters and municipal wastewater this is not the case.

The pollutants of concern for wastewater treatment and resource recovery are mostly heavy metals, CEC, and human pathogens. They are concerning due to their persistence in the environment, bioaccumulating in living organisms and leading to adverse ecological effects, including chronic and acute disorders. They are currently found in all corners of the earth. Both in conventional and new wastewater treatment technologies they are not completely removed, and in some CEC dedicated removal technologies, such as ozone, metabolites can be generated that are more harmful than the parent compound (example is carbamazepine). In recent years, due to the increase of pharmaceuticals production and consumption, there is also an increase in the environment, leading to antibiotic resistance in bacteria and endocrine disruption (Hena et al., 2021).

Even though an increasing number of studies rationalize the use of microalgae-based technologies for contributing to the circular economy, only a few studies asses the contamination level of the algal biomass (de Wilt et al., 2016; García-Galán et al., 2020; Ye et al., 2020). From these studies, it is clear that some persistent contaminants are present in the algal biomass. Even though they are at low concentrations, it is unclear if these concentrations are harmful as there are, best to our knowledge, no risk assessments made on microalgae biomass cultivated in wastewater.

In an increasing circular wastewater treatment systems, we need to come up with the right evaluation methodology, such as the value-belief-norm, to assess the risks involved in resource reclamation, but also its benefits (Poortvliet et al., 2018). Zero risks do not exist as we daily create new fabricated chemicals that finish up in the environment, and nature challenges us with new human pathogens, such as the coronaviruses. What we need is to put the risk on the balance against the benefits gained when reusing the recovered resources. This raises philosophical questions on humanity's social norms and values that go beyond financial values. What do we consider clean? Or clean enough for reusing? Upcoming wastewater technologies that support circularity will need to ensure that the resources recovered are safe for reuse, and policies, such as the end-to-waste criteria of the EU waste framework directive, will need to regulate. However, public needs to be correctly informed on the risks to embrace the circular products (Rice et al., 2016).

Risk assessment is an essential tool to inform decision-makers and the public. Risk quantification is done by assessing the hazardless of the contaminant and the level of exposure. These vary in type (chemical composition) and timescale, but also in the vulnerability of living organisms to it, which depends on the live cycle stage of the organism. For example, some pesticides result in poisoning, meaning acute hazard, while some heavy chemicals result in cancer or later infertility. Recently, the European Environmental Agency (EEA) published a report assessing the current pollution of hazardous chemicals in the environment and providing an outlook of the risks expected (EEA, 2019). Due to the large volume of chemicals industry uses and the potential combined toxicity of these compounds, there is enormous uncertainty of the risks they induce in the environment and public health. In addition, the vast majority of hazardous compounds are not monitored in the environment, resulting in incomplete risk assessments. To be able to use the resources recovered from wastewater in the circular economy, we will need to perform bioassays to define the level of toxicity. One way of achieving faster toxicity screening data would be in using embryo tests as high-throughput tools (Capela et al., 2020).

3.6 Mitigation approaches for decreasing risks

Several studies have shown that bioremediation through microbes can play an important role in cleaning polluted ecosystems, such as polluted soils (Dangi et al., 2019; Jin et al., 2018; Ye et al., 2017). By understanding the bioremediation mechanisms and exploiting the capabilities of soil microbes, we could use the soil microbiome as an extension of the wastewater treatment, therefore mitigating the possible contamination of the wastewater-derived biomass fertilizer.

Another strategy is to improve light dissipation in the microalgae cultivation system as it will increase biomass productivity but also enhance contaminants removal by photolysis (Ali et al., 2017).

Synergistically cultivating microalgae with heterotrophic bacteria also enhances CEC removal as the degradation of CEC is enhanced by the cometabolism between these two groups (Hena et al., 2021).

Besides improving the technology, we can also tackle the contamination risk by better selecting the microalgae technology to be applied for the different types of wastewaters. Depending on the source of the wastewater, tailor made wastewater treatment systems can be designed to enhance resource recovery and remove contaminant. For example, the nutrient-rich black water (toilet wastewater) does not have hazardous concentrations of heavy metals, but has high CEC and human pathogens. Brewery and dairy wastewaters on the other hand do not contain CEC and human pathogens and do have high nutrients.

Microalgae-based technologies seem promising for CEC removal; however, there are various unknowns due to the limited number of compounds tested, needed conditions for the removal, and knowledge on the removal mechanisms. Moreover, limited information is available on the formed transformation products and whether these intermediates are biologically active or inactive (Hom-Diaz et al., 2015). By understanding the mechanisms (photolysis, adsorption, biodegradation, oxidation), we can design a microalgae photobioreactor (PBR) focused on the removal of CEC and its transformation products. Combining microalgae technology with other contaminants removal technologies, such as filtration or ozone, could also offer many advantages in terms of energy savings, reduce the formation of harmful metabolites, and provide an ecologically balanced final effluent.

For the recovered microalgae biomass, we need to perform good risk assessments and develop complementary contaminants mitigation solutions for the new recycled products, thus ensuring that wastewater circularity will not compromise public health. This information will then need to reach the various players in society for public acceptance.

4. Conclusions and perspectives

Microalgae will certainly be part of our future circular economy, supplying raw materials to the food, feed, energy, and other industrial sectors, as depicted in Fig. 4.2. This will be possible as microalgae can easily grow in regions that terrestrial crops do not grow and

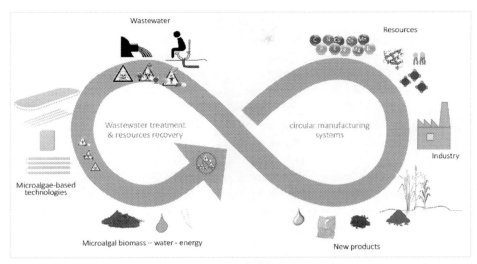

Figure 4.2 Schematic representation of circular wastewater treatment, where the resources recovered are used for manufacturing new products.

due to their fast growth rate, allowing for continuous harvest. The way of growing microalgae is as diverse as the wide range of compounds they offer. They can be cultivated in high-tech closed and fully automated systems, as in low-tech uncontrolled systems depending on the goal of the targeted function (high-value product or wastewater treatment). Whatever the goal behind the use of microalgae technology use is, more research is needed to show (1) the potential of the microalgae functions and products; (2) the best cultivation strategies and reactor's design; and (3) life-cycle and risk assessments for water reclamation and resources recovery. In addition, we need to test the lab-scale results in large-scale application, and these need to be accompanied with circular economy value-based business models (Agrawala and Börkey, 2018).

For assessing the sustainability of the new technologies, we need to perform early stage LCA (also known as Ex-Ante LCA) where the technology that is still in the R&D phase can be fairly compared with existing technologies (van der Giesen et al., 2020). In this new approach, methods for dealing with the high uncertainty and therefore high risk involving early-stage technologies are suggested.

Finally, and equally important, we need to diversify our teams when we want to establish microalgae-based technologies in the circular societies of the future. Besides the wastewater and bioprocess engineers, we need to include aquatic and microbial ecologists for understanding the complexity of the microalgae—bacteria interactions and explore its functions for targeted products. We need to work with social scientist to understand and change public perception and acceptance of the centuries-long bad image of wastewater. We need to work with urban planners, architects, and civil engineers to design our technologies in a way they fit the ambitions of the future sustainable cities where we realize that greener and biodiverse cities enhance welfare. We need to work with industries to, at an early stage, sketch the product development state.

The biggest challenge in microalgae-based circular wastewater treatment will not be the cleaning of the water, as that, when insufficient, can be coupled to other existing technologies, such as Ozone, UV, or filtration. The biggest challenge will be producing a microalgal biomass free of contaminants, such as CEC and HM that can accumulate in the biomass, therefore hindering the resource recovery feasibility.

The growing population coupled with the rising consumption patterns of synthetic chemicals and natural resources drives humanity closer and faster to Earth systems tipping points (Lenton et al., 2019). The growing risk of reaching irreversible climate changes due to human-induced pressures that exceed the carrying capacity of the planet must accelerate resource recovery circularity and replacement of synthetic chemicals by bio-based chemicals.

References

Acién Fernández, F.G., Fernández Sevilla, J.M., Molina Grima, E., 2019. Chapter 21 - costs analysis of microalgae production. In: Pandey, A., Chang, J.-S., Soccol, C.R., Lee, D.-J., Chisti, Y. (Eds.), Biofuels from Algae, second ed. Elsevier, Biomass, Biofuels, Biochemicals, pp. 551–566. https://doi.org/10.1016/B978-0-444-64192-2.00021-4.

Adey, W.H., Kangas, P.C., Mulbry, W., 2011. Algal turf scrubbing: cleaning surface waters with solar energy while producing a biofuel. Bioscience 61, 434–441. https://doi.org/10.1525/bio.2011.61.6.5.

Agrawala, S., Börkey, P., 2018. Business Models for the Circular Economy—Opportunities and Challenges from a Policy Perspective (OECD Policy Highlights). Paris, France.

Aguilar-Hernandez, G.A., Dias Rodrigues, J.F., Tukker, A., 2021. Macroeconomic, social and environmental impacts of a circular economy up to 2050: a meta-analysis of prospective studies. J. Clean. Prod. 278, 123421. https://doi.org/10.1016/j.jclepro.2020.123421.

Aiello, A.E., Larson, E.L., Sedlak, R., 2008. Hidden heroes of the health revolution. Sanitation and personal hygiene. Am. J. Infect. Control 36, S128–S151. https://doi.org/10.1016/j.ajic.2008.09.008.

Al-Gheethi, A.A., Efaq, A.N., Bala, J.D., Norli, I., Abdel-Monem, M.O., Ab Kadir, M.O., 2018. Removal of pathogenic bacteria from sewage-treated effluent and biosolids for agricultural purposes. Appl Water Sci. 8, 74. https://doi.org/10.1007/s13201-018-0698-6.

Ali, A.M.M., Kallenborn, R., Sydnes, L.K., Rønning, H.T., Alarif, W.M., Al-Lihaibi, S., 2017. Photolysis of pharmaceuticals and personal care products in the marine environment under simulated sunlight conditions: irradiation and identification. Environ. Sci. Pollut. Res. 24, 14657–14668. https://doi.org/10.1007/s11356-017-8930-8.

Amann, A., Zoboli, O., Krampe, J., Rechberger, H., Zessner, M., Egle, L., 2018. Environmental impacts of phosphorus recovery from municipal wastewater. Resour. Conserv. Recycl. 130, 127–139. https://doi.org/10.1016/j.resconrec.2017.11.002.

Amenorfenyo, D.K., Huang, X., Zhang, Y., Zeng, Q., Zhang, N., Ren, J., Huang, Q., 2019. Microalgae brewery wastewater treatment: potentials, benefits and the challenges. Int. J. Environ. Res. Publ. Health 16, 1910.

Angelakis, A.N., Asano, T., Bahri, A., Jimenez, B.E., Tchobanoglous, G., 2018. Water reuse: from ancient to modern times and the future. Front. Environ. Sci. 6. https://doi.org/10.3389/fenvs.2018.00026.

Arashiro, L.T., Ferrer, I., Pániker, C.C., Gómez-Pinchetti, J.L., Rousseau, D.P.L., Van Hulle, S.W.H., Garfí, M., 2020. Natural pigments and biogas recovery from microalgae grown in wastewater. ACS Sustain. Chem. Eng. 8, 10691–10701. https://doi.org/10.1021/acssuschemeng.0c01106.

Azua-Bustos, A., González-Silva, C., Arenas-Fajardo, C., Vicuña, R., 2012. Extreme environments as potential drivers of convergent evolution by exaptation: the Atacama Desert Coastal Range case. Front. Microbiol. 3, 1—9. https://doi.org/10.3389/fmicb.2012.00426.

Barott, K.L., Rodriguez-Mueller, B., Youle, M., Marhaver, K.L., Vermeij, M.J.A., Smith, J.E., Rohwer, F.L., 2012. Microbial to reef scale interactions between the reef-building coral Montastraea annularis and benthic algae. Proc. Biol. Sci. 279, 1655—1664. https://doi.org/10.1098/rspb.2011.2155.

Bartrons, M., Peñuelas, J., 2017. Pharmaceuticals and personal-care products in plants. Trends Plant Sci. 22, 194—203. https://doi.org/10.1016/j.tplants.2016.12.010.

Blanken, W., Magalhães, A., Sebestyén, P., Rinzema, A., Wijffels, R.H., Janssen, M., 2017. Microalgal biofilm growth under day-night cycles. Algal Res. 21, 16—26. https://doi.org/10.1016/j.algal.2016.11.006.

Boelee, N.C., Janssen, M., Temmink, H., Taparavičiūtė, L., Khiewwijit, R., Jánoska, Á., Buisman, C.J.N., Wijffels, R.H., 2013. The effect of harvesting on biomass production and nutrient removal in phototrophic biofilm reactors for effluent polishing. J. Appl. Phycol. 1—14. https://doi.org/10.1007/s10811-013-0178-1.

Borowitzka, M.A., Beardall, J., Raven, J.A., 2016. The physiology of microalgae. J. Appl. Phycol. https://doi.org/10.3354/ame01390.

Bosma, R., 2014. Design and construction of the microalgal pilot facility AlgaePARC. Algal Res. Biomass Biofuels Bioprod. 6.

Brockmann, D., Gérand, Y., Park, C., Milferstedt, K., Hélias, A., Hamelin, J., 2021. Wastewater treatment using oxygenic photogranule-based process has lower environmental impact than conventional activated sludge process. Bioresour. Technol. 319, 124204. https://doi.org/10.1016/j.biortech.2020.124204.

Bruno, I., Lobo, G., Covino, B.V., Donarelli, A., Marchetti, V., Panni, A.S., Molinari, F., 2020. Technology readiness revisited: a proposal for extending the scope of impact assessment of European public services. In: Proceedings of the 13th International Conference on Theory and Practice of Electronic Governance. Presented at the ICEGOV 2020: 13th International Conference on Theory and Practice of Electronic Governance. ACM, Athens Greece, pp. 369—380. https://doi.org/10.1145/3428502.3428552.

Bulgariu, L., Bulgariu, D., 2020. Bioremediation of toxic heavy metals using marine algae biomass. In: Naushad, M., Lichtfouse, E. (Eds.), Green Materials for Wastewater Treatment, Environmental Chemistry for a Sustainable World. Springer International Publishing, Cham, pp. 69—98. https://doi.org/10.1007/978-3-030-17724-9_4.

Burris, J.E., 1977. Photosynthesis, photorespiration, and dark respiration in eight species of algae. Mar. Biol. 39, 371—379. https://doi.org/10.1007/BF00391940.

Campello, L.D., Barros, R.M., Tiago Filho, G.L., dos Santos, I.F.S., 2021. Analysis of the economic viability of the use of biogas produced in wastewater treatment plants to generate electrical energy. Environ. Dev. Sustain. 23, 2614—2629. https://doi.org/10.1007/s10668-020-00689-y.

Capela, R., Garric, J., Castro, L.F.C., Santos, M.M., 2020. Embryo bioassays with aquatic animals for toxicity testing and hazard assessment of emerging pollutants: a review. Sci. Total Environ. 705, 135740. https://doi.org/10.1016/j.scitotenv.2019.135740.

Capodaglio, A.G., Olsson, G., 2020. Energy issues in sustainable urban wastewater management: use, demand reduction and recovery in the urban water cycle. Sustainability 12, 266. https://doi.org/10.3390/su12010266.

Cervera Sardá, R., Vicente, C.A., 2016. Case studies on the architectural integration of photobioreactors in building façades. In: Pacheco Torgal, F., Buratti, C., Kalaiselvam, S., Granqvist, C.-G., Ivanov, V. (Eds.), Nano and Biotech Based Materials for Energy Building Efficiency. Springer International Publishing, Cham, pp. 457—484. https://doi.org/10.1007/978-3-319-27505-5_17.

Chai, W.S., Tan, W.G., Halimatul Munawaroh, H.S., Gupta, V.K., Ho, S.-H., Show, P.L., 2021. Multifaceted roles of microalgae in the application of wastewater biotreatment: a review. Environ. Pollut. 269, 116236. https://doi.org/10.1016/j.envpol.2020.116236.

Choudhary, P., Malik, A., Pant, K.K., 2017. Algal biofilm systems: an answer to algal biofuel dilemma. In: Gupta, S.K., Malik, A., Bux, F. (Eds.), Algal Biofuels: Recent Advances and Future Prospects. Springer International Publishing, Cham, pp. 77—96. https://doi.org/10.1007/978-3-319-51010-1_4.

Chrispim, M.C., Scholz, M., Nolasco, M.A., 2019. Phosphorus recovery from municipal wastewater treatment: critical review of challenges and opportunities for developing countries. J. Environ. Manag. 248, 109268. https://doi.org/10.1016/j.jenvman.2019.109268.

Corradini, F., Meza, P., Eguiluz, R., Casado, F., Huerta-Lwanga, E., Geissen, V., 2019. Evidence of microplastic accumulation in agricultural soils from sewage sludge disposal. Sci. Total Environ. 671, 411—420. https://doi.org/10.1016/j.scitotenv.2019.03.368.

Croft, M.T., Lawrence, A.D., Raux-Deery, E., Warren, M.J., Smith, A.G., 2005. Algae acquire vitamin B12through a symbiotic relationship with bacteria. Nature 438, 90—93. https://doi.org/10.1038/nature04056.

Dangi, A.K., Sharma, B., Hill, R.T., Shukla, P., 2019. Bioremediation through microbes: systems biology and metabolic engineering approach. Crit. Rev. Biotechnol. 39, 79—98. https://doi.org/10.1080/07388551.2018.1500997.

Dar, R.A., Sharma, N., Kaur, K., Phutela, U.G., 2019. Feasibility of microalgal technologies in pathogen removal from wastewater. In: Application of Microalgae in Wastewater Treatment, pp. 237—268. https://doi.org/10.1007/978-3-030-13913-1_12.

Day, J.G., Gong, Y., Hu, Q., 2017. Microzooplanktonic grazers — a potentially devastating threat to the commercial success of microalgal mass culture. Algal Res. 27, 356—365. https://doi.org/10.1016/j.algal.2017.08.024.

de Haes, H.A.U., Voortman, R.L., Bastein, T., Bussink, D., Rougoor, C.W., van der Weijden, W.J., 2012. Scarcity of Micronutrients in Soil, Feed, Food, and Mineral Reserves: Urgency and Policy Options. Platform Agriculture, Innovation & Society.

de Wilt, A., Butkovskyi, A., Tuantet, K., Leal, L.H., Fernandes, T.V., Langenhoff, A., Zeeman, G., 2016. Micropollutant removal in an algal treatment system fed with source separated wastewater streams. J. Hazard Mater. 304, 84—92. https://doi.org/10.1016/j.jhazmat.2015.10.033.

Decho, A.W., Gutierrez, T., 2017. Microbial extracellular polymeric substances (EPSs) in ocean systems. Front. Microbiol. 8. https://doi.org/10.3389/fmicb.2017.00922.

Di Caprio, F., Tayou Nguemna, L., Stoller, M., Giona, M., Pagnanelli, F., 2021. Microalgae cultivation by uncoupled nutrient supply in sequencing batch reactor (SBR) integrated with olive mill wastewater treatment. Chem. Eng. J. 410, 128417. https://doi.org/10.1016/j.cej.2021.128417.

Diaz-Elsayed, N., Rezaei, N., Guo, T., Mohebbi, S., Zhang, Q., 2019. Wastewater-based resource recovery technologies across scale: a review. Resour. Conserv. Recycl. 145, 94—112. https://doi.org/10.1016/j.resconrec.2018.12.035.

Dregulo, A.M., Bobylev, N.G., 2020. Heavy metals and arsenic soil contamination resulting from wastewater sludge urban landfill disposal. Pol. J. Environ. Stud. 30, 81—89. https://doi.org/10.15244/pjoes/121989.

EC, 2020. Reinforcing Europe's Resilience - Press Release [WWW Document]. European Commission - European Commission. URL. https://ec.europa.eu/commission/presscorner/detail/en/ip_20_884 (Accessed 4.11.21).

Edreva, A., 2005. Generation and scavenging of reactive oxygen species in chloroplasts: a submolecular approach. Agric. Ecosyst. Environ. 106, 119—133. https://doi.org/10.1016/j.agee.2004.10.022.

EEA, 2019. The European Environment — State and Outlook 2020 (SOER 2020) (Publication). EEA - European Environment Agency SOER, Luxembourg.

Fang, F., Yang, L., Gan, L., Guo, L., Hu, Z., Yuan, S., Chen, Q., Jiang, L., 2014. DO, pH, and Eh microprofiles in cyanobacterial granules from Lake Taihu under different environmental conditions. J. Appl. Phycol. 26, 1689—1699. https://doi.org/10.1007/s10811-013-0211-4.

Field, C.B., 1998. Primary production of the biosphere: integrating terrestrial and oceanic components. Science 281, 237—240. https://doi.org/10.1126/science.281.5374.237.

García-Galán, M.J., Arashiro, L., Santos, L.H.M.L.M., Insa, S., Rodríguez-Mozaz, S., Barceló, D., Ferrer, I., Garfí, M., 2020. Fate of priority pharmaceuticals and their main metabolites and transformation products in microalgae-based wastewater treatment systems. J. Hazard Mater. 390, 121771. https://doi.org/10.1016/j.jhazmat.2019.121771.

Giusy, L., Brown, J., 2010. Wastewater Management through the Ages: A History of Mankind | Elsevier Enhanced Reader [WWW Document]. https://doi.org/10.1016/j.scitotenv.2010.07.062.

Grivalský, T., Ranglová, K., da Câmara Manoel, J.A., Lakatos, G.E., Lhotský, R., Masojídek, J., 2019. Development of thin-layer cascades for microalgae cultivation: milestones (review). Folia Microbiol. 64, 603–614. https://doi.org/10.1007/s12223-019-00739-7.

Grosberg, R.K., Strathmann, R.R., 2007. The evolution of multicellularity: a minor major transition? Annu. Rev. Ecol. Evol. Syst. 38, 621–654. https://doi.org/10.1146/annurev.ecolsys.36.102403.114735.

Gupta, S., Pawar, S.B., Pandey, R.A., 2019. Current practices and challenges in using microalgae for treatment of nutrient rich wastewater from agro-based industries. Sci. Total Environ. 687, 1107–1126. https://doi.org/10.1016/j.scitotenv.2019.06.115.

Gupta, S.K., Bux, F. (Eds.), 2019. Application of Microalgae in Wastewater Treatment: Volume 1: Domestic and Industrial Wastewater Treatment. Springer International Publishing, Cham. https://doi.org/10.1007/978-3-030-13913-1.

Gurunathan, B., Selvakumari, I.A.E., Aiswarya, R., Renganthan, S., 2018. Bioremediation of industrial and municipal wastewater using microalgae. In: Varjani, S.J., Agarwal, A.K., Gnansounou, E., Gurunathan, B. (Eds.), Bioremediation: Applications for Environmental Protection and Management, Energy, Environment, and Sustainability. Springer Singapore, Singapore, pp. 331–357. https://doi.org/10.1007/978-981-10-7485-1_16.

Hao, X., Li, J., van Loosdrecht, M.C.M., Jiang, H., Liu, R., 2019. Energy recovery from wastewater: heat over organics. Water Res. 161, 74–77. https://doi.org/10.1016/j.watres.2019.05.106.

Hena, S., Gutierrez, L., Croué, J.-P., 2021. Removal of pharmaceutical and personal care products (PPCPs) from wastewater using microalgae: a review. J. Hazard Mater. 403, 124041. https://doi.org/10.1016/j.jhazmat.2020.124041.

Hom-Diaz, A., Llorca, M., Rodriguez-Mozaz, S., Vicent, T., Barcelo, D., Blanquez, P., 2015. Microalgae cultivation on wastewater digestate: beta-estradiol and 17alpha-ethynylestradiol degradation and transformation products identification. J. Environ. Manag. 155, 106–113. https://doi.org/10.1016/j.jenvman.2015.03.003.

Hussain, F., Shah, S.Z., Ahmad, H., Abubshait, S.A., Abubshait, H.A., Laref, A., Manikandan, A., Kusuma, H.S., Iqbal, M., 2021. Microalgae an ecofriendly and sustainable wastewater treatment option: biomass application in biofuel and bio-fertilizer production. A review. Renew. Sustain. Energy Rev. 137, 110603. https://doi.org/10.1016/j.rser.2020.110603.

Irvine-Fynn, T.D.L., Bridge, J.W., Hodson, A.J., 2010. Rapid quantification of cryoconite: granule geometry and in situ supraglacial extents, using examples from Svalbard and Greenland. J. Glaciol. 56, 297–308. https://doi.org/10.3189/002214310791968421.

Janoska, A., Andriopoulos, V., Wijffels, R.H., Janssen, M., 2018. Potential of a liquid foam-bed photobioreactor for microalgae cultivation. Algal Res. 36, 193–208. https://doi.org/10.1016/j.algal.2018.09.029.

Jaramillo, M.F., Restrepo, I., 2017. Wastewater reuse in agriculture: a review about its limitations and benefits. Sustainability 9, 1734. https://doi.org/10.3390/su9101734.

Jin, Y., Luan, Y., Ning, Y., Wang, L., 2018. Effects and mechanisms of microbial remediation of heavy metals in soil: a critical review. Appl. Sci. 8, 1336. https://doi.org/10.3390/app8081336.

Kaur, G., Wong, J.W.C., Kumar, R., Patria, R.D., Bhardwaj, A., Uisan, K., Johnravindar, D., 2020. Value addition of anaerobic digestate from biowaste: thinking beyond agriculture. Curr Sustain. Renew. Energy Rep. 7, 48–55. https://doi.org/10.1007/s40518-020-00148-2.

Kehrein, P., Loosdrecht, M. van, Osseweijer, P., Garfi, M., Dewulf, J., Posada, J., 2020. A critical review of resource recovery from municipal wastewater treatment plants — market supply potentials, technologies and bottlenecks. Environ. Sci. Water Res. Technol. 6, 877–910. https://doi.org/10.1039/C9EW00905A.

Kirst, G.O., 1990. Salinity tolerance of eukaryotic marine algae. Annu. Rev. Plant Biol. 41, 21–53. https://doi.org/10.1146/annurev pp.41.060190.000321.

Kumar, B.R., Mathimani, T., Sudhakar, M.P., Rajendran, K., Nizami, A.-S., Brindhadevi, K., Pugazhendhi, A., 2021. A state of the art review on the cultivation of algae for energy and other valuable products: application, challenges, and opportunities. Renew. Sustain. Energy Rev. 138, 110649. https://doi.org/10.1016/j.rser.2020.110649.

Kumar, S., Prasad, S., Yadav, K.K., Shrivastava, M., Gupta, N., Nagar, S., Bach, Q.-V., Kamyab, H., Khan, S.A., Yadav, S., Malav, L.C., 2019. Hazardous heavy metals contamination of vegetables and food chain: role of sustainable remediation approaches - a review. Environ. Res. 179, 108792. https://doi.org/10.1016/j.envres.2019.108792.

Kuramae, E.E., Dimitrov, M.R., da Silva, G.H.R., Lucheta, A.R., Mendes, L.W., Luz, R.L., Vet, L.E.M., Fernandes, T.V., 2021. On-site blackwater treatment fosters microbial groups and functions to efficiently and robustly recover carbon and nutrients. Microorganisms 9, 75. https://doi.org/10.3390/microorganisms9010075.

Lenton, T.M., Rockström, J., Gaffney, O., Rahmstorf, S., Richardson, K., Steffen, W., Schellnhuber, H.J., 2019. Climate tipping points — too risky to bet against. Nature 575, 592—595. https://doi.org/10.1038/d41586-019-03595-0.

Leong, Y.K., Chang, J.-S., 2020. Bioremediation of heavy metals using microalgae: recent advances and mechanisms. Bioresour. Technol. 303, 122886. https://doi.org/10.1016/j.biortech.2020.122886.

Li, K., Liu, Q., Fang, F., Luo, R., Lu, Q., Zhou, W., Huo, S., Cheng, P., Liu, J., Addy, M., Chen, P., Chen, D., Ruan, R., 2019. Microalgae-based wastewater treatment for nutrients recovery: a review. Bioresour. Technol. 291, 121934. https://doi.org/10.1016/j.biortech.2019.121934.

Liu, N., Jin, X., Feng, C., Wang, Z., Wu, F., Johnson, A.C., Xiao, H., Hollert, H., Giesy, J.P., 2020. Ecological risk assessment of fifty pharmaceuticals and personal care products (PPCPs) in Chinese surface waters: a proposed multiple-level system. Environ. Int. 136, 105454. https://doi.org/10.1016/j.envint.2019.105454.

Lopes, B.C., Rodrigues, H.F., Costa, T.E.R., Batista, A.M.M., Filho, C.R.M., Araújo, J.C., Matos, A.T. de, 2020. Air-drying bed as an alternative treatment for UASB sludge under tropical conditions. J. Water Sanit. Hyg. Dev. 10, 458—470. https://doi.org/10.2166/washdev.2020.041.

Lutzoni, F., Miadlikowska, J., 2009. Lichens. Curr. Biol. 19, 502—503. https://doi.org/10.1016/j.cub.2009.04.034.

Mahdy, A., Mendez, L., Ballesteros, M., González-Fernández, C., 2015. Algaculture integration in conventional wastewater treatment plants: anaerobic digestion comparison of primary and secondary sludge with microalgae biomass. Bioresour. Technol. 184, 236—244. https://doi.org/10.1016/j.biortech.2014.09.145.

Mannina, G., Presti, D., Montiel-Jarillo, G., Carrera, J., Suárez-Ojeda, M.E., 2020. Recovery of polyhydroxyalkanoates (PHAs) from wastewater: a review. Bioresour. Technol. 297, 122478. https://doi.org/10.1016/j.biortech.2019.122478.

Marketwatch, 2019. microalgae-market-2019-industry-research-share-trend-global-industry-size-price-analysis-regional-outlook-to-2025-research-report [WWW Document]. https://www.marketwatch.com/press-release.

Maryjoseph, S., Ketheesan, B., 2020. Microalgae based wastewater treatment for the removal of emerging contaminants: a review of challenges and opportunities. Case Stud. Chem. Environ. Eng. 2, 100046. https://doi.org/10.1016/j.cscee.2020.100046.

Matamoros, V., Gutiérrez, R., Ferrer, I., García, J., Bayona, J.M., 2015. Capability of microalgae-based wastewater treatment systems to remove emerging organic contaminants: a pilot-scale study. J. Hazard Mater. 288, 34—42. https://doi.org/10.1016/j.jhazmat.2015.02.002.

Miranda, A.F., Ramkumar, N., Andriotis, C., Höltkemeier, T., Yasmin, A., Rochfort, S., Wlodkowic, D., Morrison, P., Roddick, F., Spangenberg, G., Lal, B., Subudhi, S., Mouradov, A., 2017. Applications of microalgal biofilms for wastewater treatment and bioenergy production. Biotechnol. Biofuels 10, 120. https://doi.org/10.1186/s13068-017-0798-9.

Mohammad, A., Zaki, M., Khallaf, E., Abbas, H., 2015. Use of fish as bio-indicator of the effects of heavy metals pollution. J Aquac. Res. Dev. 6. https://doi.org/10.4172/2155-9546.1000328.

Moreno-SanSegundo, J., Casado, C., Marugán, J., 2020. Enhanced numerical simulation of photocatalytic reactors with an improved solver for the radiative transfer equation. Chem. Eng. J. 388, 124183. https://doi.org/10.1016/j.cej.2020.124183.

Murray, A., Skene, K., Haynes, K., 2017. The circular economy: an interdisciplinary exploration of the concept and application in a global context. J. Bus. Ethics 140, 369—380. https://doi.org/10.1007/s10551-015-2693-2.

Muys, M., Phukan, R., Brader, G., Samad, A., Moretti, M., Haiden, B., Pluchon, S., Roest, K., Vlaeminck, S.E., Spiller, M., 2021. A systematic comparison of commercially produced struvite:

quantities, qualities and soil-maize phosphorus availability. Sci. Total Environ. 756, 143726. https://doi.org/10.1016/j.scitotenv.2020.143726.

Ngo, H.H., Vo, H.N.P., Guo, W., Bui, X.-T., Nguyen, P.D., Nguyen, T.M.H., Zhang, X., 2018. Advances of Photobioreactors in Wastewater Treatment: Engineering Aspects, Applications and Future Perspectives. Springer Singapore. https://doi.org/10.1007/978-981-13-3259-3_14.

Nilsen, E., Smalling, K.L., Ahrens, L., Gros, M., Miglioranza, K.S.B., Picó, Y., Schoenfuss, H.L., 2019. Critical review: grand challenges in assessing the adverse effects of contaminants of emerging concern on aquatic food webs. Environ. Toxicol. Chem. 38, 46—60. https://doi.org/10.1002/etc.4290.

Norsker, N.-H., Barbosa, M.J., Vermuë, M.H., Wijffels, R.H., 2011. Microalgal production — a close look at the economics. Biotechnol. Adv. 29, 24—27. https://doi.org/10.1016/j.biotechadv.2010.08.005.

O'Malley, M.A., 2008. 'Everything is everywhere: but the environment selects': ubiquitous distribution and ecological determinism in microbial biogeography. Stud. Hist. Philos. Biol. Biomed. Sci. 39, 314—325. https://doi.org/10.1016/j.shpsc.2008.06.005.

Oncel, S.S., Kose, A., Oncel, D.S., 2020. 8 - carbon sequestration in microalgae photobioreactors building integrated. In: Pacheco-Torgal, F., Rasmussen, E., Granqvist, C.-G., Ivanov, V., Kaklauskas, A., Makonin, S. (Eds.), Start-Up Creation, second ed. Woodhead Publishing Series in Civil and Structural Engineering. Woodhead Publishing, pp. 161—200. https://doi.org/10.1016/B978-0-12-819946-6.00008-4.

Oswald, W.J., 2003. My sixty years in applied algology. J. Appl. Phycol. 15, 99—106. https://doi.org/10.1023/A:1023871903434.

Oswald, W.J., Gotaas, H.B., Golueke, C.G., Kellen, W.R., Gloyna, E.F., Hermann, E.R., 1957. Algae in waste treatment [with discussion]. Sewage Ind. Waste 29, 437—457.

Pilnáček, V., Innemanová, P., Šereš, M., Michalíková, K., Stránská, Š., Wimmerová, L., Cajthaml, T., 2019. Micropollutant biodegradation and the hygienization potential of biodrying as a pretreatment method prior to the application of sewage sludge in agriculture. Ecol. Eng. 127, 212—219. https://doi.org/10.1016/j.ecoleng.2018.11.025.

Poortvliet, P.M., Sanders, L., Weijma, J., De Vries, J.R., 2018. Acceptance of new sanitation: the role of end-users' pro-environmental personal norms and risk and benefit perceptions. Water Res. 131, 90—99.

Puyol, D., Batstone, D.J., Hülsen, T., Astals, S., Peces, M., Krömer, J.O., 2017. Resource recovery from wastewater by biological technologies: opportunities, challenges, and prospects. Front. Microbiol. 7, 2106.

Quadra, G.R., Oliveira de Souza, H., Costa, R., dos, S., Fernandez, M.A., dos, S., 2017. Do pharmaceuticals reach and affect the aquatic ecosystems in Brazil? A critical review of current studies in a developing country. Environ. Sci. Pollut. Res. 24, 1200—1218. https://doi.org/10.1007/s11356-016-7789-4.

Raeisossadati, M., Vadiveloo, A., Bahri, P.A., Parlevliet, D., Moheimani, N.R., 2019. Treating anaerobically digested piggery effluent (ADPE) using microalgae in thin layer reactor and raceway pond. J. Appl. Phycol. 31, 2311—2319. https://doi.org/10.1007/s10811-019-01760-6.

Ramanan, R., Kim, B.-H., Cho, D.-H., Oh, H.-M., Kim, H.-S., 2016. Algae—bacteria interactions: evolution, ecology and emerging applications. Biotechnol. Adv. 34, 14—29. https://doi.org/10.1016/j.biotechadv.2015.12.003.

Rasmussen, B., Fletcher, I.R., Brocks, J.J., Kilburn, M.R., 2008. Reassessing the first appearance of eukaryotes and cyanobacteria. Nature 455, 1101—1104. https://doi.org/10.1038/nature07381.

Raven, J.A., Allen, J.F., 2003. Genomics and chloroplast evolution: what did cyanobacteria do for plants? Genome Biol. 4, 1—5. https://doi.org/10.1186/gb-2003-4-3-209.

Reike, D., Vermeulen, W.J.V., Witjes, S., 2018. The circular economy: new or refurbished as CE 3.0? — exploring controversies in the conceptualization of the circular economy through a focus on history and resource value retention options. Resour. Conserv. Recycl. 135, 246—264. https://doi.org/10.1016/j.resconrec.2017.08.027.

Ricart, S., Rico, A.M., Ribas, A., 2019. Risk-yuck factor nexus in reclaimed wastewater for irrigation: comparing farmers' attitudes and public perception. Water 11, 187. https://doi.org/10.3390/w11020187.

Rice, J., Wutich, A., White, D.D., Westerhoff, P., 2016. Comparing actual de facto wastewater reuse and its public acceptability: a three city case study. Sustain. Cities Soc. 27, 467—474. https://doi.org/10.1016/j.scs.2016.06.007.

Ruiz, J., Olivieri, G., Vree, J. de, Bosma, R., Willems, P., Reith, J.H., Eppink, M.H.M., Kleinegris, D.M.M., Wijffels, R.H., Barbosa, M.J., 2016. Towards industrial products from microalgae. Energy Environ. Sci. 9, 3036–3043. https://doi.org/10.1039/C6EE01493C.

Salama, E.-S., Roh, H.-S., Dev, S., Khan, M.A., Abou-Shanab, R.A.I., Chang, S.W., Jeon, B.-H., 2019. Algae as a green technology for heavy metals removal from various wastewater. World J. Microbiol. Biotechnol. 35, 75. https://doi.org/10.1007/s11274-019-2648-3.

Sánchez Zurano, A., Garrido Cárdenas, J.A., Gómez Serrano, C., Morales Amaral, M., Acién-Fernández, F.G., Fernández Sevilla, J.M., Molina Grima, E., 2020. Year-long assessment of a pilot-scale thin-layer reactor for microalgae wastewater treatment. Variation in the microalgae-bacteria consortium and the impact of environmental conditions. Algal Res. 50, 101983. https://doi.org/10.1016/j.algal.2020.101983.

Sand-Jensen, K., 2014. Ecophysiology of gelatinous Nostoc colonies: unprecedented slow growth and survival in resource-poor and harsh environments. Ann. Bot. 114, 17–33. https://doi.org/10.1093/aob/mcu085.

Schoenfuss, H.L., Wang, L.C., Korn, V.R., King, C.K., Kohno, S., Hummel, S.L., 2020. Understanding the ecological consequences of ubiquitous contaminants of emerging concern in the Laurentian Great Lakes Watershed: a continuum of evidence from the laboratory to the environment. In: Crossman, J., Weisener, C. (Eds.), Contaminants of the Great Lakes, the Handbook of Environmental Chemistry. Springer International Publishing, Cham, pp. 157–180. https://doi.org/10.1007/698_2020_491.

Schönknecht, G., Chen, W.H., Ternes, C.M., Barbier, G.G., Shrestha, R.P., Stanke, M., Bräutigam, A., Baker, B.J., Banfield, J.F., Garavito, R.M., Carr, K., Wilkerson, C., Rensing, S.A., Gagneul, D., Dickenson, N.E., Oesterhelt, C., Lercher, M.J., Weber, A.P.M., 2013. Gene transfer from bacteria and archaea facilitated evolution of an extremophilic eukaryote. Science 339, 1207–1210. https://doi.org/10.1126/science.1231707.

Shahid, A., Malik, S., Zhu, H., Xu, J., Nawaz, M.Z., Nawaz, S., Asraful Alam, M., Mehmood, M.A., 2020. Cultivating microalgae in wastewater for biomass production, pollutant removal, and atmospheric carbon mitigation; a review. Sci. Total Environ. 704, 135303. https://doi.org/10.1016/j.scitotenv.2019.135303.

Sies, H., 1997. Oxidative stress: oxidants and antioxidants. Exp. Physiol. 82, 291–295. https://doi.org/10.1113/expphysiol.1997.sp004024.

Silva, G.H.R., Sueitt, A.P.E., Haimes, S., Tripidaki, A., van Zwieten, R., Fernandes, T.V., 2019. Feasibility of closing nutrient cycles from black water by microalgae-based technology. Algal Res. 44, 101715. https://doi.org/10.1016/j.algal.2019.101715.

Singh, S.P., Singh, P., 2015. Effect of temperature and light on the growth of algae species: a review. Renew. Sustain. Energy Rev. 50, 431–444. https://doi.org/10.1016/j.rser.2015.05.024.

Solé-Bundó, M., Garfí, M., Ferrer, I., 2020. Pretreatment and co-digestion of microalgae, sludge and fat oil and grease (FOG) from microalgae-based wastewater treatment plants. Bioresour. Technol. 298, 122563. https://doi.org/10.1016/j.biortech.2019.122563.

Sukačová, K., Trtílek, M., Rataj, T., 2015. Phosphorus removal using a microalgal biofilm in a new biofilm photobioreactor for tertiary wastewater treatment. Water Res. 71, 55–63. https://doi.org/10.1016/j.watres.2014.12.049.

Suresh Kumar, K., Dahms, H.-U., Won, E.-J., Lee, J.-S., Shin, K.-H., 2015. Microalgae — a promising tool for heavy metal remediation. Ecotoxicol. Environ. Saf. 113, 329–352. https://doi.org/10.1016/j.ecoenv.2014.12.019.

Sutherland, D.L., Ralph, P.J., 2019. Microalgal bioremediation of emerging contaminants - opportunities and challenges. Water Res. 164, 114921. https://doi.org/10.1016/j.watres.2019.114921.

Takeuchi, N., Kohshima, S., Seko, K., 2001. Structure, formation, and darkening process of albedo-reducing material (cryoconite) on a Himalayan glacier: a granular algal mat growing on the glacier. Arctic Antarct. Alpine Res. 33, 115. https://doi.org/10.2307/1552211.

Tan, C.H., Koh, K.S., Xie, C., Tay, M., Zhou, Y., Williams, R., Ng, W.J., Rice, S.A., Kjelleberg, S., 2014. The role of quorum sensing signalling in EPS production and the assembly of a sludge community into aerobic granules. ISME J. 8, 1186–1197. https://doi.org/10.1038/ismej.2013.240.

Thorin, E., Olsson, J., Schwede, S., Nehrenheim, E., 2018. Co-digestion of sewage sludge and microalgae — biogas production investigations. Appl. Energy 227, 64–72. https://doi.org/10.1016/j.apenergy.2017.08.085.

Tolboom, S.N., Carrillo-Nieves, D., de Jesús Rostro-Alanis, M., de la Cruz Quiroz, R., Barceló, D., Iqbal, H.M.N., Parra-Saldivar, R., 2019. Algal-based removal strategies for hazardous contaminants from the environment — a review. Sci. Total Environ. 665, 358–366. https://doi.org/10.1016/j.scitotenv.2019.02.129.

Trebuch, L.M., Oyserman, B.O., Janssen, M., Wijffels, R.H., Vet, L.E.M., Fernandes, T.V., 2020. Impact of hydraulic retention time on community assembly and function of photogranules for wastewater treatment. Water Res. 173, 115506. https://doi.org/10.1016/j.watres.2020.115506.

United Nations, Department of Economic and Social Affairs, Population Division, 2019. World Urbanization Prospects: The 2018 Revision.

van der Giesen, C., Cucurachi, S., Guinée, J., Kramer, G.J., Tukker, A., 2020. A critical view on the current application of LCA for new technologies and recommendations for improved practice. J. Clean. Prod. 259, 120904. https://doi.org/10.1016/j.jclepro.2020.120904.

van der Hoek, J.P., de Fooij, H., Struker, A., 2016. Wastewater as a resource: strategies to recover resources from Amsterdam's wastewater. Resour. Conserv. Recycl. 113, 53–64. https://doi.org/10.1016/j.resconrec.2016.05.012.

Vardhan, K.H., Kumar, P.S., Panda, R.C., 2019. A review on heavy metal pollution, toxicity and remedial measures: current trends and future perspectives. J. Mol. Liq. 290, 111197. https://doi.org/10.1016/j.molliq.2019.111197.

Villarín, M.C., Merel, S., 2020. Paradigm shifts and current challenges in wastewater management. J. Hazard Mater. 390, 122139. https://doi.org/10.1016/j.jhazmat.2020.122139.

Wang, H., Ren, Z.J., 2014. Bioelectrochemical metal recovery from wastewater: a review. Water Res. 66, 219–232. https://doi.org/10.1016/j.watres.2014.08.013.

Weber, A.P.M., Bar-Even, A., 2019. Update: improving the efficiency of photosynthetic carbon reactions. Plant Physiol. 179, 803–812. https://doi.org/10.1104/pp.18.01521.

WHO (World Health Organization), 2006. Guidelines of the Safe Use of Wastewater, Excreta and Grey Water — Vol. 2: Wastewater Use in Agriculture. WHO, Geneva, Switzerland.

Wierzchos, J., DiRuggiero, J., Vítek, P., Artieda, O., Souza-Egipsy, V., Škaloud, P., Tisza, M.J., Davila, A.F., Vílchez, C., Garbayo, I., Ascaso, C., 2015. Adaptation strategies of endolithic chlorophototrophs to survive the hyperarid and extreme solar radiation environment of the Atacama Desert. Front. Microbiol. 6, 1–17. https://doi.org/10.3389/fmicb.2015.00934.

Wilbanks, E.G., Salman-Carvalho, V., Jaekel, U., Humphrey, P.T., Eisen, J.A., Buckley, D.H., Zinder, S.H., 2017. The green berry consortia of the Sippewissett salt marsh: millimeter-sized aggregates of diazotrophic unicellular cyanobacteria. Front. Microbiol. 8, 1–12. https://doi.org/10.3389/fmicb.2017.01623.

Winkler, M.K., Straka, L., 2019. New directions in biological nitrogen removal and recovery from wastewater. Curr. Opin. Biotechnol. 57, 50–55. https://doi.org/10.1016/j.copbio.2018.12.007.

Wollmann, F., Dietze, S., Ackermann, J.-U., Bley, T., Walther, T., Steingroewer, J., Krujatz, F., 2019. Microalgae wastewater treatment: biological and technological approaches. Eng. Life Sci. 19, 860–871. https://doi.org/10.1002/elsc.201900071.

Ye, S., Zeng, G., Wu, H., Zhang, C., Dai, J., Liang, J., Yu, J., Ren, X., Yi, H., Cheng, M., Zhang, C., 2017. Biological technologies for the remediation of co-contaminated soil. Crit. Rev. Biotechnol. 37, 1062–1076. https://doi.org/10.1080/07388551.2017.1304357.

Ye, Y., Ngo, H.H., Guo, W., Chang, S.W., Nguyen, D.D., Zhang, X., Zhang, J., Liang, S., 2020. Nutrient recovery from wastewater: from technology to economy. Bioresour. Technol. Rep. 11, 100425. https://doi.org/10.1016/j.biteb.2020.100425.

Ye, Y., Ngo, H.H., Guo, W., Liu, Y., Chang, S.W., Nguyen, D.D., Liang, H., Wang, J., 2018. A critical review on ammonium recovery from wastewater for sustainable wastewater management. Bioresour. Technol. 268, 749–758. https://doi.org/10.1016/j.biortech.2018.07.111.

Ylivainio, K., Lehti, A., Jermakka, J., Wikberg, H., Turtola, E., 2021. Predicting relative agronomic efficiency of phosphorus-rich organic residues. Sci. Total Environ. 773, 145618. https://doi.org/10.1016/j.scitotenv.2021.145618.

Yong, J.J.J.Y., Chew, K.W., Khoo, K.S., Show, P.L., Chang, J.-S., 2021. Prospects and development of algal-bacterial biotechnology in environmental management and protection. Biotechnol. Adv. 47, 107684. https://doi.org/10.1016/j.biotechadv.2020.107684.

Yoon, H.S., Hackett, J.D., Ciniglia, C., Pinto, G., Bhattacharya, D., 2004. A molecular timeline for the origin of photosynthetic eukaryotes. Mol. Biol. Evol. 21, 809–818. https://doi.org/10.1093/molbev/msh075.

Zeitoun, M.M., Mehana, E.-S.E., 2014. Impact of Water Pollution with Heavy Metals on Fish Health: Overview and Updates, vol. 14.

Zhang, M., Ji, B., Liu, Y., 2021. Microalgal-bacterial granular sludge process: a game changer of future municipal wastewater treatment? Sci. Total Environ. 752, 141957. https://doi.org/10.1016/j.scitotenv.2020.141957.

CHAPTER 5

Treatment of anaerobic digestion effluents by microalgal cultures

Nilüfer Ülgüdür[1], Tuba Hande Ergüder-Bayramoğlu[2] and Göksel N. Demirer[3]

[1]Department of Environmental Engineering, Düzce University, Düzce, Turkey; [2]Department of Environmental Engineering, Middle East Technical University, Ankara, Turkey; [3]School of Engineering and Technology & Institute for Great Lakes Research, Central Michigan University, Mount Pleasant, MI, United States

1. Introduction

Anaerobic digestion (AD) processes have been historically applied for the treatment and stabilization of high strength wastes such as animal manures, sewage sludges, municipal solid wastes, and agricultural and food residues (Ülgüdür and Demirer, 2019). Biogas produced in AD and incentives for AD plants have been major driving forces for biogas industry. However, there are challenges remaining in the management of digestate. These challenges are mainly associated with the drawbacks of the treatment and nutrient recovery processes offered so far, such as the requirement for additional chemicals, high energy, and high investment costs. Thus, land application of digestate as a fertilizer or soil conditioner has been considered as the most practical handling option of digestate (Ülgüdür and Demirer, 2019). On the other hand, the use of digestate on land has regulatory and applicability concerns (Ülgüdür et al., 2019a). Therefore, effective and low-cost methods need to be developed and applied for digestate management to retrieve the full benefits of AD (Praveen et al., 2018; Ülgüdür et al., 2019b).

In addition to residual organic and inorganic content, digestate has high nutrient (nitrogen and phosphorus) concentrations. This makes digestate a viable substrate for microalgae, since microalgae requires large amounts of nitrogen and phosphorus based on the composition of the biomass. Microalgal biomass contains around 10% and 1% of nitrogen and phosphorus, respectively, per unit dry weight (Muylaert et al., 2015). Total nitrogen (TN) content of digestates varies between 1.2 and 9.1 kg/Mg fresh matter (FM) of which the majority is ammonium (NH_4^+) (by 44%−81%). Total phosphorus (TP) is in the range of 0.4−2.6 kg/Mg FM and it comprises 25%−45% of water-soluble phosphorus (P) (Möller and Müller, 2012). The nutrient composition of digestates can be highly variable depending on the origin of the substrate and the management of AD process (Risberg et al., 2017). Additionally, organic matters compose 63.8%−75.0% of dry matter (DM) in digestates (Möller and Müller, 2012), which can act as a carbon and/or energy source under heterotrophic, mixotrophic, and photo-heterotrophic growth of microalgae (Chojnacka and Marquez-Rocha, 2004). Digestates may also contain essential

---

Integrated Wastewater Management and Valorization using Algal Cultures
ISBN 978-0-323-85859-5, https://doi.org/10.1016/B978-0-323-85859-5.00010-5

© 2022 Elsevier Inc.
All rights reserved.

113

nutrients such as potassium (K), magnesium (Mg), sulfur (S), and calcium (Ca) for microalgal growth (Möller and Müller, 2012; Markou et al., 2014).

Even though wastewater treatment using microalgae was first applied in 1950s (Oswald et al., 1953), the coupling of wastewater treatment with microalgal cultivation has been considered a novel approach. Microalgae can grow on a diverse range of wastewaters and are able to uptake inorganic nutrients leading to rapid growth rates (Cheng et al., 2015). Microalgal biomass obtained during wastewater treatment can be further valorized in the production of bio-based products and bioenergy (Chew et al., 2017; Katiyar et al., 2017; Koutra et al., 2018).

The treatment of digestates using microalgal cultures has been mostly investigated for the last two decades (Stiles et al., 2018). This approach is promising based on the potential for treatment and the further valorization opportunities from the microalgal biomass obtained. However, the full-scale application of microalgal digestate treatment has been limited by several challenges in application. This chapter provides the insight on the treatment of digestate by microalgal cultures along with microalgal biomass production potential with a specific focus on the challenges and the potential applicable remedies.

2. Treatment of digestates by microalgal cultures

Essential nutrients required for microalgal growth can be classified as macro- and micronutrients. Macronutrients are carbon (C), nitrogen (N), phosphorus (P), and potassium (K) and micronutrients are magnesium (Mg), calcium (Ca), sulfur (S), sodium (Na), chlorine (Cl), iron (Fe), zinc (Zn), copper (Cu), molybdenum (Mo), manganese (Mn), boron (B), and cobalt (Co) (Markou et al., 2014). Organic carbon (OC) and inorganic carbon (IC) sources are additional constituents of digestates. These carbon sources can be utilized by microalgae via photo-autotrophic, heterotrophic, mixotrophic, and photo-heterotrophic metabolisms (Xia and Murphy, 2016). The digestate is rich in macronutrients especially in the form of ammonium (NH_4) and phosphate (PO_4). Other essential elements such as K, Mg, S, Ca, Na, Cl, Fe, Zn, Cu, Mo, Mn, and Co are also commonly available in digestate. Vitamin B11 and B12 may additionally be present in digestates, which promotes the microalgal growth (Cheng et al., 2015). The composition of digestate, especially the abundance of nutrients, enables the use of digestate as a substrate for microalgae. While microalgae simultaneously serves for the treatment of the pollutional constituents in digestate. This section provides an overview on the treatment potential of digestate by microalgal cultures.

2.1 Carbon constituents

Carbon (C) forms about 17.5%—65.0% dry weight (DW) of microalgal biomass (Markou et al., 2014). Microalgae can utilize both IC via photo-autotrophic and mixotrophic metabolisms and OC via heterotrophic, mixotrophic, and photo-heterotrophic metabolisms (Chojnacka and Marquez-Rocha, 2004). Carbon dioxide (CO_2) content of biogas and IC content of digestates (mainly bicarbonate, HCO_3^-) are two streams that can be employed as IC sources for microalgae (Xia and Murphy, 2016). Additionally, digestate has residual biodegradable organic matter content, which is mainly due to incomplete stabilization in digesters (Ülgüdür et al., 2019b). Therefore, IC and OC in digestate, and biogas can be used as carbon sources for microalgae.

2.1.1 Inorganic carbon

Microalgae fixes carbon dioxide during photosynthesis. Carbon dioxide can be present as dissolved free CO_2, carbonic acid (H_2CO_3), bicarbonate (HCO_3^-), and carbonate (CO_3^{2-}) in aqueous systems depending on pH. Both CO_2 and HCO_3^- can be utilized by many microalgal species, whereas some species can uptake only CO_2 or HCO_3^- and some may use CO_3^{2-} in the aquatic environment (Markou et al., 2014).

Biogas is mainly a mixture of CH_4 (40%—75%) and CO_2 (25%—60%) (Serejo et al., 2015). Trace amounts of water (H_2O), hydrogen sulfur (H_2S), ammonia (NH_3), dinitrogen (N_2), oxygen (O_2), carbon monoxide (CO), halogenated hydrocarbons, and siloxanes can also be involved in biogas (Nagarajan et al., 2019). CO_2 content of biogas can be used for microalgae cultivation (Chuka-Ogwude et al., 2020), which may increase the value of biogas by increasing its purity. Utilization of one ton of CO_2 from the content of the biogas can produce 550 kg of algal biomass and 820 kg of oxygen (O_2), while 38 and 5 kg of N and P are required, respectively. Nutrient requirement of microalgae can be met from the content of digestate, which may reduce the associated operating costs of biogas upgrading processes (Angelidaki et al., 2019). Moreover, CO_2 content of biogas decreases the calorific value and density of biogas (Xiao et al., 2014; Awe et al., 2017; Sutanto et al., 2017). Purified biogas has much higher calorific value (39,000 kj/m^3) compared to that of raw biogas (25,000 kj/m^3), which makes it a potential substitute for natural gas and improve its use in a wider range and higher value applications (Xiao et al., 2014). Moreover, high CO_2 concentrations may lead to increased hydrocarbon and CO emissions during biogas combustion in environmental sense (Rodero et al., 2019).

Biogas upgrading can be defined as the removal of CO_2 and other impurities from the biogas content. Upgrading biogas using microalgae is a green way to increase the CH_4 content of biogas (Nagarajan et al., 2019) and has a potential to increase biomass productivities and specific growth rates of microalgae (Cheng et al., 2015; Prandini et al., 2016). Microalgal biogas upgrading has been reported as the only biotechnology that can simultaneously treat multiple constituents in the biogas composition such as CO_2, H_2S, and BTEX (mixture of benzene, toluene, and xylene) (Angelidaki et al., 2019). Microalgae and bacteria symbiosis is the key concept in microalgal biogas upgrading systems (Bahr et al., 2014). The role of microalgae in this symbiosis is to capture CO_2 and convert it into O_2 by photosynthesis. O_2 is then consumed by oxidizing bacteria, which transform H_2S to sulfate (SO_4^{2-}) and BTEX to CO_2 and water (H_2O) (Angelidaki et al., 2019). Additionally, sulfur is a micronutrient for microalgal species and SO_4^{2-} is the preferable form over H_2S and its dissolved sulfide species for microalgal assimilation (Ramírez-Rueda et al., 2020). Therefore, the products of oxidizing bacteria, i.e., CO_2 and SO_4^{2-}, can also be utilized by microalgal species in biogas upgrading.

Simultaneous biogas upgrading and digestate treatment using microalgae has been mostly studied using batch reactors in laboratory scale experiments in terms of digestate feeding strategy (fed once at the beginning of the experiment) (Table 5.1). Biogas has been usually injected to the headspace of the photobioreactor (PBR) once at the beginning of the experiment (batch mode) (Xu et al., 2015; Zhao et al., 2015; Wang et al., 2016, 2017; Gao et al., 2018) or each time prior to CO_2 exhaustion (fed-batch mode) (Prandini et al., 2016). Continuous biogas supply has generally been applied in pilot scale for biogas upgrading by microalgal—bacterial consortium (Bahr et al., 2014; Serejo et al., 2015) (Section 5).

Microalgal treatment of digestates combined with biogas upgrading has a potential to achieve 23.8%—98.2% CO_2 removal efficiency (Table 5.1). CH_4 content in biogas can be upgraded to a purity level of higher than 90% by microalgae (Yan and Zheng, 2014; Ouyang et al., 2015). Biomass concentrations obtained in digestate treatment with biogas upgrading can be in the range of 0.288—1.79 g/L (DW) with a maximum productivity of 0.311 g DW/L/d (Table 5.1). Specific growth rate of biomass in microalgal treatment is in the range of 0.134—0.6 d^{-1}. This biomass production scheme corresponds to the chemical oxygen demand (COD) removals from digestates mostly in the range of 45%—75%. TN and TP removals are recorded between 26.48% and 77%, and 22.78% and 88.79%, respectively, which are commonly measured after centrifugation and/or filtration of the samples (Table 5.1).

Digestates contain significant HCO_3^- concentration (939—1353 mg/L) (Xia and Murphy, 2016). Microalgal species can be enforced to utilize HCO_3^- from digestate content as an inorganic carbon source by not providing CO_2 to the microalgae. IC content of the digestate of piggery farm waste (80—90 mg/L as carbon equivalent), mostly HCO_3^-, was demonstrated to be consumed within 2 days for microalgal growth (Park et al., 2010).

2.1.2 Organic carbon

Some microalgal species can utilize OC such as monosaccharides, volatile fatty acids (VFAs), glycerol, and urea as an energy and/or carbon source via mixotrophic or heterotrophic metabolisms (Markou et al., 2014). Digestate COD mainly accounts for the organics such as crude fibers and organic acids (Cheng et al., 2015). COD removal can be attained in a large range of 6.0%—93.6% in digestate treatment by microalgae (Tables 5.1 and 5.2). Low COD removals are probably due to the presence of more challenging organics in terms of biodegradability remaining in the digestate for microalgal uptake (Tan et al., 2014).

Table 5.1 Studies on simultaneous digestate treatment and biogas upgrading with real digestate.

Microalgae	Substrate of digester	Reactor	Initial biogas composition	Biogas upgrading performance			Removals/removal rates from digestate content			Biomass	References
				CO_2 removal (%)	H_2S removal (%)	Final composition (%)	COD	N	P	[i]Concentration [ii]Productivity [iii]Growth rate	
Mixed culture dominated by *Scenedesmus* spp.	Raw swine manure (diluted)	Batch	CH_4: 68.7%–72.1% CO_2: 21.6%–22.5% O_2: 1.2%–1.7% H_2S:1237–1950 ppm,		>99%	CH_4: 50.4%–64.7% CO_2: 1.2%–7.5% O_2: 17.8%–21.6% H_2S: 0.4–0.5 ppmv		NH_3–N: 14.1–21.2 ± 1.2 mg/L/d	PO_4–P: 3.5 ± 2.5 mg/L/d	[i]0.9–1.1 g DW/L [ii]0.089–0.142 g DW/L/d [iii]0.5–0.6 d⁻¹	Prandini et al. (2016)
Scenedesmus obliquus	Piggery wastewater (settled, filtered, autoclaved, diluted)	Batch (7 d)	CH_4: 58.67% ± 3.45% CO_2: 37.54% ± 2.93% O_2: 0.79% ± 0.06% H_2O: 3.01% ± 0.34% H_2S < 50 ppm	54.26%–78.81%	CA[a]	CH_4: 79.28%–88.25%	61.6%–75.3% (400–3200)[b]	TN: 58.4%–74.6%	TP: 70.1%–88.8%	[ii]0.124–0.311 g DW/L/d [iii]0.334–0.486 d⁻¹	Xu et al. (2015)
Scenedesmus obliquus, *Chlorella vulgaris*, *Nitzschia palea*, *Selenastrum capricornutum*, *Anabaena spiroides*	Undefined (filtered, UV sterilized)	Batch (7 d)	CH_4: 61.38% ± 4.26% CO_2: 32.57% ± 1.98% O_2: 0.54% ± 0.02% H_2O: 5.52% ± 0.19% H_2S<0.005%		CA[a]	CH_4: 73.61%–84.28%	39.0%–68.1%	TN: 26.5%–59.1%	TP: 22.8%–60.0%	[ii]0.069–0.151 g DW/L/d [iii]0.297–0.401 d⁻¹	Wang et al. (2016)
Chlorella vulgaris, *Scenedesmus obliquus*, *Selenastrum capricornutum*, *Nitzschia palea*, *Anabaena spiroides*[S]	Undefined (filtered, UV sterilized)	Batch (10 d)	CH_4: 67.21% ± 3.72% CO_2: 29.63% ± 2.19% O_2: 0.21% ± 0.08% H_2O: 2.95% ± 0.31% H_2S < 100 ppm		CA[a]	CH_4: 79.35%–85.03%	61.0%–70.3%	TN: 59.7%–69.4%	TP: 56.1%–64.4%	[ii]0.097–0.156 g DW/L/d [iii]0.301–0.387 d⁻¹	Wang et al. (2017)
Scenedesmus obliquus, *Chlorella* sp., *Selenastrum bibraianum*	Undefined (filtered, UV sterilized)	Batch (7 d)	CH_4: 61.75% ± 2.94% CO_2: 35.28% ± 1.86% O_2: 0.31% ± 0.03% H_2O: 2.68% ± 0.29% H_2S < 50ppm	32.65%–60.11%	CA[a]	CH_4: 94.41% ± 3.16% (max)	51.9%–70.7%	TN: 47.1%–66.9%	TP: 47.6%–67.2%	[i]0.394–0.457 g DW/L (max)	Ouyang et al. (2015)

Continued

Table 5.1 Studies on simultaneous digestate treatment and biogas upgrading with real digestate.—cont'd

Microalgae	Substrate of digester	Reactor	Initial biogas composition	Biogas upgrading performance CO$_2$ removal (%)	H$_2$S removal (%)	Final composition (%)	Removals/removal rates from digestate content COD	N	P	Biomass [i]Concentration [ii]Productivity [iii]Growth rate	References
Chlorella vulgaris, Scenedesmus obliquus, Neochloris oleoabundans	Undefined (UV sterilized, filtered)	Batch (6 d)	CH$_4$: 61.32% ± 5.74% CO$_2$: 34.45% ± 3.48% O$_2$: 0.62% ± 0.05% H$_2$O: 3.65% ± 0.39% H$_2$S < 0.005%	40.25%–62.31%	CA[a]	CH$_4$: 78.53%–82.79% O$_2$: 0.19%–0.62% H$_2$O: 1.54%–3.65% (Scenedesmus obliquus)	45.3%–63.1% 78.5–109.2 mg/L/d	TN: 46.4%–59.9%	TP: 40.3%–63.2% 0.81–1.26 mg/L/d	[ii]0.014–0.217 g DW/L/d [iii]0.134–0.451 d^{-1}	Zhao et al. (2015)
Chlorella sp.	Undefined (UV sterilized, filtered)	Batch (6 d)	CH$_4$: 64.21% ± 5.82% CO$_2$: 31.38% ± 3.46% O$_2$: 0.68% ± 0.04% H$_2$O: 3.79% ± 0.45% H$_2$S < 0.005%	46.77%–57.73%	CA[a]	CH$_4$: 89.76%–93.68% O$_2$: 0.68%–1.04% H$_2$O: 3.64%–3.84%	47.3%–56.1%	TN: 40.6%–52.9%	TP: 39.8%–48.5%	[i]0.288–0.494 g DW/L	Yan and Zheng (2014)
Chlorella vulgaris, Scenedesmus obliquus[c]	Piggery wastewater (diluted)	Batch (10 d)	CH$_4$: 61.23% ± 3.68% CO$_2$: 34.68% ± 2.46% O$_2$: 0.84% ± 0.06% H$_2$O: 3.22% ± 0.29% H$_2$S < 100 ppm	42.32%–58.27%	CA[a]	CH$_4$: ~75%–81%	~61%–74%	TN: ~59%–77%	TP: ~72%–82%	[ii]0.049–0.090 g DW/L/d [iii]0.159–0.203 d^{-1}	Gao et al. (2018)
Scenedesmus sp.	GAC[d] treated desizing textile wastewater (diluted, centrifuged)	Continuous	CH$_4$: 65.2%–70.1% CO$_2$: 21%–23%	23.8%–98.2%		CH$_4$: 60.4%–66.3% CO$_2$: 0.4%–16%	2.4%–69.1% (3000–7000)[b]			[i]0.32–1.79 g DW/L [ii]−0.009–0.18 g DW/L/d [iii]0.18 d^{-1}	Nguyen et al. (2019)

[a] CA: Pretreated by chemical adsorption.
[b] The numbers indicate initial concentration in mg/L.
[c] Only the data on monoculture study is given.
[d] GAC: Granular activated carbon.

Table 5.2 Microalgal treatment of digestates.

Microalgae	Substrate of digester	Reactor	COD	N	P	[i]Concentration [ii]Productivity [iii]Growth rate	References
			Removals/removal rates from digestate content			**Biomass**	
Chlorella sp.	Sludge (diluted)	Batch (7—8 d)	47.5%—86.3% (454—3780)[a]	TN: 4.4—42.6 mg/L/d (221—1210)[a]	TP: 0.25—4.10 mg/L/d, (5—28)[a]	[i]0.96—2.11 g TSS/L [ii]0.12—0.45 g TSS/L/d	Åkerström et al. (2014)
Nannochloropsis salina	Municipal wastewater (diluted, salinity adjustment)	Batch (10 d)		TAN[b]: 100% (68—546)[a] TN: 87%—100% (80—640)[a]	TP: 99%—100% (11.43—91.44)[a]	[i]0.92 g DW/L(max) [ii]0.068—0.092 g DW/L/d [iii]0.334—0.645 d[−1]	Cai et al. (2013a)
		Semicontinuous		TN: 53.5%—89.0% (160)[a] 13.4—56.5 mg/L/d	TP: 22.9%—82.8% (22.86)[a], 2.3—4.3 mg/L/d	[ii]0.087—0.155 g/L/d	
Nannochloropsis salina, Synechocystis sp.	Municipal wastewater (diluted, salinity adjustment)	Batch (10 d)		TN: 71.2%—100% (80—2667)[a] TAN[b]: 82.5%—100% (68—2276)[a]	TP: 83.6%—100% (11.43—381)[a]	[ii]0.041—0.151 g DW/L/d	Cai et al. (2013b)
		Semicontinuous (18 d)		TN: 100% (80)[a] (max)	TP: 100% (11.43)[a] (max)	[ii]~0.125—0.212 g DW/L/d	
Chlorella PY-ZU1	Mixture of swine manure and sewage (centrifuged, autoclaved)	Batch (8—13 d)	~49%—79%	NH₃—N: ~52%—73%	TP: ~75%—99.6%	[i]4.81 g DW/L (max) [ii]0.601 g DW/L/d (max)	Cheng et al. (2015)

Continued

Table 5.2 Microalgal treatment of digestates.—cont'd

Microalgae	Substrate of digester	Reactor	Removals/removal rates from digestate content			Biomass [i]Concentration [ii]Productivity [iii]Growth rate	References
			COD	N	P		
Chlorella sp.	Dairy manure (diluted, filtered)	Batch (21 d)	27.4%—38.4%	TKN[c]: 75.7%—82.5% NH$_3$-N: 100% (81—178)[a]	PO$_4$: 62.5%—74.7%	[i]1.71 g DW/L [iii]0.282—0.409 d^{-1}	Wang et al. (2010a)
Chlorella vulgaris	Dairy manure (Diluted)	Semicontinuous	41.2%—55.4%	TN: 56.3%—93.6% NH$_4$-N: 58.3%—100%	TP: 82.4%—89.2%	[i]1.35 g DW/L	Wang et al. (2010b)
Chlorella pyrenoidosa	Starch wastewater (Settled, filtered, sterilized mixed with alcohol wastewater)	Batch (9 d)	7.1%—75.8% 95—629 mg/L/d	TN: 19.3%—91.6% (265—310)[a]	TP: 39%—98.9% (28.5—29.5)[a]	[i]0.89—3.01 g DW/L [ii]0.06—0.58 g DW/L/d [iii]0.18—0.56 d^{-1}	Yang et al. (2015)
Chroococcus sp.	*Chroococcus* sp. (settled, filtered, diluted)	Batch (12 d)	70% (578.25)[a]	TAN[b]: 85.2% (58.98)[a] NO$_3$-N[d]: 77.3% (13.90)[a]	TDP[e]: 89.3% (13.56)[a]	[i]~0.33—1.42 g DW/L	Prajapati et al. (2014)
Chlorella minutissima, Chlorella sorokiniana, Scenedesmus bijuga	Poultry litter (centrifuged, diluted)	Batch (8—12 d)		TN: 60%	TP: 80%	[i]0.612 g DW/L (max) [i]14 mg Chl-a/L (max) [ii]0.076 g/L/d (max)	Singh et al. (2011)
Desmodesmus sp.	Pig farm waste (filtered, diluted)	Batch (14 d)		TN: 75.5%—100%, 1.671—4.542 mg/L/d NH$_4$-N: 100%, 2.257—5.284 mg/L/d	TP: 100%, 0.235—0.326 mg/L/d PO$_4$-P: 100%, 0.290—0.300 mg/L/d	[i]0.272—0.412 g DW/L [ii]0.019—0.029 g DW/L/d	Ji et al. (2014)

Microalgae	Substrate	Culture mode	Nutrient removal (N)	Nutrient removal (P)	Biomass productivity	Reference
Desmodesmus sp.	Pig farm waste (filtered, diluted)	Batch (10 d)	TN: 75.6%–82.5% (25–77.4)[a], 3.333 mg/L/d NH$_4$–N: 92.7%–100% (22.4–70.8)[a], 3.723 mg/L/d	PO$_4$–P: 100% (0.81–3.12)[a], 0.132 mg/L/d	[i]0.269–0.324 g DW/L	Ji et al. (2015)
Chlorella vulgaris, Neochloris oleoabundans, Scenedesmus obliquus	Mixture of cattle slurry and raw cheese whey (centrifuged, diluted)	Fed-batch (40 d)	TN: 94.2%, 5.904 mg/L/d NH$_4$–N: 91.1%, 6.706 mg/L/d	PO$_4$–P: 88.7%, 0.161 mg/L/d	[i]1.039 g DW/L (max)	Franchino et al. (2013)
		Batch (21 d)	NH$_4$–N: 83.7%–99.9%, 3.0–7.8 mg/L/d	PO$_4$–P: 94.4%–97.3%, 0.09–0.36 mg/L/d	[ii]0.20–0.26 g DW/L/d [iii]0.23–0.64 d^{-1}	
Scenedesmus acuminatus	Piggery farm waste (filtered, autoclaved, diluted)	Batch	NH$_4$–N: 5.20–18.4 mg/L/d		[ii]0.046–0.118 g DW/L/d [iii]0.038–0.091 d^{-1}	Park et al. (2010)
		Semicontinuous	NH$_4$–N: 89%, 19.2 mg/L/d		[ii]0.213 g DW/L/d	
Chlorococcum sp.	Pig manure (Filtered, centrifuged, diluted)	Batch (24–27 d)	NH$_4$–N: 5.3 mg/L/d	PO$_4$: 1.6 mg/L/d	[i]0.52–0.85 g DW/L [ii]0.017–0.029 g DW/L/d [iii]0.48 d^{-1} (max)	Montero et al. (2018)

Continued

Table 5.2 Microalgal treatment of digestates.—cont'd

Microalgae	Substrate of digester	Reactor	Removals/removal rates from digestate content			Biomass	References
			COD	N	P	[i]Concentration [ii]Productivity [iii]Growth rate	
Chlorella vulgaris	Liquid swine manure (ammonia stripped, settled, centrifuged, diluted, autoclaved, and nonautoclaved)	Batch (7d)	31.2%–78.7% (4600–9500)[a]	TN: 49.2%–70.8%, (103.1–230.0)[a] NH_4–N: 98.5%–99.9%, (42.9–124.2)[a]	TP: 22.0%–54.4% (26.5–57.0)[a]	[i]~0.77–1.30 g VSS/L [ii]0.10–0.14 g VSS/L/d	Deng et al. (2017)
Chlorella vulgaris	Sludge (autoclaved, nonautoclaved, diluted)	Batch	59.2%–93.6% (~80–1200)[a]	TN: 16.4%–99.3% NH_4–N: ~11.4%–100%, (35–38)[a]	PO_4–P: 19.0%–100%	[i]0.07–1.11 g DW/L [ii]0.008–0.093 g DW/L/d	Ge et al. (2018)
Chlorella vulgaris, Scenedesmus acuminatus	Pulp and paper industry biosludge and wastewater treatment plant sludge (centrifuged, filtered, diluted)	Batch (11–14 d)	27.6%–55.4% (600–1259)[a]	NH_4–N: 23.8%–99.9% (70–840)[a], 17.1 mg/L/d	PO_4–P: >96.9% (0.2–16)[a]	[i]0.6–9.4 g VSS/L	Tao et al. (2017)

Chlorella vulgaris, Scenedesmus obliquus, Chlorella sorokiniana	Black water	Batch (13—14 d)		NH$_4$—N: 26%—98.4%, 13—121.9 mg/L/d	PO$_4$—P: 22%—100%, 1.4—16.8 mg/L/d	[i]3.7—12.1 g DW/L	Fernandes et al. (2015)
Mixed culture dominated by *Chlorella* and *Scenedesmus* spp.	Farm digester (centrifuged, diluted)	Batch (14 d)		NH$_4$—N: ~43% (190)[a], 6.3 mg/L/d	PO$_4$—P: 15%—100% (3—67)[a], 0.6—2.0 mg/L/d	[i]0.854 ± 0.027 g TSS/L [iii]0.78 d^{-1}	Marcilhac et al. (2015)
Desmodesmus sp.	Piggery wastes	Batch (7d)		TN: 12.9%—50.4% NH$_4$—N: 17.6%—54.1%, (374.35)[a]	TP: 35.5%—100%	[i]5.47 mg/L Chl-total (max)	M. Wang et al. (2020)
Chlorella vulgaris	Sewage sludge (centrifuged, filtered, autoclaved, diluted)	Batch	76.13%—90.06%	NH$_4$—N: 19.8%—31.8%	PO$_4$—P: 41.4%—90.9%	[i]0.67—4.23 g DW/L [ii]0.172—0.433 g DW/L/d	Cho et al. (2015)
Mixed culture dominated by Chlorella vulgaris	Mixture of laying hen (90%), cattle (10%) manures	Batch		NH$_4$—N: 47.6%—93.7%	TP: 95.6%—97.8% PO$_4$—P: 99.4%—100%	[i]2.1—4.72 g DW/L [i]6.4—15.5 mg/L Chl-a [ii]0.15—0.22 g DW/L/d	Ülgüdür et al. (2019a)

Continued

Table 5.2 Microalgal treatment of digestates.—cont'd

Microalgae	Substrate of digester	Reactor	Removals/removal rates from digestate content			Biomass	References
			COD	N	P	[i]Concentration [ii]Productivity [iii]Growth rate	
Scenedesmus sp.	Grass and molasses wastewater individually (filtered, centrifuged, diluted)	Batch (16 d)	6%—49%	TN: 20%—92% NH_3-N: 33%—100% NO_3-N[d]: 10%—74%	TP: 7%—98%	[i]1.4—4.3 g DW/L	Yang et al. (2017)
Scenedesmus sp.	Mixture of swine manure and microalgae (diluted, autoclaved)	Batch then continuous		NH_3-N: 27.2 —41.2 mg/L/d, (38.8—58.8)[a]	PO_4-P: 3.96 —6.69 mg/L/ d (5.66—9.55)[a]	[ii]0.58—0.67 g DW/L/d [iii]1.82—1.94 d^{-1} (max)	Dickinson et al. (2014)

[a] The numbers indicate initial concentration in mg/L.
[b] TAN: Total ammoniacal/ammonia nitrogen.
[c] TKN: Total kjeldahl nitrogen.
[d] NO_3-N: Nitrate nitrogen.
[e] TDP: Total dissolved phosphorus.

Digestates may contain VFAs such as lactate, acetate, propionate, butyrate, valerate, isovalerate, and formate (Nagarajan et al., 2019). VFAs can constitute a large fraction of soluble COD content of digestates (around 75%) (Cho et al., 2015). Removal of VFAs has been recorded to be over 90% during the treatment of anaerobically digested starch wastewater in an upflow anaerobic sludge blanket (UASB) reactor (initial concentration of approximately 60 mg/L) (Yang et al., 2015). Almost complete VFA removal has also been achieved in microalgal (*Chlorella vulgaris*) treatment of anaerobically digested sewage sludge (ATSS) obtained from a laboratory scale continuously stirred tank reactor operated at a pH of 6.5 ± 0.1 (Cho et al., 2015). The digestate used in the treatment ATSS using *Chlorella vulgaris* had a VFA concentration of approximately 10,000 mg/L (calculated from the given data of COD equivalent of VFA (13.73 g/L) and 75% fraction of VFA in COD). The build-up of VFA may lead to a decrease in pH and corresponding acidification in anaerobic digesters, which may develop toxic environmental conditions for methanogens (Franke-Whittle et al., 2014). Methanogens can tolerate to high VFA concentrations as long as pH is kept in an optimum range (Parkin and Owen, 1986) as applied by Cho et al. (2015). However, VFAs can leave the digester depending on the digester performance (Uggetti et al., 2014) as measured by Cho et al. (2015). Microalgal treatment of VFAs in digestate content can constitute a valorization pathway for VFAs. VFAs can be utilized as a carbon source for microalgal biomass based on the high VFA removal efficiencies obtained in digestate treatment (Cho et al., 2015).

Potential removal mechanisms of VFAs by microalgal biomass can be adsorption or assimilation (Yang et al., 2015). On the other hand, the composition of VFAs also carries a significant aspect, since not all VFA components can be equally tolerated by microalgae. For example, acetate tolerance $(1-10 \text{ g/L})$ has been reported to be higher than that of butyrate $(0.1-1.0 \text{ g/L})$ for most of the microalgal species (Chen et al., 2018).

2.2 Nitrogen constituents

Nitrogen composes 1%–14% of microalgal biomass (DW) and it is involved in the synthesis of nucleic acids, amino acids, and pigments as chlorophylls and phycocyanin. Microalgae can assimilate NH_4^+, nitrate (NO_3^-), nitrite (NO_2^-), and nitric oxide (NO) as inorganic nitrogen sources. Some species may also be capable of using organic nitrogen forms like amino acids (Markou et al., 2014).

NH_4-N has been indicated as the easily utilizable form of nitrogen by microalgae (Wang et al., 2020b), since the uptake of ammonium requires less energy than other nitrogen sources (Markou et al., 2014). Oxidized forms of nitrogen, NO_2^-, and NO_3^- can also be utilized by microalgal species (Marcilhac et al., 2015) via reduction of NO_3^- to NO_2^- and then NO_2^- to NH_4^+ which necessitates more energy (Wang et al., 2019). Digestate does not contain considerable amounts of NO_2^- and NO_3^- as compared to NH_4^+ as reported by many researchers (Botheju et al., 2010; Bahr et al., 2014; Cho et al., 2015; Marjakangas et al., 2015; Viruela et al., 2016; Tao et al., 2017; Ghyselbrecht et al.,

2019). Aerobic conditions are required for conversion of NH_4^+ into NO_2^- and NO_3^- via nitrification, which can be observed to occur during storage or can be applied as a treatment option for digestates. However, microalgae prefers to utilize NH_4-N under the presence of multiple inorganic nitrogen sources of NH_4^+, NO_2^-, and NO_3^- (Li et al., 2019).

Nitrogen removal by microalgae from digestate has been mostly reported in terms of TN after the separation of microalgal biomass via centrifugation or filtration or both. TN removal is in the range of 12.9%—100.0%, while the removal of ammonium related parameters such as NH_4-N, ammonia nitrogen (NH_3-N), and total ammoniacal/ammonia nitrogen {TAN = $[NH_4-N]$ + $[NH_3-N]$} was reported to be 11.4%—100.0% in batch studies (Table 5.2). TN removal varied between 53.5% and 100.0% in fed-batch, semicontinuous, and continuous studies (Table 5.2). Main nitrogen removal mechanisms in a digestate treatment process using microalgal cultures are microalgal and bacterial uptake, volatilization (stripping), and nitrification/denitrification (Ülgüdür et al., 2019a). Thus, nitrogen removal is required to be reported specific to the removal mechanisms involved, as there may be chemical and biological interactions in the removal apart from the microalgal assimilation.

2.3 Phosphorus constituents

Microalgal biomass contains 0.05%—3.30% phosphorus (P) of dry weight. P is present in the form of orthophosphates, polyphosphates, pyrophosphates, metaphosphates, and their organic forms in wastewaters (Markou et al., 2014). Among these P forms, orthophosphate is the most available form for microalgae. Digestate treatment by microalgae has a potential to reduce total phosphorus (TP) content by 7%—100% and that of phosphate phosphorus (PO_4-P) by 15%—100% (Tables 5.1 and 5.2). Phosphorus can also be removed by uptake of microalgal—bacterial consortium, surface adsorption (Åkerström et al., 2014), and precipitation (Li et al., 2011). Adsorption of P on the cell wall can be as high as 90% of TP removed from the system, whereas the cellular content may correspond to only a small portion (0.4%) (Åkerström et al., 2014). Hydroxyapatite and struvite like precipitates can be formed in a pH range of 8—10 depending on availability of ions such as Ca, Mg, NH_4, and PO_4 (Tan et al., 2014). Precipitate formation can be regarded as phosphorus removal mechanism in microalgal systems. However, precipitation may end up with deposits within the reactors and pipelines creating operational problems in long-term operation of facilities (Tan et al., 2014; Yang et al., 2015). Additionally, phosphorus precipitate formation is not considered as a sustainable phosphorus recycling option in microalgal systems since precipitate formation converts water-soluble P into solid form (Singh et al., 2011), which may hinder its utilization by microalgae.

2.4 Heavy metals and metalloids

Digestates may contain cadmium (Cd), chromium (Cr), cobalt (Co), copper (Cu), manganese (Mn), lead (Pb), zinc (Zn), and nickel (Ni) at concentrations of 7.1—19.0,

120—500, 16—74, 650—1800, 140—13,000, 70—240, 2800—10,000, and 74—210 µg/L, respectively (Markou et al., 2018). Arsenic (As) content of digestate has also been reported in the range of 28—690 µg/L (Cheng et al., 2015; Gao et al., 2018).

Heavy metals and metalloids (HMMs) may either promote algal growth via acting as an essential component of metabolic functions or stimulating, or conversely, may show inhibitory effects on growth (Miazek et al., 2015; Zhang et al., 2013). Metals can be components of photosynthetic electron transport proteins (Cu, iron (Fe)), photosynthetic water oxidizing centers (Mn), or vitamins (Co). They can also function as cofactors for enzymes involved in the fixation of CO_2 (Zn in carbonic anhydrase), transcription of DNA (Zn in RNA polymerase), phosphorus uptake (Zn in alkaline phosphatase), and N_2 assimilation (molybdenum (Mo), Fe, vanadium (V) in nitroglycerin). Ni can also be a cofactor in urease enzyme of *Phaeodactylum tricornutum, Cyclotella cryptica, Thalassiosira weissflogii*, and *Thalassiosira pseudonana*, when urea is the sole source of nitrogen (Miazek et al., 2015). Additionally, the stimulatory effect of HMMs such as Pb, aluminum (Al), Co, As, and Cd on microalgal growth has been reported (Miazek et al., 2015). Even 39%—57% higher growth rate has obtained for *Ostreococcus tauri* when exposed to 10 and 30 µM arsenate (As(V)) and arsenite (As(III)) concentrations which also demonstrates the stimulatory effect of As (Zhang et al., 2013). Nevertheless, high concentrations of HMMs may have inhibitory effects on microalgae, as high concentrations may damage photosynthetic mechanism, prevent cell division, and enzyme activity (Miazek et al., 2015). The potential inhibitory effects of HMMs at high concentrations require attention in microalgal treatment systems.

The removal of HMMs from digestate content has been reported to be in the range of 16%—97% in microalgal treatment systems (Table 5.3). Microalgal biomass can be effective on the removal of cationic pollutants since their cell wall carry negatively charged functional groups (Åkerström et al., 2014). Microalgae may have tendency to uptake or adsorb certain types of HMMs (Cheng et al., 2015). The major removal pathways for HMMs in microalgal treatment processes for digestates can be addressed as bioadsorption and bioaccumulation (Yang et al., 2017), and metal hydroxide precipitation induced at pHs over 6.5 (Kumar et al., 2015).

2.5 Micropollutants

Organic micropollutants may enter AD processes via feeding strategies applied to animals in agro-industrial practices to promote growth, feeding efficiency, and/or weight gain. They can be used as a therapeutic option as well (Markou et al., 2018). AD plants operated with treatment sludges are also subject to micropollutants, since conventional wastewater treatment systems operated with activated sludge processes are usually ineffective to remove such pollutants (Vassalle et al., 2020). AD processes can totally or partially remove, or conversely may increase the concentration of micropollutants.

Table 5.3 Heavy metal removals from digestates by microalgae.

Raw substrate of digestate	HMM	Removal, %	References
Swine manure and sewage	Pb	64.4—81.6	Cheng et al. (2015)
	As	35.7—64.3	
	Hg (mercury)	78.0—88.4	
	Cd	62.0—90.0	
Piggery wastewater[a]	Zn	45.9—53.3	
	Cu	42.4—53.6	Gao et al. (2018)
	As	35.3—38.5	
Sludge after dewatering	Al	86.0	Åkerström et al. (2014)
		93.8	
		94.9	
		91.3	
	Total of Fe, Mn, Cu, Zn	74.0	
		83.3	
		89.5	
		73.0	
Grass	Fe	54—87	
	Co	29—66	
	Ni	28—57	
	Cu	32—90	
	Zn	41—67	
	Mn	73—97	
	Ba (barium)	25—57	Yang et al. (2017)
	Mo	22—95	
	As	33—44	
	Cd	67—75	
	Pb	28—62	
	Cr	35—72	
Molasses ww	Fe	48—76	Yang et al. (2017)
	Co	18—69	
	Ni	23—82	
	Cu	60—82	
	Zn	36—77	
	Mn	51—75	
	Ba	16—53	
	Mo	27—33	
	As	39—42	
	Cd	66—70	
	Pb	49—63	
	Cr	32—63	

[a] Only the data on monoculture study is given.

Even though the related studies are scarce in literature, coupling of AD with microalgal nutrient removal process can offer an advantage of reducing the concentration of micropollutants. Pharmaceuticals (ibuprofen, naproxen, diclofenac, paracetamol, and gemfibrozil), estrogens (estrone, estradiol, ethinylestradiol, and estriol), and xenoestrogens (nonylphenol and bisphenol-A) have been recently tracked in such a coupled treatment system including UASB reactor and high-rate algal pond (HRAP) for treatment of raw sewage. Micropollutants received by HRAP (after UASB reactor) could be reduced by 40%—70% for pharmaceuticals, 88%—95% for estrogens, and 44%—70% for xenoestrogens in HRAP (Vassalle et al., 2020). The removal of hormones and pharmaceuticals has also been tested via external supply to digestate content, which ended up by 30%—100% removal (Hom-Diaz et al., 2015; Wilt et al., 2016). The major removal pathways for micropollutants in microalgal systems can be defined as biodegradation, adsorption, photodegradation, volatilization, and transformation (Markou et al., 2018).

3. Microalgal growth in digestates

Microalgal growth in digestate indicates the uptake of C, N, P, metals and metalloids, and micropollutants from the environment, which may decrease the pollutant load of the digestate as a result (Diniz et al., 2017). Microalgae may exert photo-autotrophic, heterotrophic, mixotrophic, and/or photo-heterotrophic metabolisms during digestate treatment depending on the changes in environmental conditions. Mixotrophic metabolism enables higher growth rates and biomass productivities, which can reach to one or two magnitudes higher biomass productivities compared to photo-autotrophic growth (Chojnacka and Marquez-Rocha, 2004). Higher biomass productivities in turn result in increased treatment capabilities in terms of nitrogen, phosphorus, IC and OC (Xia and Murphy, 2016). Mixotrophic growth can also act as a protective agent against photo-inhibition (Chojnacka and Marquez-Rocha, 2004).

Chlorella and *Scenedesmus* sp. have been the most commonly used microalgae. These species have been dedicated as being within the top eight tolerant genera against organic pollution (Palmer, 1969). Thus, these species may present high adaptability to wastewaters with high organic content such as domestic and livestock wastewaters (Tripathi and Kumar, 2017; Viswanaathan and Sudhakar, 2019) and digestate. Biomass growth, productivities, and specific growth rates have been in the range of 0.07—12.1 g/L, 0.01—0.67 g DW/L/d, and 0.038—1.94 d^{-1}, respectively, in the cultures dominated by *Chlorella* and/or *Scenedesmus* sp. given in Tables 5.1 and 5.2.

The content of lipids, carbohydrates, and proteins can vary according to the growth phase of microalgae as well as environmental conditions and the growth metabolism established by microalgae in digestates (Montero et al., 2018). Microalgae grown in digestates can contain carbohydrates, proteins, and lipids in the range of 6.3%—60.5%, 24.3%—64.2%, and 1.0%—41.0%, respectively (Table 5.4).

Table 5.4 Contents of microalgae grown in digestates.

Inoculum	Carbohydrates	Proteins	Lipids	References
Chlorella vulgaris, Scenedesmus acuminatus	6.3%–60.5%	24.3%–41.8%	19.9%–35.9%	Tao et al. (2017)
Phaeodactylum tricornutum	12.6%–26.9%	26.8%–31.6%	14.1%–23.0%	Simonazzi et al. (2019)
Filamentous microalgae	60% ± 7%	36% ± 6%	1% ± 0%	Serejo et al. (2015)
Chlorella minutissima, Chlorella sorokiniana, Scenedesmus bijuga	18.8%–27.3%	37.3%–42.9%	8.2%–13.5%	Singh et al. (2011)
Chlorella vulgaris	8.8%–16.6%	47.1%–58.8%	17.4%–33.7%	Deng et al. (2017)
Chlorella vulgaris	8.5%–16.7%	49.8%–64.2%	5.6%–29.4%	Ge et al. (2018)
Mychonastes homosphaera, Kalina, Puncochárová	31%–46%	~28%–45%	~2%–13%	Franco-Morgado et al. (2018)
Chlorella sp., *Scenedesmus* sp., pennate diatom	17.2%–39.2%	17.6%–31.4%	26.0%–41%	Nwoba et al. (2016)

4. Challenges and potential remedies for digestate treatment by microalgae

4.1 Turbidity

Turbidity caused by dissolved and suspended materials in digestates has been addressed as a major drawback of digestate use in microalgal systems (Monlau et al., 2015). High turbidity of digestates may reduce photosynthetically active radiation as well as microalgal growth. Even though microalgae have an ability to partly remove suspended materials that cause turbidity (Wang et al., 2010a; Wang et al., 2020a), some pretreatment methods such as filtration and centrifugation can be applied. Another pretreatment method proposed with the aim of using undiluted digestate for microalgae growth is a two-step process, at which the first step is air stripping for NH_4^+ recovery and the second step is an activated carbon adsorption process for turbidity reduction (Marazzi et al., 2017). This coupled process has a potential to yield approximately 10-folds higher biomass production compared to raw digestate (Marazzi et al., 2017). Integrated flocculation—biological contact oxidation process has been recently proposed to remove turbidity as well as COD, N, and P from digestate content as a pretreatment option (Zhou et al., 2019). Nevertheless, environmental performance of the latter system needs to be improved in terms of human health, ecosystem quality, and climate change categories to reduce environmental impacts and to be comparable with those of microalgal treatment of diluted digestate (Duan et al., 2020). Even though higher microalgal biomass can be obtained when digestate is pretreated, the common concern is the associated costs of the pretreatment processes employed. Therefore, pretreatment methods developed should be feasible and financially attractive.

4.2 Ammonium/ammonia inhibition

Ammonium (NH_4^+) and ammonia (NH_3) are mostly used interchangeably with each other. However, NH_4^+ is the ionized form, while NH_3 is the unionized volatile form, so called free ammonia (FA). NH_4^+ and FA concentrations are mostly related to the pH and temperature in an aqueous solution. Dominant form is FA when pH is above pK_a, negative logarithm of acid dissociation constant (pK_a is 9.25 at 25°C). NH_4^+ is the dominant form when pH is lower than pK_a (Evangelou, 1998). The inhibitory concentrations of NH_4^+ and FA are specific to microalgal strain (Wang et al., 2019). NH_4—N concentrations above 350 mg/L are reported to be toxic to phytoplankton (Barsanti and Gualtieri, 2014). Yet, much lower (200 mg/L) or higher NH_4—N (1600 mg/L) concentrations have been used in microalgal treatment of digestates (Ayre et al., 2017; Deng et al., 2017). Indeed, microalgae could be grown at elevated NH_4—N concentrations (800—1600 mg/L) in an outdoor study (Ayre et al., 2017). Compared to NH_4—N, FA has higher toxicity for microalgae which is based on the uncharged chemistry and solubility in lipids of microalgae, so that it can diffuse across cell membrane (Wang et al., 2019). FA concentration over ~ 36 mg/L (FA-nitrogen 30 mg/L) has been reported to have an inhibitory effect on the growth of green microalgae (Fernandes et al., 2015). Even much lower FA concentrations (between 2.9 and 17.6 mg/L) have been indicated as being toxic to *Scenedesmus obliquus* and *Chlorella vulgaris* (Prandini et al., 2016). Considering the fact that FA nitrogen constitutes 0.6% at pH 7%, 5% at pH 8% and 36% at pH 9 (25°C) of TAN, its inhibition is very likely at pHs above neutral. Liquid digestates may contain high concentrations of TAN in the range of 1000–3000 mg/L (Xia and Murphy, 2016). Thus, it may be advisable to keep pH at around neutral to minimize FA toxicity (Fernandes et al., 2015). Apart from the minimization of FA toxicity, such a pH management strategy can also prevent N loss from the system via volatilization.

Dilution of digestates is another common practice to avoid NH_4^+/FA inhibition and reduce turbidity (Cai et al., 2013a; Xia and Murphy, 2016). However, dilution has two main drawbacks. Firstly, it reduces nutrient concentration and thus might lower the microalgal biomass production (Xia and Murphy, 2016; Tao et al., 2017). Secondly, dilution may result in high water footprints considering high dilution factors being applied (10–20-folds) (Wang et al., 2020b). Seawater, wastewater, secondary, or tertiary wastewater can be applicable as diluents (Xia and Murphy, 2016; Praveen et al., 2018); however, availability and accessibility of such diluents may be impractical. As a response to requirement of using undiluted digestates in microalgal treatment, several laboratory and pilot scale applications have been performed (Cheng et al., 2015; Nwoba et al., 2016; Ayre et al., 2017; Raeisossadati et al., 2019) with pretreatment options such as sand filtration (Nwoba et al., 2016; Ayre et al., 2017; Raeisossadati et al., 2019) or aeration (Cheng et al., 2015).

High NH_4–N concentrations can also be managed via nitrification process. Digestates can be pretreated in an activated sludge process involving nitrification. Organics can be degraded simultaneously with the oxidation of NH_4^+ to NO_3^- in such pretreatment applications (Praveen et al., 2018). Nitrification and microalgal nutrient removal can take place simultaneously within the same reactor under unsterilized conditions (Ülgüdür et al., 2019a). Nitrification of digestates and desulfurization of biogas can also be simultaneously achieved as a pretreatment step for both AD effluents, i.e. NH_4^+, and H_2S (Sekine et al., 2020).

Other NH_4^+/NH_3 removal processes such as air stripping or struvite precipitation can be applied as a pretreatment option for a microalgal process (Marazzi et al., 2017; Sayedin et al., 2020). Air stripping has a potential to reduce ammonium concentrations to an applicable level (200–500 mg N/L) for microalgal treatment processes (Marazzi et al., 2017). NH_4^+ and PO_4 concentrations decrease due to precipitation in struvite formation, but N/P ratio can be close to an optimum value for microalgal growth after struvite is precipitated (Sayedin et al., 2020).

4.3 Phosphorus limitation

Phosphorus (P) limitation can be observed in the cases of excessive dilutions (Åkerström et al., 2014), and P removal via pretreatment of digestates such as centrifugation and filtration (Ülgüdür et al., 2019a). Several digestates may also be deficit in P such as those obtained from a digester treating wastewater sludge, especially if the wastewater treatment plant has a chemical precipitation unit (Tao et al., 2017) which converts soluble phosphorus into solid form. Phosphorus can be supplemented from commercial chemicals such as potassium dihydrogen phosphate (Cheng et al., 2015). However, it is not a sustainable approach depending on limited phosphorus reserves in the world and the associated cost of chemical supply. Phosphorus can be bound to components such as humic substances and/or precipitated with polyvalent cations (Ca, Mg), which reduces the bioavailability of P (Markou et al., 2014). Bounded P can be solubilized for microalgal utilization (Fernandes et al., 2015; Ülgüdür et al., 2019a). Simultaneous dissolution and uptake of P has a potential to increase microalgal biomass by approximately two folds, avoiding P limited conditions (Ülgüdür et al., 2019a).

4.4 Carbon limitation

Optimum C/N ratio of microalgal species by mass ranges between 4 and 8, and C content of the digestates may not be sufficient in some cases. Additional IC and OC sources may be supplied externally (Xia and Murphy, 2016). Apart from the external supply of carbon, microalgal systems can be operated in continuous or semicontinuous mode as a potential solution for carbon limitation (Nwoba et al., 2016).

4.5 Dominance of other communities

Nonsterile aerobic conditions have a potential to be dominated by fast-growing heterotrophic microorganisms. Microalgae might starve and even be a feed source for other microorganisms under such conditions (Praveen et al., 2018). On the other hand, microalgal cultures with high diversity (multiculture) have been reported as being less susceptible to invasion by other communities compared to monoculture microalgae (Chuka-Ogwude et al., 2020). The resistance/resilience to invasion may depend on several factors such as variable sizes of microalgae, the existence of silica frustules (diatoms), the specificity of predator—prey dynamics, and the production of compounds that are inhibitory for other species (Newby et al., 2016). Moreover, some strains may show better tolerance to bacterial contamination such as *Chlorella vulgaris, Chlorella sorokiniana, Scenedesmus dimorphus*, and *Parachlorella kessleri* (Koutra et al., 2018). The optimization and adjustment of environmental parameters such as light, temperature, and operational parameters like hydraulic and biomass retention time, nutrient concentrations, and pH can reduce the probability of invasion by other communities (Xia and Murphy, 2016). Adding growth inhibitors is an additional practice for elimination of undesired communities such as allylthiourea addition to prevent nitrification (Viruela et al., 2016) and *Bacillus thuringiensis* subspecies *israeliensis* addition to control grazer populations of chironomid larvae (Mulbry et al., 2008).

4.6 Limitations regarding biogas upgrading coupled with digestate treatment

One drawback can be regarded as high H_2S composition of biogas (0—10,000 ppm) since H_2S has an inhibitory effect on microalgal growth (Nagarajan et al., 2019). The dissolution of H_2S in digestate may lead to acidic growth environment for microalgae. Furthermore, sulfide inhibits the microalgal photosynthesis via reduction of electron flow between Photosystem I and Photosystem II (Nagarajan et al., 2019). H_2S tolerance of microalgae is strain specific. H_2S concentration above 100 ppm has been previously reported as being inhibitory to a mutant of *Chlorella* sp. (Kao et al., 2012). The inhibitory concentrations of H_2S in terms of lethal dose concentration of 50% (LD50) have been indicated as 1.87 and 112.2 mg/L for *Scenedesmus vacuolatus* and *Chlamydomonas* sp., respectively. Moreover, the impairment of CO_2 photoreduction and Photosystem II for *Spirulina labyrinthiformis* has been recorded during the exposure to 20—41 mg/L H_2S concentration (Prandini et al., 2016). Thus, the selection of microalgal strains that are tolerant to high H_2S concentrations is important in microalgal biogas upgrading (Nagarajan et al., 2019). On the other hand, more than 99% H_2S removal from biogas content could be achieved at an initial concentration of 2000—3000 ppm_v via biotic and abiotic removal pathways in simultaneous biogas upgrading and digestate treatment using microalgae (Figueroa, 2016; Prandini et al., 2016). If biogas is supplied to microalgal treatment system of digestate, H_2S removal follows two major pathways (biotic and abiotic). Biotic removal involves the oxidization of H_2S to SO_4^{2-} by chemolitotrophic bacteria (Prandini et al., 2016). Abiotic removal is a dissolution—chemical oxidation process. H_2S dissolves in digestate owing to its high solubility (Henry constant 20°C: 5.15×10^2 atm) and then chemically oxidizes to SO_4^{2-} by high O_2 concentration produced in microalgal digestate treatment (Figueroa, 2016). Prandini et al. (2016) reported an increase in biomass growth by 0.7—1.3-folds and NH_3—N removal by 23%—64% at H_2S concentrations up to 3000 ppm. The increase in biomass growth and improved efficiency in the removal of NH_3—N can be a consequence of the rapid conversion of H_2S to SO_4^{2-} by biotic and abiotic transformations and its subsequent conversion into microalgal biomass. The synergy of concurrent biotic and abiotic H_2S removal in PBRs treating digestate and biogas (nonsterile, high oxidative environment) allows rapid oxidation of H_2S to SO_4^{2-}, which prevents H_2S inhibition of microalgae (Figueroa, 2016). SO_4^{2-} can be utilized by microalgae, since sulfur is a micronutrient for microalgal species (Ramírez-Rueda et al., 2020). However, the proportion of SO_4^{2-} consumed for microalgal metabolism can be low compared to the SO_4^{2-} produced. The sulfate—sulfur proportion has been previously recorded to be 92% of H_2S-sulfur in biogas, whereas only 8% H_2S-sulfur is utilized by microalgae (Figueroa, 2016).

Elevated O_2 concentration in biogas is to be avoided to prevent explosion and fire. Countries set specific limits for O_2 content in biogas composition which can be as low as 0.5% as regulated in Brazil, Germany, Austria, and the Netherlands (Bahr et al., 2014; Prandini et al., 2016). Microalgal growth produces O_2 as a product of photosynthesis. Theoretically, 1 mole of CO_2 consumed in photosynthesis produces 1 mole O_2, which may significantly alter O_2 content in the biogas composition. Final O_2 content in biogas may reach to 17.8%–21.6% after microalgal treatment (Prandini et al., 2016). Various applications to decrease O_2 content of biogas have been proposed. For example, acetate (a carbon source) can be added to simulate the growth of heterotrophic bacteria to consume O_2 (Prandini et al., 2016). Biogas can be fed into the microalgal reactor when dissolved oxygen is low (dark period) in microalgal treatment coupled with adsorption columns. Cocurrent flow in the adsorption column, decreasing the liquid/biogas ratio in recirculation, and feeding the digestate into the adsorption column are the proposed methods specific to the treatment systems including adsorption columns. However, these approaches require further research and improvement (Franco-Morgado et al., 2018).

CH_4 content of biogas may decrease during microalgal upgrading (by 5.4%–26.6%) (Prandini et al., 2016; Nguyen et al., 2019). Decrease in CH_4 content can be associated with consumption of CH_4 by methanotrophic bacteria (Prandini et al., 2016). Conversely, most studies recorded an increase in CH_4 content during microalgal biogas upgrading (Table 5.1). The reasons behind the variations in CH_4 content are yet to be identified in microalgal biogas upgrading.

Other limitations for digestate treatment with biogas upgrading using microalgae can be addressed as high installation costs, limited mass transfer of CO_2 to digestate, and selection of the species tolerant to both digestate and biogas (Nagarajan et al., 2019). These limitations reveal that more research is to be undertaken to improve the applicability of biogas upgrading simultaneously with digestate treatment by microalgae.

5. Pilot scale plants

Pilot scale plants on microalgal treatment of digestates have included indoor and outdoor studies as well as those studied in greenhouses (Table 5.5). Outdoor processes for microalgal digestate treatment are subject to negative conditions such as culture contamination, high temperature variance over the seasons, and light utilization at lower levels (Tan et al., 2014). Raceways or HRAPs (shallow raceways) and PBRs have been commonly used cultivation systems for microalgae (Table 5.5). A biofilm reactor, algal turf scrubber (Mulbry et al., 2008), has also been tested in pilot scale. The most inexpensive and easy-to-build cultivation systems can be regarded as open systems like open ponds and

Table 5.5 Pilot scale applications for digestate treatment by microalgal cultures.

Inoculum	Substrate of digester	Operation	COD/Carbon	N	P	[i]Concentration [ii]Productivity [iii]Growth rate	References
				Removals from digestate content		**Biomass**	
Dominated by *Scenedesmus* sp.	Sewage (filtered)	Outdoor PBR (4×550 L), semicontinuous, HRT: 8–14 d, September–December		NH_4-N: 84.1% (max) $(40-50)^a$, 5.84 mg/L/d	PO_4-P: 95.1% (max) $(6-7)^a$, 0.85 mg/L/d	[ii]0.052 g VSS/L/d (max)	Viruela et al. (2016)
Chlorella sorokiniana, Chlorococcum sp.	Black water (filtered)	PBR (211 L), in green house, continuous, HRT: 5 d, February–November		DIN^b: 0.1%–78.0% $(1290)^a$, 28–62 mg/L/d	DIP^c: 2.4%–68.7% $(68)^a$, 2.3–5.4 mg/L/d	[i]0.67–1.81 g DW/L [ii]0.13–0.36 g DW/L/d	Silva et al. (2019)
Chlorella pyrenoidosa	Activated sludge (settled, filtered, diluted)	Outdoor PBR (160 L), batch, operated each season for 2–3 weeks	Dissolved OC: 65.7%, 105.6 mg/L/d	NH_4-N: 35.82 mg/L/d (max)		[i]1.86 g VSS/L [iii]1.06 d^{-1}	Tan et al. (2015)
Chlorella pyrenoidosa	Wheat starch processing ww (settled, filtered)	Outdoor airlift PBR (890 L), operated every season in batch for 14 d	COD: 66% (max)	TN: 83.1% (max)	TP: 97% (max)	[i]0.50–2.05 g DW/L [ii]0.06–0.37 g DW/L/d [iii]0.19–1.02 d^{-1}	Tan et al. (2014)
Filamentous green algae	Dairy manure	Outdoor algal turf scrubber raceways (4×30 m^2), biofilm, 3 years for 270 d		N(recovered): 51%–83%	P(recovered): 62%–91%	[ii]2.5–25 g DW/m^2/d	Mulbry et al. (2008)

Dominated by *Chlorella* and *Scenedesmus* spp.	Piggery effluent (sand filtered)	Outdoor thin layer reactor (11 m^2, 0.5 cm depth), conventional raceway pond (11 m^2, 15 cm depth), 1.5 months, batch, semicontinuous	COD: 39%–44%, 39–41 mg/L/d	NH_4-N: 69%–98% (110 ± 10)[a], 13–19 mg/L/d	0.03–0.75 mg/L/d	[ii]0.031–0.061 g DW/L/d [ii]1.9–6.2 g DW/m^2/d	Raeisossadati et al. (2019)
Dominated by *Chlorella*, *Scenedesmus* spp. and pennate diatom	Piggery effluent (sand filtered)	Outdoor raceway pond (1 m^2), June (winter)–September (spring)		NH_4-N: 33.2–63.7 mg/L/d (800–1600)[a]		[ii]−0.04–0.095 g DW/L/d	Ayre et al. (2017)
Dominated by *Chlorella*, *Scenedesmus* spp. and pennate diatom	Piggery effluent (sand filtered)	Outdoor helical tubular PBR (40 L) and open raceway pond (160 L), June (winter)-September (spring), batch, semicontinuous		NH_4-N: 13.3–39.2 mg/L/d (893 ± 17)[a]		[i]1–6.15 mg/L Chl-a [ii]0.008–0.047 g DW/L/d	Nwoba et al. (2016)
Scenedesmus dimorphus	Mixture of food waste and animal manure (diluted)	Indoor open raceway pond (100 L), ∼8–12 d in batch	COD: 78%–82% (272–509)[a]	TN: 65%–72% (53–109)[a] NH_4-N: 100% (36.6–78.8)[a]	TP: 63%–100% (4.2–10.8)[a]	[i]0.159–0.446 g TSS/L [ii]0.09–0.068 g TSS/L/d	Hajar (2017)
Dominated by *Chloroidium saccharophilums*, *Pseudanabaena* sp.	Sludge (centrifuged, IC supply)	Outdoor HRAP (180 L), adsorption column, 92 d (June–October)		TN: 80%–87% NH_4-N: 100% N(recovered): 54%–76%	PO_4-P: 84%–92% P(recovered): 100%	[i]0.660–1.078 g TSS/L [ii]15 g TSS/m^2/d	Posadas et al. (2017)

Continued

Table 5.5 Pilot scale applications for digestate treatment by microalgal cultures.—cont'd

Inoculum	Substrate of digester	Operation	Removals from digestate content			Biomass	References
			COD/Carbon	N	P	[i]Concentration [ii]Productivity [iii]Growth rate	
Leptolyngbya lagerheimii, Chlorella vulgaris, Parachlorella kessleri, Tetradesmus obliquus, Mychonastes homosphaera	Sludge (centrifuged, IC supply)	Outdoor HRAP (180 L), adsorption column, 1 year	C(recovered): 47%—100%	N(recovered): 0%—99%	P(recovered): 0%—100%	[ii]0—22.5 g TSS/m^2/d	Marín et al. (2018)
Mychonastes homosphaera, Kalina, Puncochárová	Sludge	Indoor HRAP (180 L), adsorption column, 171 d	COD: 81%—93% Total C: 43%—82%	TN: 97.1%—98.9%	TP: 90%—99%	[i]0.67—1.21 g TSS/L [ii]8.3—15 g TSS/m^2/d	Franco-Morgado et al. (2018)
Dominated by filamentous microalgae	Vinasse (diluted)	Indoor HRAP (170 L), adsorption column, HRT ~7.4 d, 175 d operation	COD: 31%—51% Total OC: 24%—57% IC:66%—78% Total C: 50%—73%	TN: 1%—37% NH$_4$: 100%	TP: 36%—86%	[i]0.13—0.60 g TSS/L [ii]2.5—11.8 g TSS/m^2/d	Serejo et al. (2015)
Dominated by microalgal cyanobacterial population	Sludge (centrifuged, diluted)	Indoor HRAP (180 L), absorption column, 518 d				[i]0.6 ± 0.2 g TSS/L	Bahr et al. (2014)

[a] The numbers indicate initial concentration in mg/L.
[b] DIN: Dissolved inorganic nitrogen.
[c] DIP: Dissolved inorganic phosphorus.

raceways (Åkerström et al., 2014). These systems can easily be scaled-up (Hajar et al., 2017). However, open systems are susceptible to biotic pollution by undesired species and offer less microalgal biomass productivity. Light availability can also be a problem due to dark nature of digestates (Nwoba et al., 2016). Physical, biological, and chemical parameters such as biological contamination, temperature, hydrodynamics, culture conditions, and light penetration can be more effectively controlled and managed in PBRs (Nwoba et al., 2016), leading to higher biomass productivities (Chuka-Ogwude et al., 2020). Nevertheless, these adverse conditions cannot be fully avoided. Various applications have been performed to decrease biological contamination in PBRs. Breeding of most protozoa and mini metazoan species can be inhibited via application of high CO_2 concentrations. Membranes can be installed to aeration and gas exchange locations to prevent contamination. Cutting CO_2 supply for a short period (to increase the pH of the system) can inhibit bacteria and microalgae grazers, while microalgae can survive. Medium can be filtered to eliminate large grazers (Tan et al., 2014). Excessive foaming, wall growth, and increase in O_2 concentration (inhibitory to photosynthesis) can be accounted as other problems associated with closed systems (Nwoba et al., 2016).

TN and TP removal efficiency ranges between 1%—98.9% and 36%—100%, respectively, in pilot scale applications (Table 5.5). The observed biomass concentration is in the range of 0.50—2.05 and 0.13—1.21 g/L in terms of DW and total suspended solids (TSS), respectively. Biomass productivity can be achieved in the range of (−0.04)—0.37 g DW/L/d (negative sign represents a decrease in biomass productivity). Specific growth rates have rarely been reported (0.19—1.06 d^{-1}). Biomass concentrations and productivities obtained in pilot scale so far have been lower than the ones obtained in laboratory scale studies (0.07—12.1 g DW/L, 0.01—0.67 g DW/L/d). Biomass production cannot be maintained throughout a year due to seasonal variations (especially during winter) in outdoor pilot scale applications. Despite the lower treatment efficiencies and biomass productivities, regulatory nutrient concentration limits can be achievable (Viruela et al., 2016).

Several pilot scale applications on digestate treatment with synthetic biogas upgrading using bubble column adsorption also exemplify that it is possible to obtain CH_4 content in the range of 70.5%—99.1% when inflow CH_4 is 70%. H_2S component of biogas can be almost completely removed (91%—100%). CO_2 removal efficiencies of 40%—99.5% from biogas can be obtained (Bahr et al., 2014; Serejo et al., 2015; Posadas et al., 2017; Franco-Morgado et al., 2018; Marín et al., 2018). Moreover, it has been demonstrated that volatile organics can be removed at significant levels in a pilot plant study (Franco-Morgado et al., 2018). In this study, externally supplied volatile organics (hexane, toluene) and sulfur (methyl mercaptan-MeSH) compounds were removed by 7%—11%, 97%—98%, and 59%—66% from the synthetic biogas, respectively.

Various suggestions have been made for pilot scale applications to improve the processes. Biomass retention and hydraulic retention times can be uncoupled from each other to obtain higher nutrient removal efficiencies, and thus higher biomass growth rates (Viruela et al., 2016). If the species are not tolerant to high temperature fluctuations, a cooling or heating device may be necessary for summer and/or winter conditions, respectively. However, such an installation on air conditioning may bring economic burden for facilities (Tan et al., 2014). Modular systems can be employed that are designed for the availability of light, which may potentially result in higher areal demand. Artificial light can also be supplied to increase biomass productivity. In addition, microalgal treatment of digestates can be coupled with another nutrient removal process as a supplementary process for the times of ineffective biomass production. A microalgal—bacterial community that is effective at low light intensities can be employed for low light intensity periods which requires further scientific research (Silva et al., 2019).

6. Outcomes of techno-economic and life cycle assessment analysis

Microalgal biomass obtained in digestate treatment can further be valorized owing to valuable cellular components of microalgae. However, there has been no clear evidence that a downstream processing option can be a common approach for valorization both from techno-economic and environmental perspectives. Even biorefinery approach is yet to be proven successful and techno-economic (Perez-Garcia and Bashan, 2015). Absence of full-scale industrial plants and an assigned technology create variable cost estimates for microalgal production (Serejo et al., 2015) as well as theoretical assessment of life cycle (Allen et al., 2018). This variability makes such techno-economic and life cycle assessment (LCA) studies hardly comparable with each other.

On the other hand, microalgal treatment may present a cost-effective management opportunity for digestates via valorization of biomass depending on the promising results obtained from different studies. For example, it was demonstrated that microalgal biomass obtained from vinasse digestate (60% ± 7% carbohydrate) in HRAP had 102 L/ton bioethanol production potential which was higher than that of obtained from sugar cane (70 L/ton) (Serejo et al., 2015). Land requirement for digestate treatment in outdoor raceway ponds was also calculated to be 3% of the total area required for land application of digestate. The energy content of biomass was found to be 6% of the biogas energy for a 2 MW_e agricultural biogas plant in a case study (Xia and Murphy, 2016). A closed loop system including algal growth using the digestate obtained from the digestion of algal biomass had a potential to treat 2600 L/d algal digestate diluted with rural sector wastewater (Prajapati et al., 2014). Microalgal treatment was also found to be more economic than upgrading the wastewater treatment plants existing in sensitive basins based on the nitrogen removal costs without considering the valorization opportunities (Mulbry et al., 2008).

Even though promising results are obtained, techno-economic feasibility is required to be more thoroughly examined. Studies conducted up to now have demonstrated that microalgal productivity, areal requirement, recyclable water, total capital and operational expenses, and the cost of microalgal production are important parameters for determination of the economic feasibility of large-scale applications. Open pond systems have been dedicated as being more economical than closed systems, and thus areal productivity may represent the most critical parameter. Moreover, strain selection in terms of tolerance, process performance and optimization are necessary applications to develop economically feasible processes (Chuka-Ogwude et al., 2020). Biofuel pathway for valorization from microalgal biomass becomes an attractive option, considering large amounts of digestates produced from AD plants (10,000 tons/year for a 500 kW plant) (Koutra et al., 2018). Nevertheless, microalgal biomass is required to be refined for high value compounds to increase the total product value, which would make the biofuel option compatible with traditional fossil fuels (Perez-Garcia and Bashan, 2015; Koutra et al., 2018). A techno-economic analysis on integrated biorefinery of microalgal biomass coupled with AD process of microalgal cake demonstrated that biodiesel production cost could be reduced to $0.48/kg diesel via production of astaxanthin and poly(hydroxybutyrate) byproducts along with biodiesel (Prieto et al., 2017).

From the environmental perspective, chemical use in pretreatment of digestates, digestate storage and transportation and composting the solid fraction of digestate are the major factors affecting the environmental performance. The use of diluted liquid digestate in microalgal treatment has been demonstrated to have less effect on human health, ecosystem quality, and climate change terms compared to the treatment option including ammonia stripping and struvite precipitation. The coupled process (AD + microalgae) may potentially contribute to the energy recovery and save water resources (Duan et al., 2020). Direct application of pig manure digestate on farmlands (DA) without oversupply and additional transportation cost was also compared with the case of using of diluted liquid fraction of the digestate in a microalgal treatment process and composting the solid fraction of digestate (DLF). Both DA and DLF applications had close human health effect. DA was found to be better than DLF application in climate change and resources damage categories. Ecosystem quality can be preserved better via DLF application compared to DA (Duan et al., 2020). Liquid digestate application on farmlands has a potential to gain environmental credits since fertilizer use and spreading would be avoided. On the other hand, short distances between AD plants and applications fields, and the economic preference for larger AD plants may lead to over application of liquid digestate, which may eventually lead to low efficiency of nutrient use and poor environmental balance (Stiles et al., 2018).

The infrastructure, energy, and water requirement for microalgal digestate treatment process and biomass valorization still create burdens associated with global warming, abiotic, and fossil resource depletion. Further LCA studies is to look for the answer whether these burdens can be overbalanced by environmental credits from the high-value products (Stiles et al., 2018).

7. Conclusions

Stricter regulations impose biogas industry to manage digestates with more environmentally sound and low-cost processing options. Microalgal treatment of AD effluents is a promising approach in this context. However, the coupled AD and microalgae processes are yet to be applied in full scale. Several suggestions can be made for further improvements. Commonly applied pretreatment options may be reevaluated to improve the feasibility and environmental aspects of the processes. Microalgal biomass productivity needs to be improved for the field studies especially for the seasons of lower productivity. Innovative and low-cost reactor designs or operational procedures may enhance the applicability in large scale. Increasing the value of biomass is another opportunity via introduction of new valorization pathways. Techno-economic and LCA studies are limited and required to be improved to be more informative for further applications and investments.

References

Åkerström, A.M., Mortensen, L.M., Rusten, B., Gislerød, H.R., 2014. Biomass production and nutrient removal by *Chlorella* sp. as affected by sludge liquor concentration. J. Environ. Manag. 144, 118—124. https://doi.org/10.1016/j.jenvman.2014.05.015.

Allen, J., Unlu, S., Demirel, Y., Black, P., Riekhof, W., 2018. Integration of biology, ecology and engineering for sustainable algal-based biofuel and bioproduct biorefinery. Bioresour. Bioprocess. 5 (47). https://doi.org/10.1186/s40643-018-0233-5.

Angelidaki, I., Xie, L., Luo, G., Zhang, Y., Oechsner, H., Lemmer, A., Munoz, R., Kougias, P.G., 2019. Biogas upgrading: current and emerging technologies. In: Pandey, A., Larroche, C., Dussap, C.-G., Gnansounou, E., Khanal, S.K., Ricke, S. (Eds.), Biomass, Biofuels, Biochemicals: Biofuels: Alternative Feedstocks and Conversion Processes for the Production of Liquid and Gaseous Biofuels, second ed. Academic Press, pp. 817—843. https://doi.org/10.1016/B978-0-12-816856-1.00033-6.

Awe, O.W., Zhao, Y., Nzihou, A., Minh, D.P., Lyczko, N., 2017. A Review of biogas utilisation, purification and upgrading technologies. Waste Biomass Valorization 8, 267—283. https://doi.org/10.1007/s12649-016-9826-4.

Ayre, J.M., Moheimani, N.R., Borowitzka, M.A., 2017. Growth of microalgae on undiluted anaerobic digestate of piggery effluent with high ammonium concentrations. Algal Res. 24, 218—226. https://doi.org/10.1016/j.algal.2017.03.023.

Bahr, M., Díaz, I., Dominguez, A., González Sánchez, A., Muñoz, R., 2014. Microalgal-biotechnology as a platform for an integral biogas upgrading and nutrient removal from anaerobic effluents. Environ. Sci. Technol. 48, 573—581. https://doi.org/10.1021/es403596m.

Barsanti, L., Gualtieri, P., 2014. Algae: Anatomy, Biochemistry, and Biotechnology. CRC Press, ISBN 9781439867327.

Botheju, D., Svalheim, O., Bakke, R., 2010. Digestate nitrification for nutrient recovery. Open Waste Manag. J. 3, 1—12. https://doi.org/10.2174/1876400201003010001.

Cai, T., Park, S.Y., Racharaks, R., Li, Y., 2013a. Cultivation of *Nannochloropsis salina* using anaerobic digestion effluent as a nutrient source for biofuel production. Appl. Energy 108, 486—492. https://doi.org/10.1016/j.apenergy.2013.03.056.

Cai, T., Ge, X., Park, S.Y., Li, Y., 2013b. Comparison of *Synechocystis* sp. PCC6803 and *Nannochloropsis salina* for lipid production using artificial seawater and nutrients from anaerobic digestion effluent. Bioresour. Technol. 144, 255—260. https://doi.org/10.1016/j.biortech.2013.06.101.

Chen, Y., Li, S., Ho, S.-H., Wang, C., Lin, Y.-C., Nagarajan, D., Chang, J.-S., Ren, N.-Q., 2018. Integration of sludge digestion and microalgae cultivation for enhancing bioenergy and biorefinery. Renew. Sustain. Energy Rev. 96, 76—90. https://doi.org/10.1016/j.rser.2018.07.028.

Cheng, J., Xu, J., Huang, Y., Li, Y., Zhou, J., Cen, K., 2015. Growth optimisation of microalga mutant at high CO_2 concentration to purify undiluted anaerobic digestion effluent of swine manure. Bioresour. Technol. 177, 240—246. https://doi.org/10.1016/j.biortech.2014.11.099.

Chew, K.W., Yap, J.Y., Show, P.L., Suan, N.H., Juan, J.C., Ling, T.C., Lee, D.J., Chang, J.S., 2017. Microalgae biorefinery: high value products perspectives. Bioresour. Technol. 229, 53—62. https://doi.org/10.1016/j.biortech.2017.01.006.

Cho, H.U., Kim, Y.M., Choi, Y., Xu, X., Shin, D.Y., Park, J.M., 2015. Effects of pH control and concentration on microbial oil production from *Chlorella vulgaris* cultivated in the effluent of a low-cost organic waste fermentation system producing volatile fatty acids. Bioresour. Technol. 184, 245—250. https://doi.org/10.1016/j.biortech.2014.09.069.

Chojnacka, K., Marquez-Rocha, F.-J., 2004. Kinetic and stoichiometric relationships of the energy and carbon metabolism in the culture of microalgae. Biotechnology 3, 21—34. https://doi.org/10.3923/biotech.2004.21.34.

Chuka-Ogwude, D., Ogbonna, J., Moheimani, N.R., 2020. A review on microalgal culture to treat anaerobic digestate food waste effluent. Algal Res. 47, 101841. https://doi.org/10.1016/j.algal.2020.101841.

Deng, X.Y., Gao, K., Zhang, R.C., Addy, M., Lu, Q., Ren, H.Y., Chen, P., Liu, Y.H., Ruan, R., 2017. Growing *Chlorella vulgaris* on thermophilic anaerobic digestion swine manure for nutrient removal and biomass production. Bioresour. Technol. 243, 417—425. https://doi.org/10.1016/j.biortech.2017.06.141.

Dickinson, K.E., Bjornsson, W.J., Garrison, L.L., Whitney, C.G., Park, K.C., Banskota, A.H., Mcginn, P.J., 2014. Simultaneous remediation of nutrients from liquid anaerobic digestate and municipal wastewater by the microalga *Scenedesmus* sp. AMDD grown in continuous chemostats. J. Appl. Microbiol. 118, 75—83. https://doi.org/10.1111/jam.12681.

Diniz, G.S., Silva, A.F., Araújo, O.Q.F., Chaloub, R.M., 2017. The potential of microalgal biomass production for biotechnological purposes using wastewater resources. J. Appl. Phycol. 29 (2), 821—832. https://doi.org/10.1007/s10811-016-0976-3.

Duan, N., Khoshnevisan, B., Lin, C., Liu, Z., Liu, H., 2020. Life cycle assessment of anaerobic digestion of pig manure coupled with different digestate treatment technologies. Environ. Int. 137, 105522. https://doi.org/10.1016/j.envint.2020.105522.

Evangelou, V.P., 1998. Environmental Soil and Water Chemistry: Principles and Applications. John Wiley and Sons.

Fernandes, T.V., Shrestha, R., Sui, Y., Papini, G., Zeeman, G., Vet, L.E.M., Wijffels, R.H., Lamers, P., 2015. Closing domestic nutrient cycles using microalgae. Environ. Sci. Technol. 49 (20), 12450—12456. https://doi.org/10.1021/acs.est.5b02858.

Figueroa, L.M.M., 2016. Biogas Upgrading Using Microalgae. Universidad de la Frontera.

Franchino, M., Comino, E., Bona, F., Riggio, V.A., 2013. Growth of three microalgae strains and nutrient removal from an agro-zootechnical digestate. Chemosphere 92, 738—744. https://doi.org/10.1016/j.chemosphere.2013.04.023.

Franco-Morgado, M., Toledo-Cervantes, A., González-Sánchez, A., Lebrero, R., Muñoz, R., 2018. Integral (VOCs, CO_2, mercaptans and H_2S) photosynthetic biogas upgrading using innovative biogas and digestate supply strategies. Chem. Eng. J. 354, 363—369. https://doi.org/10.1016/j.cej.2018.08.026.

Franke-Whittle, I.H., Walter, A., Ebner, C., Insam, H., 2014. Investigation into the effect of high concentrations of volatile fatty acids in anaerobic digestion on methanogenic communities. Waste Manag. 34, 2080—2089. https://doi.org/10.1016/j.wasman.2014.07.020.

Gao, S., Hu, C., Sun, S., Xu, J., Zhao, Y., Zhang, H., 2018. Performance of piggery wastewater treatment and biogas upgrading by three microalgal cultivation technologies under different initial COD concentration. Energy 165, 360—369. https://doi.org/10.1016/j.energy.2018.09.190.

Ge, S., Qiu, S., Tremblay, D., Viner, K., Champagne, P., Jessop, P.G., 2018. Centrate wastewater treatment with *Chlorella vulgaris*: simultaneous enhancement of nutrient removal, biomass and lipid production. Chem. Eng. J. 342, 310—320. https://doi.org/10.1016/j.cej.2018.02.058.

Ghyselbrecht, K., Monballiu, A., Somers, M.H., Sigurnjak, I., Meers, E., Appels, L., Meesschaert, B., 2019. The fate of nitrite and nitrate during anaerobic digestion. Environ. Technol. 40 (8), 1013–1026. https://doi.org/10.1080/09593330.2017.1415380.

Hajar, H.A.A., Riefler, R.G., Stuart, B.J., 2017. Cultivation of *Scenedesmus dimorphus* using anaerobic digestate as a nutrient medium. Bioproc. Biosyst. Eng. 40, 1197–1207. https://doi.org/10.1007/s00449-017-1780-4.

Hom-Diaz, A., Llorca, M., Rodríguez-Mozas, S., Vicent, T., Barcelo, D., Blanquez, P., 2015. Microalgae cultivation on wastewater digestate: β-estradiol and 17α-ethynylestradiol degradation and transformation products identification. J. Environ. Manag. 155, 106–113. https://doi.org/10.1016/j.jenvman.2015.03.003.

Ji, F., Liu, Y., Hao, R., Li, G., Zhou, Y., Dong, R., 2014. Biomass production and nutrients removal by a new microalgae strain *Desmodesmus* sp. in anaerobic digestion wastewater. Bioresour. Technol. 161, 200–207. https://doi.org/10.1016/j.biortech.2014.03.034.

Ji, F., Zhou, Y., Pang, A., Ning, L., Rodgers, K., Liu, Y., Dong, R., 2015. Fed-batch cultivation of *Desmodesmus* sp. in anaerobic digestion wastewater for improved nutrient removal and biodiesel production. Bioresour. Technol. 184, 116–122. https://doi.org/10.1016/j.biortech.2014.09.144.

Kao, C.Y., Chiu, S.Y., Huang, T.T., Dai, L., Wang, G.H., Tseng, C.P., Chen, C.H., Lin, C.S., 2012. A mutant strain of microalga *Chlorella* sp. for the carbon dioxide capture from biogas. Biomass Bioenergy 36, 132–140. https://doi.org/10.1016/j.biombioe.2011.10.046.

Katiyar, R., Gurjar, B.R., Biswas, S., Pruthi, V., Kumar, N., Kumar, P., 2017. Microalgae: an emerging source of energy based bio-products and a solution for environmental issues. Renew. Sustain. Energy Rev. 72, 1083–1093. https://doi.org/10.1016/j.rser.2016.10.028.

Koutra, E., Economou, C.N., Tsafrakidou, P., Kornaros, M., 2018. Bio-based products from microalgae cultivated in digestates. Trends Biotechnol. 36, 819–833. https://doi.org/10.1016/j.tibtech.2018.02.015.

Kumar, K.S., Dahms, H.-U., Won, E.-J., Lee, J.-S., Shin, K.-H., 2015. Microalgae- A promising tool for heavy metal remediation. Ecotoxicol. Environ. Saf. 113, 329–352. https://doi.org/10.1016/j.ecoenv.2014.12.019.

Li, G., Bai, X., Li, H., Lu, Z., Zhou, Y., Wang, Y., Cao, J., Huang, Z., 2019. Nutrients removal and biomass production from anaerobic digested effluent by microalgae: a review. Int. J. Agric. Biol. Eng. 12 (5), 8–13. https://doi.org/10.25165/j.ijabe.20191205.3630.

Li, Y., Chen, Y.F., Chen, P., Min, M., Zhou, W., Martinez, B., Zhu, J., Ruan, R., 2011. Characterization of a microalga *Chlorella* sp. well adapted to highly concentrated municipal wastewater for nutrient removal and biodiesel production. Bioresour. Technol. 102, 5138–5144. https://doi.org/10.1016/j.biortech.2011.01.091.

Marazzi, F., Sambusiti, C., Monlau, F., Cecere, S.E., Scaglione, D., Barakat, A., Mezzanotte, V., Ficara, E., 2017. A novel option for reducing the optical density of liquid digestate to achieve a more productive microalgal culturing. Algal Res. 24, 19–28. https://doi.org/10.1016/j.algal.2017.03.014.

Marcilhac, C., Sialve, B., Pourcher, A.M., Ziebal, C., Bernet, N., Béline, F., 2015. Control of nitrogen behaviour by phosphate concentration during microalgal-bacterial cultivation using digestate. Bioresour. Technol. 175, 224–230. https://doi.org/10.1016/j.biortech.2014.10.022.

Marín, D., Posadas, E., Cano, P., Pérez, V., Blanco, S., Lebrero, R., Muñoz, R., 2018. Seasonal variation of biogas upgrading coupled with digestate treatment in an outdoors pilot scale algal-bacterial photobioreactor. Bioresour. Technol. 263, 58–66. https://doi.org/10.1016/j.biortech.2018.04.117.

Marjakangas, J.M., Chen, C., Lakaniemi, A., Puhakka, J.A., Whang, L., Chang, J., 2015. Simultaneous nutrient removal and lipid production with *Chlorella vulgaris* on sterilized and non-sterilized anaerobically pretreated piggery wastewater. Biochem. Eng. J. 103, 177–184. https://doi.org/10.1016/j.bej.2015.07.011.

Markou, G., Vandamme, D., Muylaert, K., 2014. Microalgal and cyanobacterial cultivation: the supply of nutrients. Water Res. 65, 186–202. https://doi.org/10.1016/j.watres.2014.07.025.

Markou, G., Wang, L., Ye, J., Unc, A., 2018. Using agro-industrial wastes for the cultivation of microalgae and duckweeds: contamination risks and biomass safety concerns. Biotechnol. Adv. J. 36, 1238–1254. https://doi.org/10.1016/j.biotechadv.2018.04.003.

Miazek, K., Iwanek, W., Remacle, C., Richel, A., Goffin, D., 2015. Effect of metals, metalloids and metallic nanoparticles on microalgae growth and industrial product biosynthesis: a review. Int. J. Mol. Sci. 16, 23929–23969. https://doi.org/10.3390/ijms161023929.

Möller, K., Müller, T., 2012. Effects of anaerobic digestion on digestate nutrient availability and crop growth: a review. Eng. Life Sci. 12 (3), 242–257. https://doi.org/10.1002/elsc.201100085.

Monlau, F., Sambusiti, C., Ficara, E., Aboulkas, A., Barakat, A., Carrère, H., 2015. New opportunities for agricultural digestate valorization: current situation and perspectives. Energy Environ. Sci. 8 (9), 2600–2621. https://doi.org/10.1039/c5ee01633a.

Montero, E., Olguín, E.J., De Philippis, R., Reverchon, F., 2018. Mixotrophic cultivation of *Chlorococcum* sp. under non-controlled conditions using a digestate from pig manure within a biorefinery. J. Appl. Phycol. 30, 2847–2857. https://doi.org/10.1007/s10811-018-1467-5.

Mulbry, W., Kondrad, S., Pizarro, C., Kebede-Westhead, E., 2008. Treatment of dairy manure effluent using freshwater algae: algal productivity and recovery of manure nutrients using pilot-scale algal turf scrubbers. Bioresour. Technol. 99, 8137–8142. https://doi.org/10.1016/j.biortech.2008.03.073.

Muylaert, K., Beuckels, A., Depraetere, O., Foubert, I., Markou, G., Vandamme, D., 2015. Wastewater as a source of nutrients for microalgae biomass production. In: Moheimani, N., McHenry, M., de Boer, K., P., B. (Eds.), Biomass and Biofuels from Microalgae, vol. 2. Springer, Cham, pp. 75–94. https://doi.org/10.1007/978-3-319-16640-7_5.

Nagarajan, D., Lee, D.J., Chang, J.S., 2019. Integration of anaerobic digestion and microalgal cultivation for digestate bioremediation and biogas upgrading. Bioresour. Technol. 290, 121804. https://doi.org/10.1016/j.biortech.2019.121804.

Newby, D.T., Mathews, T.J., Pate, R.C., Huesemann, M.H., Lane, T.D., Wahlen, B.D., Mandal, S., Engler, R.K., Feris, K.P., Shurin, J.B., 2016. Assessing the potential of polyculture to accelerate algal biofuel production. Algal Res. 19, 264–277. https://doi.org/10.1016/j.algal.2016.09.004.

Nguyen, M.T., Lin, C., Lay, C., 2019. Microalgae cultivation using biogas and digestate carbon sources. Biomass Bioenergy 122, 426–432. https://doi.org/10.1016/j.biombioe.2019.01.050.

Nwoba, E.G., Ayre, J.M., Moheimani, N.R., Ubi, B.E., Ogbonna, J.C., 2016. Growth comparison of microalgae in tubular photobioreactor and open pond for treating anaerobic digestion piggery effluent. Algal Res. 17, 268–276. https://doi.org/10.1016/j.algal.2016.05.022.

Oswald, W.J., Gotaas, H.B., Ludwig, H.F., Lynch, V., 1953. Algae symbiosis in oxidation ponds: III. Photosynthetic oxygenation. Sewage Ind. Wastes 25 (6), 692–705.

Ouyang, Y., Zhao, Y., Sun, S., Hu, C., Ping, L., 2015. Effect of light intensity on the capability of different microalgae species for simultaneous biogas upgrading and biogas slurry nutrient reduction. Int. Biodeterior. Biodegrad. 104, 157–163. https://doi.org/10.1016/j.ibiod.2015.05.027.

Palmer, C.M., 1969. A composite rating algae tolerating organic pollution. J. Phycol. 5, 78–82. https://doi.org/10.1111/j.1529-8817.1969.tb02581.x.

Park, J., Jin, H.F., Lim, B.R., Park, K.Y., Lee, K., 2010. Ammonia removal from anaerobic digestion effluent of livestock waste using green alga Scenedesmus sp. Bioresour. Technol. 101, 8649–8657. https://doi.org/10.1016/j.biortech.2010.06.142.

Parkin, G.F., Owen, W.F., 1986. Fundamentals of anaerobic digestion of wastewater sludges. J. Environ. Eng. 112, 867–920. https://doi.org/10.1061/(asce)0733-9372(1986)112:5(867).

Perez-Garcia, O., Bashan, Y., 2015. Microalgal heterotrophic and mixotrophic culturing for bio-refining: from metabolic routes to techno-economics. In: Prokop, A., Bajpai, R., Zappi, M. (Eds.), Algal Biorefineries. Springer, Cham, pp. 61–131. https://doi.org/10.1007/978-3-319-20200-6_3.

Posadas, E., Marín, D., Blanco, S., Lebrero, R., Muñoz, R., 2017. Simultaneous biogas upgrading and centrate treatment in an outdoors pilot scale high rate algal pond. Bioresour. Technol. 232, 133–141. https://doi.org/10.1016/j.biortech.2017.01.071.

Prajapati, S.K., Kumar, P., Malik, A., Vijay, V.K., 2014. Bioconversion of algae to methane and subsequent utilization of digestate for algae cultivation: a closed loop bioenergy generation process. Bioresour. Technol. 158, 174–180. https://doi.org/10.1016/j.biortech.2014.02.023.

Prandini, J.M., da Silva, M.L.B., Mezzari, M.P., Pirolli, M., Michelon, W., Soares, H.M., 2016. Enhancement of nutrient removal from swine wastewater digestate coupled to biogas purification by microalgae *Scenedesmus* spp. Bioresour. Technol. 202, 67–75. https://doi.org/10.1016/j.biortech.2015.11.082.

Praveen, P., Guo, Y., Kang, H., Lefebvre, C., Loh, K.C., 2018. Enhancing microalgae cultivation in anaerobic digestate through nitrification. Chem. Eng. J. 354, 905—912. https://doi.org/10.1016/j.cej.2018.08.099.

Prieto, C.V.G., Ramos, F.D., Estrada, V., Villar, M.A., Diaz, M.S., 2017. Optimization of an integrated algae-based biorefinery for the production of biodiesel, astaxanthin and PHB. Energy 139, 1159—1172. https://doi.org/10.1016/j.energy.2017.08.036.

Raeisossadati, M., Vadiveloo, A., Bahri, P.A., Parlevliet, D., Moheimani, N.R., 2019. Treating anaerobically digested piggery effluent (ADPE) using microalgae in thin layer reactor and raceway pond. J. Appl. Phycol. 31, 2311—2319. https://doi.org/10.1007/s10811-019-01760-6.

Ramírez-Rueda, A., Velasco, A., González-Sánchez, A., 2020. The effect of chemical sulfide oxidation on the oxygenic activity of an alkaliphilic microalgae consortium deployed for biogas upgrading. Sustainability 12. https://doi.org/10.3390/su12166610.

Risberg, K., Cederlund, H., Pell, M., Arthurson, V., Schnürer, A., 2017. Comparative characterization of digestate versus pig slurry and cow manure — chemical composition and effects on soil microbial activity. Waste Manag. 61, 529—538. https://doi.org/10.1016/j.wasman.2016.12.016.

Rodero, M. del R., Lebrero, R., Serrano, E., Lara, E., Arbib, Z., García-Encina, P.A., Muñoz, R., 2019. Technology validation of photosynthetic biogas upgrading in a semi-industrial scale algal-bacterial photobioreactor. Bioresour. Technol. 279, 43—49. https://doi.org/10.1016/j.biortech.2019.01.110.

Sayedin, F., Kermanshahi-pour, A., He, Q.S., Tibbetts, S.M., Lalonde, C.G.E., Brar, S.K., 2020. Microalgae cultivation in thin stillage anaerobic digestate for nutrient recovery and bioproduct production. Algal Res. 47, 101867. https://doi.org/10.1016/j.algal.2020.101867. July 2019.

Sekine, M., Yoshida, A., Akizuki, S., Kishi, M., Toda, T., 2020. Microalgae cultivation using undiluted anaerobic digestate by introducing aerobic nitrification-desulfurization treatment. Water Sci. Technol. 1—11. https://doi.org/10.2166/wst.2020.153.

Serejo, M.L., Posadas, E., Boncz, M.A., Blanco, S., García-Encina, P., Muñoz, R., 2015. Influence of biogas flow rate on biomass composition during the optimization of biogas upgrading in microalgal-bacterial processes. Environ. Sci. Technol. 49, 3228—3236. https://doi.org/10.1021/es5056116.

Silva, G.H.R., Sueitt, A.P.E., Haimes, S., Tripidaki, A., Zwieten, R. Van, Fernandes, T.V., 2019. Feasibility of closing nutrient cycles from black water by microalgae-based technology. Algal Res. 44, 101715. https://doi.org/10.1016/j.algal.2019.101715.

Simonazzi, M., Pezzolesi, L., Guerrini, F., Vanucci, S., Samorì, C., Pistocchi, R., 2019. Use of waste carbon dioxide and pre-treated liquid digestate from biogas process for Phaeodactylum tricornutum cultivation in photobioreactors and open ponds. Bioresour. Technol. 292, 121921. https://doi.org/10.1016/j.biortech.2019.121921.

Singh, M., Reynolds, D.L., Das, K.C., 2011. Microalgal system for treatment of effluent from poultry litter anaerobic digestion. Bioresour. Technol. 102, 10841—10848. https://doi.org/10.1016/j.biortech.2011.09.037.

Stiles, W.A.V., Styles, D., Chapman, S.P., Esteves, S., Bywater, A., Melville, L., Silkina, A., Lupatsch, I., Grünewald, C.F., Lovitt, R., Chaloner, T., Bull, A., Morris, C., Llewellyn, C.A., 2018. Using microalgae in the circular economy to valorise anaerobic digestate: challenges and opportunities. Bioresour. Technol. 267, 732—742. https://doi.org/10.1016/j.biortech.2018.07.100.

Sutanto, S., Dijkstra, J.W., Pieterse, J.A.Z., Boon, J., Hauwert, P., Brilman, D.W.F., 2017. CO_2 removal from biogas with supported amine sorbents: first technical evaluation based on experimental data. Separ. Purif. Technol. 184, 12—25. https://doi.org/10.1016/j.seppur.2017.04.030.

Tan, X.-B., Yang, L., Zhang, Y., Zhao, F., Chu, H., Guo, J., 2015. *Chlorella pyrenoidosa* cultivation in outdoors using the diluted anaerobically digested activated sludge. Bioresour. Technol. 198, 340—350. https://doi.org/10.1016/j.biortech.2015.09.025.

Tan, X.-B., Chu, H., Zhang, Y., Yang, L., Zhao, F., Zhou, X., 2014. *Chlorella pyrenoidosa* cultivation using anaerobic digested starch processing wastewater in an airlift circulation photobioreactor. Bioresour. Technol. 170, 538—548. https://doi.org/10.1016/j.biortech.2014.07.086.

Tao, R., Kinnunen, V., Praveenkumar, R., Lakaniemi, A., Rintala, J.A., 2017. Comparison of *Scenedesmus acuminatus* and *Chlorella vulgaris* cultivation in liquid digestates from anaerobic digestion of pulp and paper industry and municipal wastewater treatment sludge. J. Appl. Phycol. 29, 2845—2856. https://doi.org/10.1007/s10811-017-1175-6.

Tripathi, B.N., Kumar, D., 2017. Prospects and Challenges in Algal Biotechnology. Springer. https://doi.org/10.1007/978-981-10-1950-0.

Uggetti, E., Sialve, B., Trably, E., Steyer, J.P., 2014. Integrating microalgae production with anaerobic digestion: a biorefinery approach. Biofuel. Bioprod. Biorefin. 8, 516—529. https://doi.org/10.1002/bbb.1469.

Ülgüdür, N., Demirer, G.N., 2019. Anaerobic treatability and residual biogas potential of the effluent stream of anaerobic digestion processes. Water Environ. Res. 91, 259—268. https://doi.org/10.1002/wer.1048.

Ülgüdür, N., Ergüder, T.H., Demirer, G.N., 2019a. Simultaneous dissolution and uptake of nutrients in microalgal treatment of the secondarily treated digestate. Algal Res. 43, 101633. https://doi.org/10.1016/j.algal.2019.101633.

Ülgüdür, N., Ergüder, T.H., Uludağ-Demirer, S., Demirer, G.N., 2019b. High-rate anaerobic treatment of digestate using fixed film reactors. Environ. Pollut. 252, 1622—1632. https://doi.org/10.1016/j.envpol.2019.06.115.

Vassalle, L., García-Galán, M.J., Aquino, S.F., Afonso, R.J. de C.F., Ferrer, I., Passos, F., Mota, C.R., 2020. Can high rate algal ponds be used as post-treatment of UASB reactors to remove micropollutants? Chemosphere 248, 125969. https://doi.org/10.1016/j.chemosphere.2020.125969.

Viruela, A., Murgui, M., Gómez-Gil, T., Durán, F., Robles, Á., Ruano, M.V., Ferrer, J., Seco, A., 2016. Water resource recovery by means of microalgae cultivation in outdoor photobioreactors using the effluent from an anaerobic membrane bioreactor fed with pre-treated sewage. Bioresour. Technol. 218, 447—454. https://doi.org/10.1016/j.biortech.2016.06.116.

Viswanaathan, S., Sudhakar, M.P., 2019. Microalgae: potential agents for CO_2 mitigation and bioremediation of wastewaters. In: Singh, J.S. (Ed.), New and Future Developments in Microbial Biotechnology and Bioengineering: Microbes in Soil, Crop and Environmental Sustainability. Elsevier, pp. 129—148. https://doi.org/10.1016/B978-0-12-818258-1.00008-X.

Wang, J., Zhou, W., Chen, H., Zhan, J., He, C., Wang, Q., 2019. Ammonium nitrogen tolerant *Chlorella* strain screening and its damaging effects on photosynthesis. Front. Microbiol. 9, 1—13. https://doi.org/10.3389/fmicb.2018.03250.

Wang, L., Li, Y., Chen, P., Min, M., Chen, Y., Zhu, J., Ruan, R.R., 2010a. Anaerobic digested dairy manure as a nutrient supplement for cultivation of oil-rich green microalgae *Chlorella* sp. Bioresour. Technol. 101, 2623—2628. https://doi.org/10.1016/j.biortech.2009.10.062.

Wang, L., Wang, Y., Chen, P., Ruan, R., 2010b. Semi-continuous cultivation of *Chlorella vulgaris* for treating undigested and digested dairy manures. Appl. Biochem. Biotechnol. 162, 2324—2332. https://doi.org/10.1007/s12010-010-9005-1.

Wang, L., Chen, L., Wu, S.X., 2020a. Microalgae cultivation using screened liquid dairy manure applying different folds of dilution: nutrient reduction analysis with emphasis on phosphorus removal. Appl. Biochem. Biotechnol. https://doi.org/10.1007/s12010-020-03316-8.

Wang, M., Zhang, S.C., Tang, Q., Shi, L.D., Tao, X.M., Tian, G.M., 2020b. Organic degrading bacteria and nitrifying bacteria stimulate the nutrient removal and biomass accumulation in microalgae-based system from piggery digestate. Sci. Total Environ. 707, 134442. https://doi.org/10.1016/j.scitotenv.2019.134442.

Wang, X., Bao, K., Cao, W., Zhao, Y., Hu, C.W., 2017. Screening of microalgae for integral biogas slurry nutrient removal and biogas upgrading by different microalgae cultivation technology. Sci. Rep. 7, 1—12. https://doi.org/10.1038/s41598-017-05841-9.

Wang, Z., Zhao, Y., Ge, Z., Zhang, H., Sun, S., 2016. Selection of microalgae for simultaneous biogas upgrading and biogas slurry nutrient reduction under various photoperiods. J. Chem. Technol. Biotechnol. 91 (7), 1982—1989. https://doi.org/10.1002/jctb.4788.

Wilt, A. De, Butkovskyi, A., Tuantet, K., Hernandez, L., Fernandes, T.V., Langenhoff, A., Zeeman, G., 2016. Micropollutant removal in an algal treatment system fed with source separated wastewater streams. J. Hazard. Mater. 304, 84—92.

Xia, A., Murphy, J.D., 2016. Microalgal cultivation in treating liquid digestate from biogas systems. Trends Biotechnol. 34, 264–275. https://doi.org/10.1016/j.tibtech.2015.12.010.

Xiao, Y., Yuan, H., Pang, Y., Chen, S., Zhu, B., Zou, D., Ma, J., Yu, L., Li, X., 2014. CO_2 removal from biogas bywater washing system. Chin. J. Chem. Eng. 22 (8), 950–953. https://doi.org/10.1016/j.cjche.2014.06.001.

Xu, J., Zhao, Y., Zhao, G., Zhang, H., 2015. Nutrient removal and biogas upgrading by integrating freshwater algae cultivation with piggery anaerobic digestate liquid treatment. Appl. Microbiol. Biotechnol. 99, 6493–6501. https://doi.org/10.1007/s00253-015-6537-x.

Yan, C., Zheng, Z., 2014. Performance of mixed LED light wavelengths on biogas upgrade and biogas fluid removal by microalga *Chlorella* sp. Appl. Energy 113, 1008–1014. https://doi.org/10.1016/j.apenergy.2013.07.012.

Yang, L., Tan, X., Li, D., Chu, H., Zhou, X., Zhang, Y., Yu, H., 2015. Nutrients removal and lipids production by *Chlorella pyrenoidosa* cultivation using anaerobic digested starch wastewater and alcohol wastewater. Bioresour. Technol. 181, 54–61. https://doi.org/10.1016/j.biortech.2015.01.043.

Yang, S., Xu, J., Wang, Z.M., Bao, L.J., Zeng, E.Y., 2017. Cultivation of oleaginous microalgae for removal of nutrients and heavy metals from biogas digestates. J. Clean. Prod. 164, 793–803. https://doi.org/10.1016/j.jclepro.2017.06.221.

Zhang, S.Y., Sun, G.X., Yin, X.X., Rensing, C., Zhu, Y.G., 2013. Biomethylation and volatilization of arsenic by the marine microalgae *Ostreococcus tauri*. Chemosphere 93 (1), 47–53. https://doi.org/10.1016/j.chemosphere.2013.04.063.

Zhao, Y., Sun, S., Hu, C., Zhang, H., Xu, J., Ping, L., 2015. Performance of three microalgal strains in biogas slurry purification and biogas upgrade in response to various mixed light-emitting diode light wavelengths. Bioresour. Technol. 187, 338–345. https://doi.org/10.1016/j.biortech.2015.03.130.

Zhou, J., Wu, Y., Pan, J., Zhang, Y., Liu, Z., Lu, H., Duan, N., 2019. Pretreatment of pig manure liquid digestate for microalgae cultivation via innovative flocculation-biological contact oxidation approach. Sci. Total Environ. 694, 133720. https://doi.org/10.1016/j.scitotenv.2019.133720.

CHAPTER 6

Techno-economic analysis and life cycle assessment of algal cultivation on liquid anaerobic digestion effluent for algal biomass production and wastewater treatment

Sibel Uludag-Demirer[3], Mauricio Bustamante[2], Yan Liu[1] and Wei Liao[1]

[1]Biosystems Engineering, Michigan State University, East Lansing, MI, United States; [2]Biosystems Engineering, University of Costa Rica, San Pedro, San José, Costa Rica; [3]Biosystems and Agricultural Engineering Department & The Anaerobic Digestion Research and Education Center (ADREC), Michigan State University, East Lansing, MI, United States

1. Introduction

Organic rich wastes produced from food and agricultural industry present great opportunities for reuse and generation of valuable chemicals and materials especially during their management. The common approach for rendering the strength of organic rich agricultural and food waste typically starts with anaerobic digestion (AD) during which the complex organic compounds are broken down to simpler organic compounds as precursors for biomethane (CH_4) production (Peng et al., 2020; Linville et al., 2015; Ward et al., 2008). The effluent from AD process (digestate) is typically more homogeneous than the substrate in its structure, but it is still loaded with solids, organic carbon, and nutrients (NH_3-N and PO_4-P) (Marcato et al., 2008; Mangwandi et al., 2013; Shi et al., 2018). Current digestate management practices are very much based on land applications. However, the use of liquid digestate as fertilizer and/or nutrient rich irrigation water is far from meeting water and fertilizer needs of agricultural lands, and its overapplication frequently results in negative environmental impacts of nutrient runoff into soil and ground and surface water (Koszel and Lorencowicz, 2015; Styles et al., 2018; Panuccio et al., 2019). Therefore, recovering nutrients from the digestate is an urgent need to address the challenges regarding disposal of digestates. Several biorefinery approaches have been developed for utilization of digestate for value-added product generation. Centrifugation, belt press filter, and settling are often used to separate solid and liquid fractions of digestates (Moeller and Mueller, 2012; Tampio et al., 2016; Bolzonella et al., 2018). The solid digestate can be used for biofuel and chemical production (Yue et al., 2010; Zhong et al., 2016; Liu et al., 2016), animal bedding (Raul Pelaez-Samaniego et al., 2017), and biochar production (Streubel et al., 2012), while the liquid portion of the digestate can be used as the nutrients as well as carbon source to produce

Integrated Wastewater Management and Valorization using Algal Cultures
ISBN 978-0-323-85859-5, https://doi.org/10.1016/B978-0-323-85859-5.00009-9

© 2022 Elsevier Inc.
All rights reserved.

bioethanol (Sambusiti et al., 2016; Gao and Li, 2011), polyhydroxyalkanoate (Passanha et al., 2013), and carboxylic acids (Villegas-Rodríguez and Buitrón, 2020; Xiong et al., 2015). Recently, there has been an increasing interest in using liquid digestate as a nutrient source to grow protein-rich microalgae for value-added chemical production (Zuliani et al., 2016; Xu et al., 2015; Stiles et al., 2018). It is a promising method for harvesting nutrients in microalgal biomass while reducing the carbon footprint of protein production and alleviating environmental impacts of liquid digestate disposal.

Liquid digestate contains essential nutrients, mainly NH_3-N and PO_4-P, required by microalgae for growth. However, N:P mole ratio of the liquid digestate often exceeds the preferred ratio (16:1) of microalgae growth (Redfield, 1958). In addition, the presence of humic substances and solids (making the digestate not transparent) prevents light penetration and inhibits algal growth. Therefore, liquid digestate needs to be pretreated and conditioned before being used to culture microalgae. Several techniques have been investigated to treat liquid digestate and adjust N:P ratio such as membrane filtration (Gerardo et al., 2014), and chemical and electrochemical coagulation (Ritigala et al., 2021; Chen et al., 2012, 2016). Our previous studies successfully demonstrated that conventional coagulation and electrocoagulation (EC) remove both solids and nutrients in the liquid digestate and adjust N:P ratio that enable algal growth without need for additional nutrients and water (Chen et al., 2012).

This chapter presents a thorough techno-economic analysis and life cycle assessment of algal cultivation on AD liquid digestate with different pretreatment methods (conventional chemical coagulation and EC). Mass and energy analysis of the cultivations on chemical coagulation or EC-treated liquid digestate are conducted to delineate mass and energy flows of both processes. An economic analysis is then carried out to determine capital expenditure (CapEx) and operational expenditure (OpEx) of the processes and then conclude their economic performance. A life cycle impact assessment is then conducted to explore their environmental impacts.

2. Material and methods

2.1 Chemical and EC treatment of liquid digestate

Two methods of chemical coagulation and EC treatment were applied to prepare the liquid digestate for algal cultivation (Chen et al., 2012). The digestate used in the conventional coagulation was obtained from a commercial digester in a private diary farm (3000 cows) in Michigan. Dairy manure and food wastes were the feeds of the digester. Total solids (TS), volatile solids (VS), and chemical oxygen demand (COD) of the mixed feed are 11.4%, 9.6%, and 1.1 g/g TS, respectively. The digester is a continuously stirred tank reactor (CSTR) of an effective volume of 3500 m^3. The digestion temperature was 35°C and the hydraulic retention time was 30 days. Two 5.5 kW FAN screw press separators with a screen (opening size of 2 mm) were used to separate liquid and solid digestates. The liquid digestate was used for the chemical treatment. The chemical coagulation process was conducted as follows: The original digestate was first diluted with same amount of tap water; then $Ca(OH)_2$ and $AlK(SO_4)_2 \cdot 12H_2O$ were added to reach the concentration of 7.5 kg m^{-3} and 1 kg m^{-3}, respectively. After centrifugation, the supernatant was collected as the chemical treated digestate for algal cultivation (Chen et al., 2012).

The liquid digestate for the EC treatment was collected from the commercial anaerobic digester at Michigan State University. The digester is a CSTR and has an effective volume of 1800 m^3. Feedstock of the digester consisted of roughly 60% dairy manure and 40% food waste (wet mass). TS, VS, and COD of the mixed feed are 9.9%, 8.3%, and 1.05 g/g TS, respectively. The digester was operated at 35°C with a hydraulic retention time (HRT) of 25 days, and the digestate was separated into liquid and solid digestates using a screw press with a screen (opening size of 2 mm). The liquid digestate was used for the EC treatment with iron (Fe) as anode and cathode. The digestate was first diluted 5.5 times using tap water and then treated in a 50 L column EC reactor with the anode surface area/volume ratio of 12.4 m^2/m^3. The electrocoagulation was carried out at 20A and 8V for 4 h. After centrifugation of the EC-treated effluent, the supernatant was collected as the EC treated digestate for algal cultivation (Chen et al., 2016).

The characteristics of the original and treated liquid digestates are shown in Table 6.1.

2.2 Algal cultivation systems

Algae were collected from a local pond near Michigan State University (MSU) campus and maintained as the inoculum seeds (Chen et al., 2012, 2016). Algal cultivations on pretreated digestates (chemical and EC treated) were carried out using bench cultivation units. The total nitrogen concentration of chemical-treated digestate was adjusted to 200 mg/L by dilution using tap water. The raceway pond containing 19.8 L of chemical-treated liquid digestate was inoculated with 0.2 L of algal seed and cultured

Table 6.1 Chemical analysis of original and treated liquid digestates[a].

	Treatment	TS (w/w)	TN (mg/L)	TP (mg/L)	COD (g/L)	Total iron (mg/L)	pH
EC treatment	Original liquid digestate	4.80%	3100	1500	21.5	—	8.0
	EC-treated liquid digestate	0.03%	350	25.4	0.907	5.41	8.5
Chemical treatment	Original liquid digestate	4.37%	3200	1190	30.1	—	8.1
	Chemical treated liquid digestate	—	475	23.4	—	—	8.5

[a]Data are average of two replicates.

at 18°C ± 2°C under continuous illumination of 100 μmol m^{-2}s^{-1}. 0.3 L of the culture broth was harvested each day and chemical-treated AD liquid digestate and water were added to keep the total volume constant at 20 L (Chen et al., 2012). Semicontinuous algal culture on EC-treated AD liquid digestate was conducted in 2-L Erlenmeyer flasks including 60 mL algal seeds and 1 L EC treated digestate. The flasks were placed on an orbital shaker with speed of 150 rpm. The culture temperature was 22°C±2°C under the continuous illumination of 100 μmol m^{-2}s^{-1} (Chen et al., 2016). When algae reached its maximum growth rate, 100 mL of broth was harvested from each flask and 100 mL of EC-treated digestate was added every day (Chen et al., 2016).

2.3 Mass and energy balance analysis

Detailed mass and energy balance were carried out to generate data for an economic analysis and life cycle assessment of both cultivation systems. Overall process flowchart of the algal cultivation systems was shown in Fig. 6.1. The algal cultivation system includes liquid digestate pretreatment unit, race-way open pond algal cultivation, and algal biomass harvesting and drying. The liquid digestate from the screw filter is treated in the pretreatment unit by either chemicals or EC. The treated digestate is then used for continuous algal cultivation. An aliquot of culture broth is collected daily from algal pond, and the same amount of treated digestate mixed with reclaimed water is added to the pond. A centrifugation unit is applied to separate algal biomass from liquid broth. The supernatant liquid with less nutrients can be used as the reclaimed water or as the dilution water for EC or chemical pretreatments. The solid algal biomass is dried and

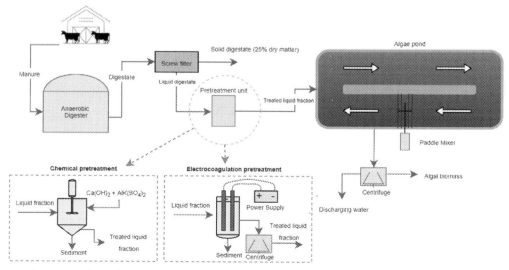

Figure 6.1 Flow chart of algal cultivations on chemical and EC-treated AD liquid digestates.

packed as a protein-rich feedstock for value-added chemical production. Mass and energy flows through the systems were concluded and used for the following economic analysis and life cycle assessment.

2.4 Economic analysis

An economic analysis was conducted to elucidate the feasibility of algal cultivation on treated digestate (Chen et al., 2012, 2016). The analysis is based on a commercial digester unit with a feeding rate of 100 m^3 per day. It generates 97.5 m^3 of liquid digestate. The size of pretreatment, cultivation, and drying units was based on the amount of liquid digestate and treatment capacity of individual systems. Capital expenditure (CapEx) and operational expenditure (OpEx) of individual unit operations were calculated using linear scaling of reference numbers (Davis et al., 2016). The Modified Accelerated Cost Recovery System (MACRS) (the current tax depreciation system in the United States) was used to calculate the annual depreciation of CapEx. An annual inflation of 3% was set for OpEx and revenues based on the 5-year average inflation rate in the United States (from 2016 to 2020). Break-even price of algal biomass production of the two systems was estimated based on CapEx, OpEx, and a payback period of 20 years.

2.5 Life cycle assessment

Besides economic feasibility, environmental impact is another important aspect of the studied systems. A life cycle assessment was then carried out to elucidate such impacts. The digester with a feeding rate of 100 m^3 per day was used for life cycle assessment as well. The direct land application of liquid digestate was used as the control. Mass and energy flows from the mass and energy balance analysis are used as life cycle inventory data. Two impact categories were chosen to run life cycle impact assessment: global warming potential (GWP) and water eutrophication potential. The classification of each category is defined by the US Environmental Protection Agency (US EPA). The analysis was conducted using the data from US EPA's TRACI-2 characterization factors (Bare, 2011). Contribution analysis was performed to interpret the factors that influence each impact category.

3. Results and discussion

3.1 Mass and energy balance of the studied systems

Mass and energy balance analysis based on a commercial digester with a feeding rate of 100 m^3 per day was conducted on two systems for algal biomass production and water reclamation (Fig. 6.2 and Table 6.2). After liquid and solid separation, the liquid digestate (97.5 m^3) is treated with either conventional chemical coagulation or EC.

For the treatment via conventional coagulation (Fig. 6.2A), the liquid digestate was diluted with the same amount of reclaimed water obtained from the clarifier used to

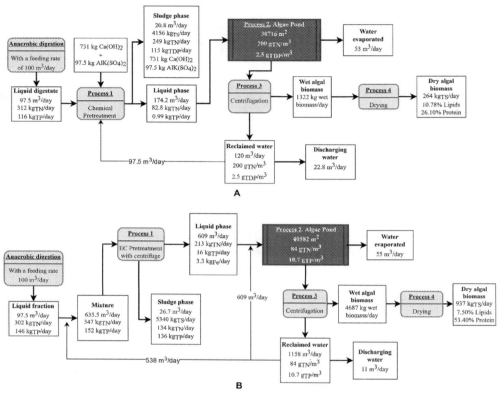

Figure 6.2 Mass balance of algal cultivation systems based on an anaerobic digestion system with a feeding rate of 100 m³/day. A. The cultivation system with chemical coagulation treatment; B. The cultivation system with EC treatment.

settle the flocs formed following the addition of 731 kg/day of $Ca(OH)_2$ and 97.5 kg/day of $AlK(SO_4)_2 \cdot 12H_2O$ to treat the liquid digestate overnight. After centrifugation, the supernatant (174.2 m³ with 82.8 kg of TN and 0.99 kg of TDP) was pumped to the algal pond (38,716 m² of surface area and 0.3 m of depth) as nutrients for algal cultivation. The sludge of 20.8 m³ including added chemicals and majority of nutrients (75% TN and 99% TDP) in the digestate can be used as a nutrient-rich solid fertilizer. It was calculated that 53 m³/day of water is evaporated from the race-way open pond cultivation based on water evaporation data for East Lansing, MI. The daily harvesting volume is 121.4 m³. Another centrifugation step is used to separate the algal biomass and less-nutrient reclaimed water. The wet algal biomass of 1322 kg per day with 20% total solids is dried by a triple-pass rotary dryer to produce 264.4 kg/day of dry algal biomass. The liquid digestate is diluted by pumping 97.5 m³ of the reclaimed water into the chemical coagulation process and the rest of the reclaimed water (22.8 m³) is discharged.

Table 6.2 Energy demand of algal cultivation on liquid anaerobic digestion effluent for algal biomass production[a].

Energy demand	Energy (kWh-e/kg dry algae biomass produced)	
	The culture with the chemical coagulation treatment	The culture with the EC treatment
Process 1 — Treatment	0.66[b]	10.34[c]
Process 2 — Open pond algal cultivation	1.17[d]	0.37[e]
Process 3 — Centrifugation of biomass collection	0.40[f]	1.11[g]
Process 4 — Drying	2.91[h]	2.91[i]
Overall energy demand	5.12	14.73

[a]Energy demand calculation is based on dry algae biomass of 1 kg.

[b]The chemical coagulation treatment uses a 1.1-kW centrifugal pump to transfer 195 m³ liquid to 11-kW disc stack centrifuge to do liquid and sludge separation. The energy demands for the centrifugal pump and centrifuge are 0.06 and 0.59 kWh/kg dry algal biomass, respectively. The chemicals are mixed during the pumping.

[c]The EC treatment uses a batch EC unit with a current of 20 A and voltage of 8 V. Two 3.73-kW centrifugal pumps are used to transfer 635 m³ to two 11-kW disc stack centrifuges to do liquid and sludge separation. The energy demands for the EC unit, centrifugal pumps, and centrifuges are 9.76, 0.04, and 0.54 kWh/kg dry algal biomass, respectively.

[d]A 1.1-kW centrifugal pump was used to transfer 174 m³ chemical treated liquid to the pond. The mixing of the pond is achieved by paddle wheels. The energy demands for the pump and paddle wheels are 0.04 and 1.13 kWh/kg dry algal biomass, respectively (Davis et al., 2016).

[e]A 3.73-kW centrifugal pump was used to transfer 609 m³ EC-treated liquid to the pond. The mixing of the pond is achieved by paddle wheels. The energy demands for the pump and paddle wheels are 0.04 and 0.33 kWh/kg dry algal biomass, respectively (Davis et al., 2016).

[f]A 1.1-kW centrifugal pump was used to transfer 121 m³ algal solution to 11-kW disc stack centrifuge to collect the algal biomass. The energy demands for the centrifugal pump and centrifuge are 0.04 and 0.36 kWh/kg dry algal biomass, respectively (data from a pilot operation).

[g]Two 3.73-kW centrifugal pumps were used to transfer 967 m³ algal solution to two 11-kW disc stack centrifuges to collect the algal biomass. The energy demands for the centrifugal pumps and centrifuges are 0.08 and 1.03 kWh/kg dry algal biomass, respectively (data from a pilot operation).

[h]A triple-pass rotary dryer was used to dry 1320 kg wet algal biomass. The heat and electricity required to dry the algal biomass are 2.90 and 0.01 kWh-e/kg dry algal biomass, respectively. The energy consumptions were calculated using heat equations. The parameters are as follows: initial biomass temperature is 20°C, drying temperature is 100°C, specific heat capacity of water is 2.18 kJ/kg/K, specific heat capacity of dry biomass is 1.48 kJ/kg/K, and latent heat of water at 100°C is 2244 kJ/kg.

[i]A triple-pass rotary dryer was used to dry 4685 kg wet algal biomass. The heat and electricity required to dry the algal biomass are 2.90 and 0.01 kWh-e/kg dry algal biomass, respectively. The energy consumptions were also calculated using hat equations. The parameters are the same with h.

Compared to the cultivation with chemically treated digestate, the EC treatment requires a high dilution (5.5X) of the liquid digestate for the treatment (Fig. 6.2B). After diluting 5.5 times, 635 m³/day of the diluted liquid digestate is treated by the EC. After 4 h of the EC treatment, the treated digestate is centrifuged, which generated 26.7 m³ of the nutrient-rich sludge containing 134 kg of TN and 136 kg of TP, and 609 m³ of the liquid fraction. The liquid is pumped into an algae pond with a surface area of 40,582 m² and a depth of 0.3 m. The volume of water evaporation from the pond is 55 m³/day, which is again calculated using the local water evaporation data. 1162 m³ of the algal solution is harvested daily. Another centrifugation dedicated for algal biomass harvesting generates 4687 kg/day of wet algal biomass and 1158 m³/day of reclaimed water. The wet algal biomass is dried by a triple-pass rotary dryer to produce 937 kg/day of dry algal biomass. 1147 m³ of the reclaimed water is pumped back to the EC treatment unit to dilute the liquid digestate and the liquid phase before algae pond, and rest of the reclaimed water (11.6 m³) is discharged. It is apparent that the cultivation with EC treated digestate (937 kg/day) has much better biomass production than the cultivation using chemically coagulated digestate (264 kg/day). The difference on biomass production between two systems is mainly caused by concentrations of available nutrient. The cultivation with chemical treatment has a much higher TN content (200 g/m³) than the cultivation with EC treatment (84 g/m³), which results in slow algae growth due to inhibition caused by high TN (mainly ammonia).

Energy balance analysis further delineated the energy consumption of individual unit operations (processes) in both systems (Table 6.2). The drying process (Process 4, Fig. 6.2) is an energy-intensive process for both systems, which requires 2.91 kWh-e/kg dry biomass. The drying process requires the largest energy input among four processes of the cultivation with chemical treatment. Energy demands of the chemical treatment with centrifuge (Process 1, Fig. 6.2), open-pond cultivation (Process 2, Fig. 6.2), and biomass collection (Process 3, Fig. 6.2) are 0.66, 1.17, and 0.40 kWh-e/kg dry algal biomass, respectively. For the algae cultivation in the liquid digestate treated by EC, the most energy-intensive unit operation is the EC process since it requires high dilution and high electricity demand (Process 1, Fig. 6.2). The energy consumption of the EC unit is calculated as 10.34 kWh-e/kg dry biomass, which is much higher than other three processes (0.37, 1.11, and 2.91 kWh-e/kg dry biomass for open-pond cultivation, biomass collection, and drying, respectively). Due to the high energy consumption of the EC, the total energy consumption of the cultivation system with the EC treatment (14.73 kWh-e/kg dry biomass) is nearly three times higher than that of the cultivation system with the chemical treatment (5.12 kWh-e/kg dry biomass).

3.2 Economic analysis

Based on the mass and energy balance data for a culturing period of 240 days, 63 and 225 metric tons of dry algal biomass per year can be produced using digestate treated by chemical coagulation and EC respectively. CapEx, OpEx, and the payback period of 20 years are used as the parameters to estimate the break-even algal biomass prices of the algal cultivation systems with two treatment methods investigated in this study (Table 6.3). The total CapEx includes direct cost of all equipment required for different processes, land cost, and indirect cost of proratable expenses, field expenses, office and construction, project contingency, and other costs. The total OpEx includes annual costs of chemicals, electricity, and labor.

Table 6.3 Economic performance of the algal cultivation with chemical and EC treatment for a digestion unit with a feeding rate of 100 m^3/day.

	The culture on chemical treated digestate			The culture on EC-treated digestate		
	Unit cost	Unit	Cost	Unit cost	Unit	Cost
Capital expenditure (CapEx) − Direct cost						
Chemical treatment vessel[a]	$34.3/$m^3$ vessel volume	200 m^3	$6860			
EC treatment reactor[b]				$16,667/$m^3$/hr	27 m^3/h	$450,009
Pump to transfer treated digestate[c]	$2000	1	$2000	$3000	1	$3000
Centrifuge for threated digestate[d]	$50,000	1	$50,000	$50,000	2	$100,000
Algae ponds with paddle wheels[e]	$6.13/$m^2$ pond	38,716 m^2	$237,259	$6.13/$m^2$ pond	40,582 m^2	$248,768
Pump to transfer algal solution[e]	$2000	1	$2000	$3000	2	$6000
Centrifuge for algal solution[d]	$50,000	1	$50,000	$50,000	2	$100,000
Dryer[f]	$22/kg/hr water removed	53 kg/h water removed	$1166	$22/kg/hr water removed	187 kg/h water removed	$4114
Total direct CapEx			$349,285			$908,891
Capital expenditure (CapEx) − Indirect cost						
Proratable expenses[g]			$17,464			$45,445
Field expenses[h]			$17,464			$45,445
Office and construction[i]			$34,929			$90,889
Project contingency[j]			$34,929			$90,889
Other costs[k]			$17,464			$45,445
Total indirect CapEx			$122,250			$318,112
Land cost[l]			$58,074			$60,873

Continued

Table 6.3 Economic performance of the algal cultivation with chemical and EC treatment for a digestion unit with a feeding rate of 100 m^3/day.—cont'd

	The culture on chemical treated digestate			The culture on EC-treated digestate		
	Unit cost	Unit	Cost	Unit cost	Unit	Cost
Total CapEx [m]			**$514,510**			**$1,272,048**
Operational expenditure (OpEx)						
Ca(OH)$_2$	$0.011/kg	351,000 kg/year	$3773			
AlK(SO$_4$)$_2$	$0.2/kg	46,800 kg/year	$9360			
Electricity for chemical or EC treatment operation[n]	$0.1/kWh	40,800 kWh/year	$4080/year	$0.1/kWh	2,325,360 kWh/year	$232,536/year
Electricity for algal pond cultivation operation[n]	$0.1/kWh	73,920 kWh/year	$7392/year	$0.1/kWh	84,240 kWh/year	$8424/year
Electricity for algal solution centrifuge[n]	$0.1/kWh	25,440 kWh/year	$2544/year	$0.1/kWh	250,800 kWh/year	$25,080/year
Drying[o]	$0.1/kWh	480 kWh/year	$48/year	$0.1/kWh	1200 kWh/year	$120/year
Maintenance[p]			$10,479/year			$27,267/year
Labor cost						
Plant engineer[q]	$80,000/employee/year	1 employee	$80,000/year	$80,000/employee/year	1 employee	$80,000/year
Operators[q]	$40,000/employee/year	2 employees	$80,000/year	$40,000/employee/year	3 employees	$120,000/year

Labor burden[r]			$160,000/ year		$200,000/ year
Total labor cost			$144,000/ year		$180,000/ year
Total OpEX [s]			**$341,676/ year**		**$673,427/ year**
Break-even price of algal biomass production[t]			**$5.78/kg dry biomass**		**$3.26/kg dry biomass**

[a]The cost of the vessel for the chemical treatment is based on the unit cost ($34.3/m^3) of a vessel for the reference (Davis et al., 2016).
[b]The unit cost of the EC treatment is $27/m^3/hr (Davis et al., 2016). The flow rate of 27 m^3/h is needed to process 635 m^3/day of the liquid using the EC unit.
[c]The costs of the centrifugal pumps are obtained from a local pump supplier.
[d]The cost of the stack disc centrifuge is obtained from a centrifuge manufacturer.
[e]The unit price of the algae pond is based on the MicroBio system reported in the reference (Davis et al., 2016).
[f]The cost of the triple-pass rotary dryer is calculated based on the capital cost of $22/kg water removed/hr.
[g]Proratable expenses including fringe benefits, burdens, and insurance of the construction contractor are set at 5% of the total direct CapEx.
[h]Field expenses including consumables, small tool and equipment rental, field services, temporary construction facilities, and field construction supervision are set at 5% of the total direct CapEx.
[i]Office and construction costs including engineering plus incidentals, purchasing, and construction are set at 10% of the total direct CapEx.
[j]Project contingency including extra cash on hand for unforeseen issues during construction is set at 10% of the total direct CapEx.
[k]Other costs including start-up and commissioning costs; land, rights-of-way, permits, surveys, and fees; piling, soil compaction/dewatering, and unusual foundations; sales, use, and other taxes; freight, insurance in transit and import duties on equipment, piping, steel, and instrumentation; overtime pay during construction; field insurance; project team; and transportation equipment are set at 5% of the CapEx.
[l]The land cost for purchasing the land for the system is set at $0.74/m^2 according to the current price of land in Michigan.
[m]Total CapEx is the sum of total direct CapEx, total indirect CapEx, and land cost.
[n]Electricity cost for pumps and centrifuges in process 1, 2, and 3 is based on a current electricity cost of $0.1/kWh and 240 days of operation time.
[o]The heat demand for the drying is from the digestion operation without charge. The electricity usage for the dryer is considered in the cost based on a current electricity cost of $0.1/kWh and 240 days of operation time.
[p]The maintenance cost is set at 3% of the total direct CapEx.
[q]The labor cost is estimated based on a similar operation in Michigan.
[r]The labor burden is set at 90% of the total salary.
[s]The total OpEx is the sum of electricity cost, maintenance, and labor cost.
[t]The break-even price of algal biomass is calculated based on both CapEx and OpEx. The life time of the system is set at 20 years. The 5-year average local inflation of 3% is used as the inflation rate. The depreciation is just on CapEx. The annual depreciation rates from MARCRS are 0.100, 0.188, 0.144, 0.115, 0.092, 0.074, 0.066, 0.066, 0.065, 0.065, 0.033, and 0.033 (after 10 years).

160 Integrated Wastewater Management and Valorization using Algal Cultures

The data show that the total CapEx for establishing the cultivation system with EC treatment is $1,272,048, which is 2.5 times higher than the CapEx for the cultivation system with chemical treatment ($514,510). The high CapEx is mainly attributed to the high dilution (5X) of EC. Among individual unit operations, the EC unit is the most expensive one ($450,009) in the cultivation system with EC treatment, and the algal pond is the most expensive one ($237,259) in the cultivation system with chemical treatment. Correspondingly, since the cultivation system with EC treatment deals with much larger water flow (1163 m^3 algal solution/day from the pond) compared to 121 m^3 algal solution/day of the cultivation with chemical treatment, the OpEx of the system with EC treatment (673,427/year) is also much higher than that ($341,677/year) of the latter system. The electricity consumption for the EC in the system with EC treatment and the labor cost in the system with chemical treatment are $232,536/year and $144,000/year, which are the highest expenditures in corresponding systems. Based on the data of CapEx and OpEx, the local inflation rate, and MACRS depreciation rate, and the break-even prices of algal biomass are $5.78/kg dry biomass and $3.26/kg dry biomass for the system with chemical treatment and the system with EC treatment, respectively, for a system lifetime of 20 years. It has been reported that current selling price of algal biomass using similar open pond raceway on freshwater is in the range of $3.92—4.19/kg dry biomass (Davis et al., 2016). The studied system with EC treatment clearly has a better economic performance than the reported cultivation on freshwater. It is interesting to conclude that the cultivation system with EC treatment leads to a better break-even price with much higher CapEx and OpEx than the other system and current commercial system on freshwater, which is mainly attributed to much higher biomass production of the system with EC (225 metric ton/year). Therefore, improving algal biomass production is key to enhance economic performance of algal cultivation on liquid digestate, even considering using more expensive equipment and energy-intensive operations.

3.3 Life cycle impact assessment

Life cycle assessment is mainly focused on GWP and water eutrophication potential. Corresponding impact categories are analyzed using contribution analysis (Chen et al., 2015) (Fig. 6.3). The GWP is the amount of greenhouse gases that is potentially released in a year. During direct land application of the liquid digestate, the high nutrient contents of COD and TN could contribute to high emissions of methane (CH_4) and nitrous oxide (N_2O). According to the IPCC emission conversion factors (IPCC, 2006), COD has a methane conversion factor of 0.282 kg CH_4/kg COD, and TN has a N_2O conversion factor of 0.005 kg N_2O/kg TN. Emissions of CH_4 and N_2O are normalized to a metric ton of CO_2 equivalent (CO_2-e) based on the following conversions: 1 kg CH_4 = 25 kg CO_2-e and 1 kg N_2O = 300 kg CO_2-e (Chen et al., 2015). To sum up both emissions

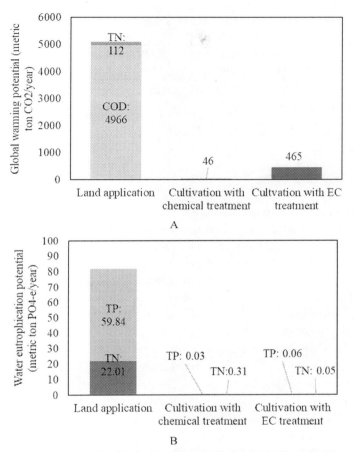

Figure 6.3 Contribution analysis of individual impact categories for land application and the culture with treatment. A. Global warming potential; B. Water eutrophication potential.

from both COD and TN, the direct land application of liquid digestate has an overall GWP of 5077 metric ton CO_2-e/year (Table 6.4 and Fig. 6.3A). Compared to the land application, both cultivation systems have much lower GWP. The GWPs for the cultivation system with chemical coagulation and with EC are 46 metric ton CO_2-e/year and 465 metric ton CO_2-e/year respectively (Table 6.4). Between the two algae cultivation systems, the cultivation with chemical coagulation has 10 times less GWP than the one with EC.

Water eutrophication potential is calculated using the total amount of TN and TP discharged into the environment. The unit for water eutrophication potential is metric ton PO_4-e/year. The conversion factors for TN and TP are 0.42 kg PO_4-e/kg TN

Table 6.4 Comparison of the life cycle impact assessment between land application and the cultivation with chemical or EC treatment.

	Land application of liquid digestate	Cultivation with chemical treatment	Cultivation with EC treatment
Global warming potential (metric ton CO_2-e/year)	5077	46	465
Water eutrophication potential (metric ton PO_4-e/year)	59.84	0.03	0.06

and 3.07 kg PO_4-e/kg TP, respectively. Since liquid digestate has relatively high TN and TP (Table 6.1), direct land application has the largest water eutrophication potential of 59.84 metric ton PO_4-e/year (Table 6.4 and Fig. 6.3B). Meanwhile, the cultivation systems with chemical and EC treatment have much lower water eutrophication potential due to lower TN and TP contents in the reclaimed water as well as small amount of discharging (Fig. 6.2). Most of the nutrients are reclaimed in the sludge as fertilizer and converted into value-added algal biomass as protein and cell components. Water eutrophication potentials of the systems with chemical treatment and EC treatment are 0.03 and 0.06 metric ton PO_4-e/year, respectively (Table 6.4 and Fig. 6.3B).

The life cycle impact assessment demonstrates the advantages of algal cultivation using the anaerobic digestate treated with chemical coagulation or EC over the current practice of direct land application. The cultivation with either treatment significantly reduces both GWP and water eutrophication potential, which indicates that algal cultivation combined with pretreatment technologies, such as coagulation, could play a critical role on digestate management to further reduce emissions and protect the environment.

4. Conclusions

This study comprehensively analyzed techno–economic performance and environmental impacts of two algal cultivation systems using liquid AD effluent as the growth medium. Based on a digestion system with a feeding rate of 100 m^3/day, the cultivation system with EC treatment can produce 225 metric tons of algal biomass per year, which is about 3.5 times the production of the cultivation with chemical pretreatment (63 metric tons/year). The techno–economic analysis revealed that even with high CapEx and OpEx, the cultivation system with EC had a better break-even price of algal biomass production ($3.25/kg dry biomass) than the cultivation system with chemical treatment ($5.78/kg dry biomass). The life cycle assessment further concludes that both systems greatly reduced GWP and water eutrophication potential compared to the current practices of direct land application; while compared with EC treatment, the algal cultivation on chemical treated liquid digestate had less negative environmental impacts and performed better on emission reduction. This study clearly shows that pretreatment must be applied to enable the digestate amenable for algal cultivation. In addition, comprehensive TEA and LCA are critical to conclude technically feasible, economically sound, and environmentally friendly cultivation systems for both wastewater treatment and algal biorefining.

Acknowledgments

The authors would like to thank the AgBioResearch at Michigan State University for funding this work through faculty salaries.

References

Bare, J., 2011. Traci 2.0: the tool for the reduction and assessment of chemical and other environmental impacts 2.0. Clean Technol. Environ. Policy 13, 687–696.

Bolzonella, D., Fatone, F., Gottardo, M., Frison, N., 2018. Nutrients recovery from anaerobic digestate of agro-waste: techno-economic assessment of full scale applications. J. Environ. Manag. 216, 111–119.

Chen, R., Li, R., Deitz, L., Liu, Y., Stevenson, R.J., Liao, W., 2012. Freshwater algal cultivation with animal waste for nutrient removal and biomass production. Biomass Bioenergy 39, 128–138.

Chen, R., Rojas-Downing, M.M., Zhong, Y., Saffron, C.M., Liao, W., 2015. Life cycle and economic assessment of anaerobic Co-digestion of dairy manure and food waste. Ind. Biotechnol. 11, 127–139.

Chen, R., Liu, Y., Liao, W., 2016. Using an environmentally friendly process combining electrocoagulation and algal cultivation to treat high-strength wastewater. Algal Res. 16, 330–337.

Davis, R., Markham, J., Kinchin, C., Grundl, N., Tan, E.C.D., 2016. Process Design and Economics for the Production of Algal Biomass: Algal Biomass Production in Open Pond Systems and Processing through Dewatering for Downstream Conversion. The U.S. Department of Energy National Renewable Energy Laboratory, Oak Ridge, TN.

Gao, T., Li, X., 2011. Using thermophilic anaerobic digestate effluent to replace freshwater for bioethanol production. Bioresour. Technol. 102, 2126–2129.

Gerardo, M.L., Oatley-Radcliffe, D.L., Lovitt, R.W., 2014. Integration of membrane technology in microalgae biorefineries. J. Membr. Sci. 464, 86–99.

IPCC, 2006. IPCC guidelines for national greenhouse gas inventories, prepared by the national greenhouse gas inventories programme. In: Eggleston, H.S., Buendia, L., Miwa, K., Ngara, T., Tanabe, K. (Eds.), I.P.o.C.C. (IPCC), Institute for Global Environmental Strategies (IGES), Hayama, Japan, 2006.

Koszel, M., Lorencowicz, E., 2015. Agricultural use of biogas digestate as a replacement fertilizers. Agric. Agric. Sci. Procedia 7, 119–124.

Linville, J.L., Shen, W.Y., M, M., Urgun-Demirtas, M., 2015. Current state of anaerobic digestion of organic wastes in North America. Curr. Sustain. Renew. Energy Rep. 2, 136–144.

Liu, Z., Liao, W., Liu, Y., 2016. A sustainable biorefinery to convert agricultural residues into value-added chemicals. Biotechnol. Biofuels 9 (17 September 2016)-(2017 September 2016).

Mangwandi, C., Liu, J., Albadarin, A.B., Allen, S.J., Walker, G.M., 2013. The variability in nutrient composition of Anaerobic Digestate granules produced from high shear granulation. Waste Manag. 33, 33–42.

Marcato, C.E., Pinelli, E., Pouech, P., Winterton, P., Guiresse, M., 2008. Particle size and metal distributions in anaerobically digested pig slurry. Bioresour. Technol. 99, 2340–2348.

Moeller, K., Mueller, T., 2012. Effects of anaerobic digestion on digestate nutrient availability and crop growth: a review. Eng. Life Sci. 12, 242–257.

Panuccio, M.R., Papalia, T., Attina, E., Giuffre, A., Muscolo, A., 2019. Use of digestate as an alternative to mineral fertilizer: effects on growth and crop quality. Arch. Agron Soil Sci. 65, 700–711.

Passanha, P., Esteves, S.R., Kedia, G., Dinsdale, R.M., Guwy, A.J., 2013. Increasing polyhydroxyalkanoate (PHA) yields from Cupriavidus necator by using filtered digestate liquors. Bioresour. Technol. 147, 345–352.

Peng, W., Lu, F., Hao, L.P., Zhang, H., Shao, L.M., He, P.J., 2020. Digestate Management for High-Solid Anaerobic Digestion of Organic Wastes: A Review. Bioresource Technology, p. 297.

Raul Pelaez-Samaniego, M., Hummel, R.L., Liao, W., Ma, J., Jensen, J., Kruger, C., Frear, C., 2017. Approaches for adding value to anaerobically digested dairy fiber. Renew. Sustain. Energy Rev. 72, 254–268.

Redfield, A.C., 1958. The biological control OF chemical factors IN the environment. Am. Sci. 46, 205–221.

Ritigala, T., Demissie, H., Chen, Y., Zheng, J., Zheng, L., Zhu, J., Fan, H., Li, J., Wang, W.D., S, K., Weerasooriya, R., 2021. Optimized pre-treatment of high strength food waste digestate by high content aluminum–nanocluster based magnetic coagulation. J. Environ. Sci. 104, 430–443.

Sambusiti, C., Monlau, F., Barakat, A., 2016. Bioethanol fermentation as alternative valorization route of agricultural digestate according to a biorefinery approach. Bioresour. Technol. 212, 289–295.

Shi, L., Simplicio, W.S., Wu, G., Hu, Z., Hu, H., Zhan, X., 2018. Nutrient recovery from digestate of anaerobic digestion of livestock manure: a review. Curr. Pollut. Rep. 4, 74–83.

Stiles, W.A.V., Styles, D., Chapman, S.P., Esteves, S., Bywater, A., Melville, L., Silkina, A., Lupatsch, I., Grunewald, C.F., Lovitt, R., Chaloner, T., Bull, A., Morris, C., Llewellyn, C.A., 2018. Using microalgae in the circular economy to valorise anaerobic digestate: challenges and opportunities. Bioresour. Technol. 267, 732–742.

Streubel, J.D., Collins, H.P., Tarara, J.M., Cochran, R.L., 2012. Biochar produced from anaerobically digested fiber reduces phosphorus in dairy lagoons. J. Environ. Qual. 41, 1166–1174.

Styles, D., Adams, P., Thelin, G., Vaneeckhaute, C., Chadwick, D., Withers, P.J.A., 2018. Life cycle assessment of biofertilizer production and use compared with conventional liquid digestate management. Environ. Sci. Technol. 52, 7468–7476.

Tampio, E., Marttinen, S., Rintala, J., 2016. Liquid fertilizer products from anaerobic digestion of food waste: mass, nutrient and energy balance of four digestate liquid treatment systems. J. Clean. Prod. 125, 22–32.

Villegas-Rodríguez, G.S., Buitrón, 2020. Performance of native open cultures (winery effluents, ruminal fluid, anaerobic sludge and digestate) for medium-chain carboxylic acid production using ethanol and acetate. J. Water Process Eng. 101784.

Ward, A.J., Hobbs, P.J., Holliman, P.J., Jones, D.L., 2008. Optimisation of the anaerobic digestion of agricultural resources. Bioresour. Technol. 99, 7928–7940.

Xiong, B., Richard, T.L., Kumar, M., 2015. Integrated acidogenic digestion and carboxylic acid separation by nanofiltration membranes for the lignocellulosic carboxylate platform. J. Membr. Sci. 489, 275–283.

Xu, J., Zhao, Y., Zhao, G., Zhang, H., 2015. Nutrient removal and biogas upgrading by integrating freshwater algae cultivation with piggery anaerobic digestate liquid treatment. Appl. Microbiol. Biotechnol. 99, 6493–6501.

Yue, Z., Teater, C., Liu, Y., MacLellan, J., Liao, W., 2010. A sustainable pathway of cellulosic ethanol production integrating anaerobic digestion with biorefining. Biotechnol. Bioeng. 105.

Zhong, Y., Liu, Z., Isaguirre, C., Liu, Y., Liao, W., 2016. Fungal fermentation on anaerobic digestate for lipid-based biofuel production. Biotechnol. Biofuels 9.

Zuliani, L., Frison, N., Jelic, A., Fatone, F., Bolzonella, D., Ballottari, M., 2016. Microalgae cultivation on anaerobic digestate of municipal wastewater, sewage sludge and agro-waste. Int. J. Mol. Sci. 17.

CHAPTER 7

Biomethane production from algae biomass cultivated in wastewater

Esteban Serrano, Maikel Fernandez, Raúl Cano, Enrique Lara, Zouhayr Arbib and Frank Rogalla
FCC Aqualia S.A. Innovation and Technology Department, Madrid, Spain

1. Introduction

This chapter explores the energy balance of low-cost secondary treatment of wastewater with High Rate Algae Ponds (HRAPs), and the methane production potential of the produced biomass. Microalgae can be an attractive solution for wastewater treatment (WWT) while producing biofuel (Arbib et al., 2013; De Godos et al., 2011). Microalgae provide oxygenation of organic matter in the culture while assimilating nitrogen and phosphorous in their cells. High-quality effluents can be generated in shallow channels exposed to sunlight, where microalgae grow in symbiosis with bacteria. This biomass can easily be converted into biogas through anaerobic digestion, while the extraction of lipids and transesterification for biodiesel is far remote from any sustainability (Craggs et al., 2011).

Typically in WWT, primary settled effluents are fed to the microalgae in HRAP, while biomass is then separated in secondary clarifiers (Park and Craggs, 2010; Canovas et al., 1996). However, to maximize energy recovery from raw effluents, anaerobic pre-treatment could result in a higher energy production per unit of volume processed. Since suspended and dissolved carbonaceous components from the raw influent can directly be converted into methane, the reduced suspended solids could favor microalgae growth in the downstream HRAPs.

Microalgae activity, which determines the oxygenation potential and biomass production rates, is driven by light availability. This direct correlation between solar radiation and the performance of HRAPs means that shorter hydraulic retention times (HRTs) could be applied during summer, and longer during winter, when low temperatures decrease cell growth. This strategy has been applied in locations with significant seasonal variations in solar radiation as Europe, California, and New Zealand (Garcia et al., 2000; Craggs et al., 2008).

To achieve reliable WWT, and to generate a biomass and biofuels thorough anaerobic digestion, efficient and low-cost harvesting must be developed. This process must overcome difficulties of microalgae to settle, due to the small cell size and strong negative surface charge, resulting in low biomass recoveries. In this study, Dissolved Air

Integrated Wastewater Management and Valorization using Algal Cultures
ISBN 978-0-323-85859-5, https://doi.org/10.1016/B978-0-323-85859-5.00005-1

© 2022 Elsevier Inc.
All rights reserved.

Flotation (DAF) is used to provide low suspended solids effluents, and to simultaneously concentrate the biomass to a level suitable for anaerobic digestion (AD).

Coupling AD with algal production was first mentioned more than 60 years ago (Golueke et al., 1957). Since then, AD of microalgal biomass has been investigated in various combinations, temperatures, reactor configurations, and pretreatment methods, and recently reviewed by several authors (Lakaniemi et al., 2013). AD of microalgae produces biogas containing mainly methane and carbon dioxide (Chisti, 2007; Ras et al., 2011), converting into energy the organic matter in the cells produced by capturing the sunlight through the photosynthetic process (Golueke and Oswald, 1959).

The higher photosynthetic efficiency and growth rate of microalgae compared to terrestrial crops are their main interest as potential for biofuel production (Vasudevan and Briggs, 2008). Methane yields from microalgae vary as a function of cellular protein, carbohydrate and lipid content, cell wall structure, as well as process parameters such as the bioreactor type and the digestion temperature. Chemical composition of microalgal biomass varies among species, and even within the same species under different growth conditions (Sheehan et al., 1998). Thus, CH_4 production from microalgae needs to be related to the different conditions to seek high growth yields and to determine optimal biomass composition for AD (Lakaniemi et al., 2013).

This work summarizes the results obtained during the preparation and operation of an industrial scale WWT coupled to bioenergy production in Spain. The experiments optimized the performance of HRAPs fed with anaerobically pretreated WW and screened raw WW in terms of biomass production and anaerobic digestibility. The variables studied include the HRT in the algae ponds, and CO_2 supply (with and without), as well as the separation of biomass with DAF. Biomethane production of microalgae at mesophilic and thermophilic range was also evaluated in laboratory and in pilot scale reactors.

2. Material and methods

2.1 Pilot and prototype plant description

Results presented in this work are obtained at two different scales of HRAP and AD:
- Pilot (200 m^2) and 5 L digesters
- Prototype (1000 m^2) with 600 L digesters

2.1.1 Pilot plant (200 m^2)

2.1.1.1 Anaerobic sludge blanket reactors (UASB) for raw wastewater pretreatment

Three UASB reactors (Fig. 7.1) with total volume of 24 m^3 each were used to supply pretreated WW to pilot HRAPs. Sludge accumulated in the UASB reactors was controlled at a bed height of 2.5 m, wasting the excess sludge from the top of the sludge bed. Dissolved methane stripping was made with a cascade reactor under vacuum (-200 mbar).

Figure 7.1 24 m³ Pilot UASB reactors.

2.1.1.2 Cultivation and harvesting

UASB effluent was fed continuously to six Pilot HRAPs that were made of concrete blocks and each had a surface area of 32 m² and a total volume of 9.6 m³ (Fig. 7.2A). They were operated over several years at the WWTP in Chiclana (Spain), as described in De Godos et al. (2016). With a six-blade paddle wheel moving at 7 rpm, a liquid velocity between 0.20 and 0.25 m/s was maintained. Two flow rectifiers in each loop of the raceways were used to decrease the dead zones, the backflow and eddies generated.

Harvesting was done by DAF unit after coagulation with PAX-18 and flocculation with Flopam FO 4800 SSH (Fig. 7.2B). The DAF was operated in batches of 3–4 h/d with

Figure 7.2 (A) 32 m² Raceway reactors and (B) complete DAF harvesting plant.

an air/solids ratio of 0.02 mg air/mg TSS, within the typical values to separate secondary solids (0.005—0.09 mg air/mg TSS according to Shammas and Bennett (2010)). Flow to the HRAP varied from 0.040 to 0.133 m^3/h (between HRT of 7 d in winter and 3 d in summer), while one DAF with a capacity of 1.67 m^3/h served all the ponds.

2.1.1.3 Anaerobic digesters

Two sets of four identical continuously stirred reactors, each with a total volume of 5 L, were operated in a field lab at Chiclana WWTP (Fig. 7.3). The digesters were constructed of PVC tube with gas-tight top and bottom plates, fitted with a 32 mm drain tube and valve to remove digestate. As illustrated in Fig. 7.3, a 60 rpm motor is mounted on the top plate, fitted with a gas outlet, a feed port sealed with a rubber bung, and a draught tube liquid seal through which a stainless steel asymmetric bar stirrer is inserted. Temperature is controlled by circulating water from a thermo-statically controlled bath through a heating coil around each digester.

2.1.2 Scale up process: prototype plant (1000 m^2)

In order to confirm the results obtained in the pilot HRAPs (6 × 32 m^2), the process was scaled up to a 2 × 500 m^2 prototype.

2.1.2.1 Cultivation area

Algae were cultivated with two parallel HRAPs of 500 m^2 each (Fig. 7.4). In order to evaluate the improvements on algae growth and energy balance, an innovative low

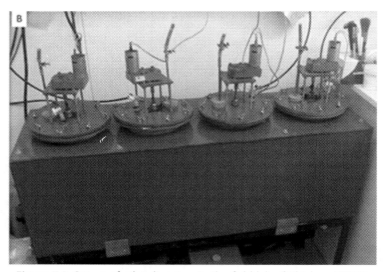

Figure 7.3 Set up of pilot digestors in the field lab of Chiclana WWTP.

Biomethane production from algae biomass cultivated in wastewater 169

Figure 7.4 2 × 500 m² prototype HRAP (left paddle wheel, right venturi mixer).

energy algae reactor patented by FCC aqualia (European Patent: EP 2875724 A1) was developed, reducing the power consumed by pond with:
- Hydraulic improvements to lower head losses along the ponds, especially those produced in the bends and sumps.
- Increasing the hydraulic efficiency of recirculation by using a submergible impeller installed in a venturi flow booster.

2.1.2.2 Harvesting by flotation
The main objective of biomass separation is to obtain an effluent that fits the most restrictive limits of the EU WWT directive 91/271/CE, and to concentrate the biomass up to 5% of TSS to feed AD simultaneously. To overcome one of the bottlenecks of microalgae biofuels production, efficient harvesting is essential. DAF combined with a low energy pressurization pump can have an energy requirement lower than 0.1 kwh/m³. Coagulation followed by flocculation precipitates phosphorus and separates efficiently the biomass from the culture media, thickening the solids for AD.

2.1.2.3 Anaerobic digesters
Three prototype ADs of 2 × 600 L and 1 × 1500 L of effective volumes were continuously fed with harvested algae from HRAP and DAF (Fig. 7.5). Algae were fed twice a day, setting a certain flow according to the desired HRT in AD (Table 7.1) for the main operational parameters. Proper mixing was achieved by mechanical vertical stirrers and internal recycling in the cylindrical reactors. A proportional—integral—derivative controller for pH and temperature, using external water jacket, and a hot/cold water circuit coming from a biomass boiler, maintained the temperature in each digester

Figure 7.5 Pilot anaerobic digestion plant.

Table 7.1 Operational parameters of the three prototype digesters.

	HRT (d)			Tª (°C)				Days of operation
Period	Ambient	Meso	Thermo	Ambient		Meso	Thermo	n
A	—	—	—	ª19.0 ± 1.2	ᵇ17.6	35	55	31
B	30	21	21	ª22.8 ± 0.4	ᵇ8.7	35	55	91
C	25	18	18	ª29.9 ± 0.7	ᵇ10.5	35	55	72

ª$X \pm IC$ = confidence interval calculated for $\alpha = 0.05$; n = n° of samples.
ᵇCV = variation coefficient (%)).

(either mesophilic or thermophilic conditions: 35 or 55°C, respectively). Three independent gas holders were used to store and quantify the biogas production, and a portable biogas analyzer COMBIMASS GA-m (BINDER) was used to measure its composition.

2.2 Experimental design

Algae growth in the smaller pilot HRAP was performed in continuous mode by using the effluent of UASB fed with screened raw WW.

The pilot experiments were started during autumn with two ponds operated at 7 d HRT and two others at 10 d, with one pond each controlled by CO_2 addition. In spring, HRT was reduced from 7 to 5 d and from 10 to 7 d, respectively. HRT was further lowered for summer (from 5 to 3 and from 7 to 5 d), corresponding to the maximum productivity of the year. At the end of the summer, HRT was increased from 3 to 5 and from 5 to 7 d.

The 500 m^2 prototype HRAP plants were operated without pH control or external CO_2 addition, and HRT was kept at constant 3 d even during wintertime, to operate the plant under the most critical parameters and exploring the limits of the system.

2.3 Analytical procedures

Samples of 300 mL from the culture broth were taken three times a week in order to evaluate biomass concentration, suspended solids, nutrients, soluble COD, and optical density. Total and soluble COD were measured in influent and effluent, but in culture broth just the soluble phase was determined (after filtration or centrifugation). Samples for microscopical identification of the phytoplankton were drawn weekly.

Biomass productivity is measured as total and volatile suspended solids (TSS and VSS), using the latter in calculations of biomass production. Settleable solids concentration was determined using Imhoff cones according to (APHA, 2005). Optical density was analyzed as absorbance in the wavelength of 550 and 750 nm.

NH_4^+ was analyzed by means of a selective electrode (Thermo Scientific), and Total Kjeldahl Nitrogen was determined by distillation and titration (VELP UDK 152) according to Lundquist et al. (2010). This parameter was used to determine the nitrogen content of the algae biomass produced in the ponds.

NO_3^- was analyzed with a selective electrode (Thermo Scientific) and also with the spectrometric method of UV absorbance (APHA, 2005). NO_2^- was determined using the colorimetric method described by APHA (2005) with sulphanilamide diazote and diclorhidrate of $N-(1-naftil)-etilendiamide$.

PO_4^+ concentration was determined using two protocols of (APHA, 2005) ascorbic method and molybdenum—vanadium method, with the latter chosen as reference for its wider range between minimum and maximum detention limits. Total phosphorous was determined by oxidation with potassium persulfate (Oxysolve) to PO_4^+ and then analyzed with the molybdenum—vanadium method. Total phosphorous with acid oxidation (APHA, 2005) was also used to validate the results found with the Oxysolve.

3. Results and discussion

3.1 Pilot plant performance

3.1.1 Wastewater treatment and biomass characteristics

All the ponds, supplied with carbon dioxide or not, showed very similar nitrogen removal percentages (average removal rates of $54 \pm 5\%$). However, assimilation into algae biomass showed sharp variations during the different seasons. While summer conditions promoted the generation of biomass and thus the assimilation of nitrogen, during winter, the assimilation showed considerably lower values ($28 \pm 7\%$ and $25 \pm 5\%$, respectively).

Removal of phosphorus was very constant ($53 \pm 5\%$) with very little seasonal influence. Although the removal was only around 50% in HRAP, low phosphorus residuals are achieved in the DAF, as the addition of aluminum precipitates soluble phosphorus together with the algae biomass resulting in very low P concentrations (>0.2 mg/L).

Table 7.2 shows biomass characteristics in three different periods of operation, while HRT of AD was varied between 28, 21, and 14 d. During the last period, feed solids to the ponds increased by 40%, as the HRT of the UASB was decreased from 18 to 12 h. The biomass characteristics to feed AD fluctuated between 3% (P2) and 4% (P1 and 3), which is within the range expected for secondary solids and DAF thickening (Metcalf et al., 1991).

3.1.2 Biomass production

Fig. 7.6 illustrates the main climate conditions (temperature and solar radiation) and its influence on the microalgae composition. The AD operating period with HRT of 28 d covers the summer, while the period with HRT of 21d covers fall and winter with a decrease of temperature and radiation, which in turn seems to reduce the percentage of VSS/TSS by about 7%, while COD biomass concentration is lowered by 15%.

As climate conditions influence microalgae growth (Sutherland and Ralph, 2020), this period (HRT = 21 d) also coincides with less productivity as described by (De Godos et al., 2016). Autumn and winter growth rates are limited ($8.4-12.2$ g/m^2/d), while in spring and summer (HRT = 14 y 28 d) they can reach values twice as high ($18.6-29.3$ g/m^2/d).

3.1.3 Biogas production: anaerobic digestion on lab scale

For microalgae grown on municipal wastewaters, dominated by *Chlorella* sp. and *Scenedesmus* sp., methane yields in literature range between 60 and 280 mL CH_4/g VS $_{fed}$ (Choudhary et al., 2020), below typical methane production obtained for wastewater solids at $240-340$ mL CH_4/g VS $_{fed}$ (Shen et al., 2015).

Fig. 7.7 provides weekly averages of methane production in pilot AD, with CH_4 concentration in the biogas remaining stable between 63% and 68%. Specific methane yield from the biomass reached values up to 156 ± 16 mL CH_4/g VS $_{fed}$ for thermophilic conditions and 28 d HRT (Fig. 7.8), about 30% higher than mesophilic AD. At lower HRT in AD of 21 d, methane production dropped by almost 30% for thermophilic and by 50% for mesophilic operation. While AD remained stable at HRT of 28 d for both temperatures, at HRT of 21 d a decrease in the pH and alkalinity PA/IA ratio (Jiang et al., 2020) was observed, and an increase in soluble COD concentration—indicating the start of an acidification process.

Table 7.2 HRAP feed conditions (UASB effluent) and microalgae biomass characteristics (harvested from DAF).

	Period	TRH (h)	Days of operation (d)	Total COD $\overline{X} \pm$ IC (mg O₂/L)	n	CV	Total solids $\overline{X} \pm$ IC (mg/L)	n	CV	Volatile solids $\overline{X} \pm$ IC (mg/L)	n	CV
UASB$_{effluent}$	P1	12 y24 (h)	85	232 ± 6.7	40	9	51.7 ± 5.1	40	31	43.6 ± 3.6	40	26
	P2	12 y24 (h)	179	277 ± 18.7	71	29	49.5 ± 4.0	72	35	40.0 ± 3.3	72	35
	P3	12 y18 (h)	109	274 ± 12.6	50	17	69.4 ± 5.0	53	27	56.3 ± 3.6	51	25
		(d)	(d)	$\overline{X} \pm$ IC (mg O₂/L)	n	CV	$\overline{X} \pm$ IC (g/kg)	n	CV	$\overline{X} \pm$ IC (g/kg)	n	CV)
DAF$_{Biomass}$	P1	28	85	49.086 ± 7,171	11	25	37.5 ± 4.9	11	22	29.0 ± 4.4	11	26
	P2	21	179	35.031 ± 3,501	23	25	30.7 ± 2.0	23	16	22.1 ± 1.3	23	15
	P3	14	109	49.077 ± 2,269	16	9	39.5 ± 2.3	16	12	32.2 ± 1.7	16	11

CV, variation coefficient (%), *IC*, confidence interval calculated for $\alpha = 0.05$; *n*, n° of samples.

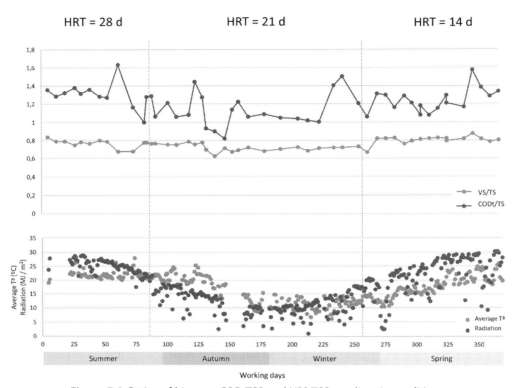

Figure 7.6 Ratios of biomass COD/TSS and VSS/TSS vs. climatic conditions.

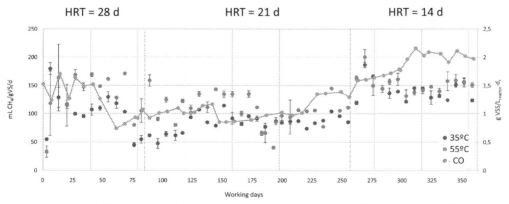

Figure 7.7 AD methane yields and organic loading rates for different digester HRT.

During the second period (HRT = 21 d), the difference in methane yield between mesophilic and thermophilic reactors was lower, reaching values between 84 and 95 mL CH$_4$/gVS $_{fed}$. These values were only half of what was previously reported

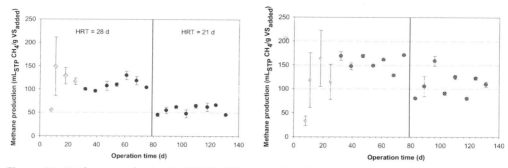

Figure 7.8 Methane production (mLCH4/g VS$_{added}$) during the operation time (d) under mesophilic conditions (left) and thermophilic conditions (right) with two HRT in AD.

(Passos et al., 2014) with 180 mL CH$_4$/gVS $_{fed}$ from mesophilic AD at HRT of 20 d on mesotrophic HRAP. In the third period (HRT = 14 d), the organic load was doubled to 1.9 gVS $_{fed}$/L/d, increasing mesophilic methane yields to between 140 and 153 mL CH$_4$/g VS $_{fed}$, with thermophilic yields tending to be 10% higher.

Comparing different HRTs (14, 21, and 28 d), methane yields increase with reduced HRT, reaching 140 mL CH$_4$/g VS $_{fed}$ at 14 d for mesophilic AD. The differences were 50% higher than obtained at twice the HRT of 28 d. These results improved the values previously reported (Passos et al., 2014), with mesophilic AD of microalgae at HRT of 15 d and yields of 100 ± 27 mL CH$_4$/g VS $_{fed}$.

Similarly, thermophilic AD had higher methane yields at HRT of 14 d with an average 153 mL CH$_4$/g VS $_{fed}$, which increased 40% compared the values observed at HRT = 28 d. Nevertheless these values are lower than observed by Ferrer et al. (2010) on secondary sludge with 230 mL CH$_4$/gVS $_{fed}$ with thermophilic reactors and similar organic loadings (1.65 ± 0.3 gVS $_{fed}$/L/d). This is related to lower biodegradability of microalgae due to the cell wall structures (Mendez et al., 2015; Cavinato et al., 2017). It has been also reported by De Godos et al. (2016) with a predominance of *Coelastrum* sp. that are known to have a three layered membrane formed of a resistant polymer (Rodríguez and Cerezo, 1996).

Solubilization of organic matter as measured by COD showed significant difference, with soluble COD in thermophilic reactors. It reached to 3466 ± 137 mg O$_2$/l (n = 168; CV = 26%; α = 0.05) or 300% of the COD in mesophilic digesters at 1118 ± 58 mg O$_2$/l (n = 178; CV = 35%; α = 0.05), reflecting better kinetics of hydrolysis of particulate matter (Arras et al., 2019). Nevertheless, not all soluble VFA is easily degradable, such as propionate (Perez, 2016), as the removal of total COD is similar with 41 ± 2% (n = 173; α = 0.05) in mesophilic reactors compared to 48 ± 2% (n = 165; α = 0.05) in thermophilic AD.

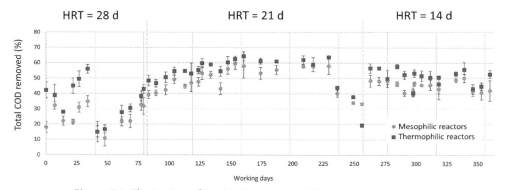

Figure 7.9 Elimination of total COD in mesophilic and thermophilic AD.

Fig. 7.9 reports the elimination of organic matter, showing initially limited removal of COD at HRT of 28 d with $24 \pm 2\%$ (n = 43; CV = 33%; $\alpha = 0.05$) in mesophilic and $35 \pm 4\%$ (n = 44; CV = 37%; $\alpha = 0.05$) in thermophilic reactors. These results are similar to those reported by Passos et al. (2014) for algae biomass with COD elimination between 31.4% and 34.2% in mesophilic AD with HRT at 20 d.

In the second period (TRH = 21 d), total COD in the mesophilic and thermophilic digesters was 40% lower but its removal increased by $48 \pm 2\%$ (n = 94; CV = 18%; $\alpha = 0.05$) for mesophilic conditions and $54 \pm 2\%$ (n = 84; CV = 15%; $\alpha = 0.05$) for the thermophilic AD. These results are in the range of digestion of secondary sludge at HRT of 20 d observed by Bolzonella et al. (2009) for mesophilic (35%) or thermophilic (45%) digesters.

Total COD removal was maintained in the last phase (HRT = 14 d) for mesophilic reactors at $45 \pm 1\%$ (n = 58; CV = 10%; $\alpha = 0.05$) and thermophilic conditions at $50 \pm 1\%$ (n = 59; CV = 10%; $\alpha = 0.05$). No significant accumulation of VFAs was observed, even with a much higher organic load, as removal of soluble COD was always around 50%.

3.2 Prototype performance

3.2.1 Wastewater treatment

Screened raw urban wastewater was fed to the prototype HRAPs to confirm the effect of the higher C/N ratio of 10. This was done in comparison to the pretreated UASB effluent of four fed to the pilots, as a lower inorganic carbon demand should affect positively microalgae and bacteria symbiotic interaction. Comparing HRAP with paddlewheel (PDW) and low energy mixing (LEAR), no significant differences in the feed concentrations of UASB effluent and raw WW are observed for the total nitrogen (TN) and soluble phosphorus (PO_4).

Nevertheless, despite the prototype HRT of 3 d being more than 2 times lower than the pilots with 7 d, the nitrogen removal efficiencies of both prototype HRAPs were at least 10% higher compared to the pilots and reached $75.9 \pm 6.6\%$ in PDW. This is due to a significantly higher biomass production, with LEAR and PDW averaging for winter period 18.4 ± 4 and 16.5 ± 3.3 g VSS/m^2/d respectively, and about twice that value in summer.

In both prototypes (LEAR and PDW), the effluent after cultivation and harvesting meets the discharge limits of the EU WWT directive 91/271/EC, achieving simultaneous removal of nitrogen, phosphorus, and organic matter in a single aerobic (day)/anoxic (night) reactor. Nutrients can therefore be removed without the need for aeration, but the biosolids production is higher than activated sludge process.

3.2.2 Biogas production: anaerobic digestion in prototype scale

On the larger prototype HRAPs, COD, nitrogen, and phosphorous concentrations were removed by 75%, 70%, and 90%, respectively. An average algae biomass of 18 g VS/m^2/d, or 1.5 kg VS/kg $_{COD\ removed}$ was produced. The species *Coelastrum* dominated in all experimental measurements of bacteria and protozoa along the year. Ambient temperatures fluctuated between springtime at $15 \pm 2.3°C$ (n = 91; CV = 16%) and summer at $22 \pm 3.2°C$ (n = 72; CV = 15%), and radiation also increasing by 50% during the summer season.

Table 7.3 summarizes the biogas production obtained in the three algal digesters which were operated as described in Table 7.1. Methane yields of 180 mL CH$_4$/g VS $_{added}$ and 280 mL CH$_4$/g VS were observed under ambient and thermophilic conditions, respectively. Thermophilic conditions improved methane yield by 55%. These methane yields at thermophilic range are much higher than the 145 mL/g VS$_{in}$ observed for waste activated sludge (Kepp et al., 2000; Bougrier et al., 2006).

The biogas production in prototype AD is also much higher than for the lab scale reactors, with the surplus reaching 100 mL CH$_4$/g VS in thermophilic conditions. The more concentrated influent from raw sewage generates more heterotrophic biomass, which in turn had a sludge age about half (HRT 3 d) in comparison to the pilot plant (from 5 to 7 d); therefore, the biomass produced was "younger" and mixed with more easily degradable primary solids.

3.3 Energy balances

From experimental results obtained, mass and energy balances for a WWT plant for 5000 m^3/d (25,000 PE.) were developed. Using the same design parameters and WW characterization, operational conditions and efficiencies as well as electric consumptions have been properly evaluated.

The main energy flows (yearly averaged) are depicted in Table 7.4 and plotted in Fig. 7.10 for both scenarios considering mesophilic and thermophilic digestion

Table 7.3 Prototype AD operating conditions and biogas characteristics.

Period (radiation)	January to April: 16 ± 6 MJ/m^2									May to July: 26 ± 3.5 MJ/m^2								
Digestor (HRT)	Ambient (30 d)			Meso (21 d)			TP (21 d)			Ambient (25 d)			Meso (18 d)			TP (18 d)		
	$\bar{X} \pm$ IC	n	CV	$\bar{X} \pm$ IC	n	CV	$\bar{X} \pm$ IC	n	CV	$\bar{X} \pm$ IC	n	CV	$\bar{X} \pm$ IC	n	CV	$\bar{X} \pm$ IC	n	CV
CH_4 (%)	62 ± 1	14	3	62 ± 1	14	2	62 ± 1	14	2	60 ± 1	7	3	58 ± 1	7	3	58 ± 3	7	8
CO_2 (%)	36 ± 1	14	6	34 ± 1	14	5	33 ± 1	14	4	33 ± 2	7	6	35 ± 3	7	12	34 ± 3	7	13
H_2S (ppm)	776 ± 130	14	32	2688 ± 291	14	21	3043 ± 429	14	27	5080 ± 2957	7	79	6975 ± 2797	7	54	7227 ± 2581	7	48
m^3 CH_4/kg VS $_{fed}$	0.18 ± 0.03	11	27	0.23 ± 0.04	11	27	0.28 ± 0.02	11	10	0.17 ± 0.02	3	13	0.17 ± 0.06	3	31	0.3 ± 0.04	3	11
CO (gVS/ L$_{digestor}$·d)	0.67 ± 0.08	11	20	0.95 ± 0.11	11	20	0.95 ± 0.11	11	20	0.88 ± 0.16	7	25	1.23 ± 0.23	7	25	1.23 ± 0.23	7	25

CO, Organic load; CV, variation coefficient (%); IC, confidence interval calculated for $\alpha = 0.05$; n, n° of samples.

Table 7.4 Energy balances per treated wastewater m³ (negative values mean consumption and positive ones, production—with an equivalence of 0.33 kwh el/kwh th).

			Mesophilic	Thermophilic	Units
Inputs	**Electricity**				
	UASB and grit removal			−0.032	kWh el/m³
	HRAPs			−0.032	kWh el/m³
	Harvesting			−0.045	kWh el/m³
	Digestion			−0.044	kWh el/m³
	Biogas upgrading			−0.053	kWh el/m³
	Electricity			−0.206	kWh el/m³
				−0.618	kWh th/m³
	Heat		−0.23	−0.52	kWh th/m³
	Total		−0.85	−1.14	kWh th/m³
Outputs	Biomethane	UASB	59.7	59.67	L CH₄/m³
		Algae			
	Digestion		59.1	87.46	L CH₄/m³
		Total	118.8	147.15	L CH₄/m³
			1.32	1.64	kWh th/m³
Net balance			0.47	0.5	kWh th/m³

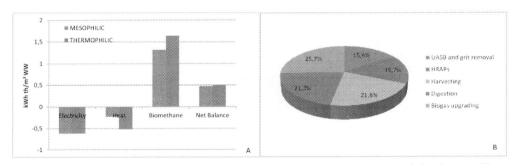

Figure 7.10 Comparison in terms of energy flows (A) and electricity demand distribution (B).

(data from the prototype, Table 7.3). All consumption terms, corresponding to electricity and biomass to feed the boiler, account for 0.85 and 1.14 kWh th/m³ of treated WW for each scenario, respectively. In addition to water for reuse, a considerable biomethane is produced by the two anaerobic processes: 1.32 and 1.64 kWh th/m³ of treated WW.

Higher biogas production during thermophilic digestion compared to mesophilic does not increase final net energy production, due to the higher heating requirements The heating requirements could be supplied by biomass boilers using low cost residual biomass such as olive pits. Considering that biomethane could be sold for cars in a petrol station, the overall process (mesophilic preferably) will result in an energy producing WWT, with a positive net energy balance of 0.5 kWh th/m³.

Compared to conventional WWT in Spain (Hernández-Sancho et al., 2011) with an average electricity consumption of 0.51 kWh el/m^3 (equivalent to 1.53 kWh th/m^3), microalgae processes not only consume half the energy (0.28 kWh el/m^3 or 0,85 kwh th/m^3), but they also generate a biofuel output of up to 150 L CH$_4$/m^3. This energy corresponds to 0.12 kg CH$_4$/m^3 of wastewater, so each m^3 would allow to run a methane fueled car around 4.2 km (consumption of VW Eco Up is 2.9 kg CH$_4$ per 100 km).

4. Conclusions

Microalgae and bacteria consortia show a great capability for complete WWT, reducing TCOD, TN, TP, and TSS below the limits of discharge of the European directive 91/271/ECC (European Commission Directive, 1998). Both raw screened WW and effluent from an anaerobic sludge blanket reactor were compatible for growing high energy content microalgae and bacteria biomass.

Microalgae—bacteria biomass can be anaerobically digested, averaging biomethane production at prototype scale above 200 mL CH$_4$/g VS$_{added}$ for mesophilic conditions or even 300 mL CH$_4$/g VS$_{added}$ at thermophilic temperatures. The biogas energy generated is higher than needed for mixing of HRAP and harvesting of algae with DAF.

The implementation of anaerobic processes in microalgae WWT (UASB for WW and AD for harvested microalgae) leads to a net energy generation of 0.5 kWh th/m^3 which can be used as transportation biofuel. This approach allows to transform residuals contained in WW into biomethane and reuse water and changes the concept of WWT to create a new paradigm of sustainability based on microalgae.

References

APHA, 2005. Standard Methods for the Examination of Water and Wastewater. American Public Health Association (APHA), Washington, DC, USA.

Arbib, Z., Ruiz, J., Álvarez-Díaz, P., Garrido-Pérez, C., Barragan, J., Perales, J.A., 2013. Effect of pH control by means of flue gas addition on three different photo-bioreactors treating urban wastewater in long-term operation. Ecol. Eng. 57, 226—235.

Arras, W., Hussain, A., Hausler, R., Guiot, S.R., 2019. Mesophilic, thermophilic and hyperthermophilic acidogenic fermentation of food waste in batch: effect of inoculum source. Waste Manag. 87, 279—287.

Bolzonella, D., Fatone, F., DI Fabio, S., Cecchi, F., 2009. Mesophilic, thermophilic and temperature phased anaerobic digestion of waste activated sludge. TVS 80, 74.

Bougrier, C., Delgenes, J.-P., Carrère, H., 2006. Combination of thermal treatments and anaerobic digestion to reduce sewage sludge quantity and improve biogas yield. Process Saf. Environ. Protect. 84, 280—284.

Canovas, S., Picot, B., Casellas, C., Zulkifi, H., Dubois, A., Bontoux, J., 1996. Seasonal development of phytoplankton and zooplankton in a high-rate algal pond. Water Sci. Technol. 33, 199—206.

Cavinato, C., Ugurlu, A., DE Godos, I., Kendir, E., Gonzalez-Fernandez, C., 2017. Biogas Production from Microalgae. Microalgae-Based Biofuels and Bioproducts. Elsevier.

Craggs, R., Heubeck, S., Lundquist, T., Benemann, J., 2011. Algal biofuels from wastewater treatment high rate algal ponds. Water Sci. Technol. 63, 660—665.

Craggs, R., Park, J., Heubeck, S., 2008. Methane emissions from anaerobic ponds on a piggery and a dairy farm in New Zealand. Aust. J. Exp. Agric. 48, 142—146.

Chisti, Y., 2007. Biodiesel from microalgae. Biotechnol. Adv. 25, 294—306.

Choudhary, P., Assemany, P.P., Naaz, F., Bhattacharya, A., De Siqueira Castro, J., DO Couto, E.D.A., Calijuri, M.L., Pant, K.K., Malik, A., 2020. A review of biochemical and thermochemical energy conversion routes of wastewater grown algal biomass. Sci. Total Environ. 137961.

De Godos, I., Arbib, Z., Lara, E., Rogalla, F., 2016. Evaluation of High Rate Algae Ponds for treatment of anaerobically digested wastewater: effect of CO_2 addition and modification of dilution rate. Bioresour. Technol. 220, 253—261.

De Godos, I., Guzman, H.O., Soto, R., García-Encina, P.A., Becares, E., Muñoz, R., Vargas, V.A., 2011. Coagulation/flocculation-based removal of algal—bacterial biomass from piggery wastewater treatment. Bioresour. Technol. 102, 923—927.

European Commission Directive, 1998. 98/15/EC of 27 February. Off. J. Eur. Commun. (https://eur-lex.europa.eu/eli/dir/1998/15/1998-03-27). OJ L 067 7.3.1998, p. 29.

Ferrer, I., Vázquez, F., Font, X., 2010. Long term operation of a thermophilic anaerobic reactor: process stability and efficiency at decreasing sludge retention time. Bioresour. Technol. 101, 2972—2980.

Garcia, J., Mujeriego, R., Hernandez-Marine, M., 2000. High rate algal pond operating strategies for urban wastewater nitrogen removal. J. Appl. Phycol. 12, 331—339.

Golueke, C.G., Oswald, W.J., 1959. Biological conversion of light energy to the chemical energy of methane. Appl. Microbiol. 7, 219.

Golueke, C.G., Oswald, W.J., Gotaas, H.B., 1957. Anaerobic digestion of algae. Appl. Microbiol. 5, 47.

Hernández-Sancho, F., Molinos-Senante, M., SALA-Garrido, R., 2011. Energy efficiency in Spanish wastewater treatment plants: a non-radial DEA approach. Sci. Total Environ. 409, 2693—2699.

Jiang, J., He, S., Kang, X., Sun, Y., Yuan, Z., Xing, T., Guo, Y., Li, L., 2020. Effect of organic loading rate and temperature on the anaerobic digestion of municipal solid waste: process performance and energy recovery. Front. Energy Res. 8.

Kepp, U., Machenbach, I., Weisz, N., Solheim, O.E., 2000. Enhanced stabilisation of sewage sludge through thermal hydrolysis-three years of experience with full scale plant. Water Sci. Technol. 42, 89—96.

Lakaniemi, A.-M., Tuovinen, O.H., Puhakka, J.A., 2013. Anaerobic conversion of microalgal biomass to sustainable energy carriers—a review. Bioresour. Technol. 135, 222—231.

Lundquist, T.J., Woertz, I.C., Quinn, N., Benemann, J.R., 2010. A Realistic Technology and Engineering Assessment of Algae Biofuel Production, vol. 1. Energy Biosciences Institute.

Mendez, L., Mahdy, A., Ballesteros, M., González-Fernández, C., 2015. Biomethane production using fresh and thermally pretreated Chlorella vulgaris biomass: a comparison of batch and semi-continuous feeding mode. Ecol. Eng. 84, 273—277.

Metcalf, L., Eddy, H.P., Tchobanoglous, G., 1991. Wastewater Engineering: Treatment, Disposal, and Reuse. McGraw-Hill, New York.

Park, J., Craggs, R., 2010. Wastewater treatment and algal production in high rate algal ponds with carbon dioxide addition. Water Sci. Technol. 633—639.

Passos, F., Hernandez-Marine, M., García, J., Ferrer, I., 2014. Long-term anaerobic digestion of microalgae grown in HRAP for wastewater treatment. Effect of microwave pretreatment. Water Res. 49, 351—359.

Perez, M., 2016. Thermophilic and mesophilic temperature phase anaerobic co-digestion (TPAcD) compared with single-stage co-digestion of sewage sludge and sugar beet pulp lixiviation. Biomass Bioenergy 93, 107—115.

Ras, M., Lardon, L., Bruno, S., Bernet, N., Steyer, J.-P., 2011. Experimental study on a coupled process of production and anaerobic digestion of Chlorella vulgaris. Bioresour. Technol. 102, 200—206.

Rodríguez, M.C., Cerezo, A.S., 1996. The resistant 'biopolymer' in cell walls of Coelastrum sphaericum. Phytochemistry 43, 731—734.

Shammas, N.K., Bennett, G.F., 2010. Principles of Air Flotation Technology. Flotation Technology. Springer.

Sheehan, J., Dunahay, T., Benemann, J., Roessler, P., 1998. Look Back at the US Department of Energy's Aquatic Species Program: Biodiesel from Algae; Close-Out Report. National Renewable Energy Lab., Golden, CO.(US).

Shen, Y., Linville, J.L., Urgun-Demirtas, M., Mintz, M.M., Snyder, S.W., 2015. An overview of biogas production and utilization at full-scale wastewater treatment plants (WWTPs) in the United States: challenges and opportunities towards energy-neutral WWTPs. Renew. Sustain. Energy Rev. 50, 346—362.

Sutherland, D.L., Ralph, P.J., 2020. 15 years of research on wastewater treatment high rate algal ponds in New Zealand: discoveries and future directions. N. Z. J. Bot. 58, 334—357.

Vasudevan, P.T., Briggs, M., 2008. Biodiesel production—current state of the art and challenges. J. Ind. Microbiol. Biotechnol. 35, 421.

CHAPTER 8

Anaerobic digester biogas upgrading using microalgae

Kaushik Venkiteshwaran[1], Tonghui Xie[2], Matthew Seib[3], Vaibhav P. Tale[4] and Daniel Zitomer[5]

[1]Department of Civil, Coastal and Environmental Engineering, University of South Alabama, Mobile, AL, United States; [2]School of Chemical Engineering, Sichuan University, Chengdu, Sichuan, China; [3]Madison Metropolitan Sewerage District, Madison, WI, United States; [4]Chemtron Riverbend Water, Saint Charles, MO, United States; [5]Department of Civil, Construction and Environmental Engineering, Marquette University, Milwaukee, WI, United States

1. Introduction

Anaerobic biotechnology is being used more frequently to convert organic feedstocks including industrial and agricultural wastes to biogas containing methane (CH_4). The CH_4 can be used as a renewable energy source for heat and electricity generation. However, energy recovery is often complicated by the presence of biogas impurities. For example, raw biogas can contain more than 40% v/v carbon dioxide (CO_2) which reduces the gas energy density to less than the minimum value of 23 MJ/kg ($0°C$, 1 atm) required by typical boilers and engine generator sets (gen-sets). Also, raw biogas can contain water vapor, hydrogen sulfide (H_2S), siloxanes, and other impurities that are corrosive or otherwise damaging to boilers, gen-sets, as well as other equipment, and must be removed.

Current physical and chemical technologies to remove biogas impurities can be expensive and may produce unwanted byproducts. Therefore, new methods to efficiently condition biogas for subsequent use would be beneficial. This chapter describes relatively new microalgae technologies to condition biogas for renewable energy generation. Specifically, the state of the art regarding application of microalgae to remove CO_2, H_2S, siloxanes, and other biogas impurities is described. Recommendations for future research are made to help guide the development of new commercially viable, microalgae biogas conditioning technologies.

2. Raw biogas characteristics

Raw biogas is a mixture of CH_4 and CO_2 and may contain H_2S, CO_2, water vapor, nitrogen (N_2), and siloxanes. Typically, CH_4 represents 50%–85% of the biogas volume, whereas the CO_2 concentration ranges from 4% to 40% (Speece, 2008). The biogas CH_4 content primarily determines its heating value, ranging from 550 to 715 Btu/ft^3 (U.S. EPA, 2011). When biogas is produced, it is saturated with water vapor. The

---

Integrated Wastewater Management and Valorization using Algal Cultures
ISBN 978-0-323-85859-5, https://doi.org/10.1016/B978-0-323-85859-5.00004-X

© 2022 Elsevier Inc.
All rights reserved.

183

liquid—vapor pressure of pure water at 20, 35, and 55°C is 0.023, 0.056, and 0.16 atm, respectively (Lide, 1999). The remaining biogas components are present in trace amounts, but some of these components such as H_2S and siloxanes pose challenges to beneficial use.

There are several typical equipment choices to harvest energy from biogas. These choices include: (1) heat generation in boilers, (2) electricity generation and heat recovery using reciprocating engines, microturbines, or gas turbines, (3) electricity generation through fuel cells, and (4) conditioning biogas to natural gas quality for injection into a natural gas pipeline network and/or use as a vehicle fuel.

Among the technologies mentioned above, those that simultaneously capture heat and electricity are more efficient than technologies that can only produce electricity. Such technologies are classified as combined heat and power (CHP) technologies. A detailed review of CHP technologies is presented in a report published by the US Environmental Protection Agency (U.S. EPA, 2017).

3. Required gas quality for heat and power equipment

The U.S. Department of Energy CHP Installation Database indicates that out of 4675 CHP installations in the USA, 824 applications use biogas or landfill gas as their primary fuel, and out of these, more than 61% utilize reciprocating engines to produce heat and electricity. Use of reciprocating engines for CHP applications is an attractive choice since it can lead to 27%—41% electric efficiency and high (77%—80%) overall energy recovery if CHP technologies are used (U.S. DOE, 2020). Several engine manufacturers supply CHP equipment for biogas, and the biogas quality requirements for CHP units differ among manufacturers (Table 8.1).

The key trace contaminants of concern in biogas are water vapor, H_2S, and siloxanes. Raw biogas may need to be cleaned to make it suitable for use in a CHP unit. Upon burning, H_2S can lead to SO_2, SO_3, or H_2SO_4 emissions from the engine. Moreover, the presence of moisture and these gaseous emissions are corrosive to the engine and its components reducing total engine life and requiring frequent maintenance.

Table 8.1 Typical range of biogas requirements for reciprocating engines.

Parameter	Limiting range
Biogas temperature	$<40°C$
Relative humidity	$<80\%$
H_2S	$<150-250$ ppmV — normal operation
	<700 ppmV — enhanced maintenance schedule
Siloxanes	$<10-28$ mg/m^3 biogas
Exhaust gas temperature limitation	$>220°C$ to avoid SO_X deposits

Another prominent contaminant present in the raw biogas is siloxanes. Siloxanes are a subgroup of silicones containing Si—O bonds with organic radicles that are components of personal care products. When exposed to higher temperatures in an engine or boiler, siloxanes leave behind hard silicone residues on piston and valve assemblies leading to engine performance deterioration and life reduction (U.S. EPA, 2017). There are several technologies available on the market to remove siloxanes from the raw biogas. Choosing a siloxanes removal technology depends on the concentration of siloxanes in the biogas and other factors. Siloxanes removal technologies research and development is progressing, and it is probable that new technologies will be available in the future. Manufacturers of siloxane removal systems provide technologies that use capture medium like activated carbon for filtering out siloxanes from biogas. For higher total siloxane quantity to be removed, often a medium regeneration technology like thermal regeneration or pressure swing adsorption (PSA) is coupled with the siloxane filter (U.S. EPA, 2020).

Different regional jurisdictions follow separate standards for biogas to be added to public natural gas pipeline systems in the U.S.A. minimum acceptable specifications are laid out by either the local natural gas utility accepting the cleaned biomethane, a relevant professional biogas association or an applicable engineering standard. Table 8.2 presents a summary of the general standards applicable in the North American markets (Audrey, 2019), whereas Table 8.3 summarizes the relevant permitted concentrations of biogas constituents for pipeline injection in Europe (Petersson and Wellinger, 2009).

In the United States, the Society of Automotive Engineers Surface Vehicle Recommended Practice J1616 for Compressed Natural Gas Vehicle Fuel (SAE J1616, 1994) is often followed for upgraded biogas for either pipeline injection or use as a vehicle fuel

Table 8.2 American Biogas Council (USA) pipeline biomethane recommendations.

Physical property	Units	Lower limit	Upper limit
Heating value	BTU/ft^3	960	1100
CO_2	mol %		2
O_2	mol %		0.4
Total inerts	mol %		5
H_2S	gr./100 ft^3		0.25
Total sulfur	gr./100 ft^3		1
Water	lbs/mmSft^3		7
Siloxanes	ppm(v)		1
Hydrocarbon dew point	Fahrenheit		−40
Temperature	Fahrenheit	50	120
Dust, particulate			Commercially free[a]
Biologicals			Commercially free[a]
Heavy metals			Commercially free[a]

[a]Commercially free is defined as equal to or less than the levels present in conventional natural gas.

Table 8.3 General requirements for pipeline quality biomethane.

Compound	Unit	France	Germany	Sweden	Switzerland	Austria	The Netherlands
CH_4 content	Vol-%			95–99	>96		>80
CO_2	Vol-%	<2	<6		<6	≤2	
O_2	Vol-%		<3		<0.5	≤0.5	
	ppmV	<100					
	Mol-%						<0.5
H_2	Vol-%	<6	≤5		<5	≤4[c]	<12
$CO_2+O_2+N_2$	Vol-%			<5			
Relative humidity					<60%		
Sulfur	mg/Nm^3	<100[a] <75[b]	<30	<23	<30	≤5	<45

[a]Maximum permitted.
[b]Average content.
[c]Mole percentage.

Table 8.4 American Biogas Council (USA) vehicle biomethane recommendations.

American biogas council recommendations - CNG vehicles			
Physical property	Units	Lower limit	Upper limit
Heating value	BTU/ft^3	940	1100
$CO_2 + N_2$	mol %		4
O_2	mol %		1
H_2	ppm(v)		300
Total sulfur	wt%		0.001
H_2S	ppm(v)		6
Siloxanes	ppm(v)		3
Liquids			Commercially free[a]
Dust, particulate			Commercially free[a]
Chlorine additives			Commercially free[a]
Heavy metals			Commercially free[a]

[a]Commercially free is defined as equal or less than the levels present in conventional natural gas.

(U.S. EPA, 2020). The SAE J1616 standard sets minimum requirements for fuel composition and properties to ensure vehicle, engine and component safety, durability, and performance. The SAE J1616 provides guidelines regarding CH_4, N_2, CO_2, water, O_2, trace amounts of lubricating oil (from compressors), and sulfur found as H_2S and other sulfur compounds as well as particulate material in the fuel. The American Biogas Council (USA) also has published typical biomethane vehicle fuel requirements (Table 8.4).

4. Current commercial biogas conditioning technologies

This section describes existing commercial technologies, other than algal systems, applied to industrial-scale biogas conditioning for the removal of CO_2, H_2S, water vapor, and trace biogas constituents such as siloxanes. Subsequently, microalgae technology to remove these constituents is presented.

4.1 CO_2 removal

Removal of CO_2 from biogas increases the gas heating value to meet the quality required for pipeline distribution or other applications (Awe et al., 2017). Current technologies that are mature and are commercially applied for industrial scale biogas CO_2 removal include water scrubbing, organic solvent scrubbing, chemical absorption, PSA, cryogenic separation, and membrane separation techniques (Ryckebosch et al., 2011; Bauer et al., 2013; Thrän et al., 2014; Muñoz et al., 2015; Awe et al., 2017). Biological methods for CO_2 removal have also been developed; however, the biotechnologies have only been assessed at lab or pilot scale.

4.1.1 Water scrubbing

CO_2 removal using water scrubbing is an established and most widely used technology accounting for approximately 41% of the global biogas conditioning market (Thrän et al., 2014). Removal of CO_2 from biogas using water scrubbing relies on the higher water solubility of CO_2 (26 times higher at 25°C) compared to that of CH_4 (Bauer et al., 2013). A schematic diagram of a biogas water scrubbing system is shown in Fig. 8.1. The raw biogas is passed through a CO_2 absorption unit from the bottom operated at 6–10 bars pressure, countercurrent to the flow of water which is introduced from the top of the absorption unit (Fig. 8.1). The water is introduced as small droplets through a nozzle, and the absorption unit typically contains random packing materials (Pall or Raschig ring) to increase the water:gas surface area and support higher gas–liquid mass transfer. Water with absorbed CO_2 is then passed through a flash tank unit which is operated at a lower pressure (2–4 bars) that results in the release of a CO_2-rich off-gas (80%–90% CO_2 and 10%–20% CH_4) that is recycled back through the first absorption unit (Fig. 8.1) (Bauer et al., 2013).

Apart from removing CO_2, water scrubbing can also remove H_2S from biogas, which is a benefit but can also cause operational problems from accumulation of sulfur and corrosion of surfaces. This technology can deal with H_2S concentrations ranging from 300 to 2500 ppm in the biogas; however, H_2S removal is often applied prior to water scrubbing to improve efficiency and reduce maintenance costs. Microbial growth and foaming in the packed absorption unit can also contribute to operational problems (Muñoz et al., 2015).

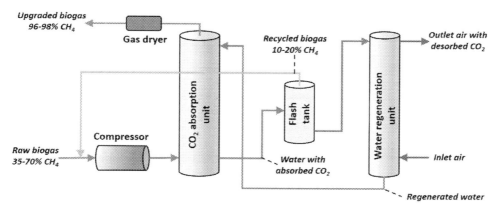

Figure 8.1 Schematic of the water scrubbing biogas upgradation system. *(Adapted from Muñoz, R., Meier, L., Diaz, I., Jeison, D., 2015. A review on the state-of-the-art of physical/chemical and biological technologies for biogas upgrading. Rev. Environ. Sci. Biotechnol. 14 (4), 727–759. https://doi.org/10.1007/s11157-015-9379-1).*

4.1.2 Organic solvent scrubbing

Organic solvent scrubbing is technically similar to a water scrubbing system (Fig. 8.1), but uses a polyethylene glycol-based absorbents such as Selexol or Genosorb, which exhibit a higher affinity (5 times higher) for CO_2 than water (Muñoz et al., 2015). In addition to the use of an organic absorbent, this system also differs from water scrubbing in requiring an initial biogas condition step to remove excess moisture and several cooling/heating steps to promote efficient absorption and desorption of CO_2 (Ryckebosch et al., 2011).

The use of a high affinity organic solvent provides advantages over water scrubbing such as a low recycling rate and a smaller system footprint; in addition, the longer life of the absorbent and its anticorrosive nature helps to reduce the investment and operational costs (Muñoz et al., 2015; Awe et al., 2017). The main operating costs derive from the electricity consumption used for biogas compression, solvent pumping, and heating costs (Muñoz et al., 2015). Despite these advantages, the organic solvent scrubbing systems only accounted for 6% of the global market in 2014 (Thrän et al., 2014).

4.1.3 Chemical absorption

Chemical absorption systems rely on similar biogas—liquid mass transfer principles as water and organic solvent scrubbing systems. The CO_2 removal is performed using absorbents such as alkanolamines (mono-ethanol amine, di-methyl ethanol amine, piperazine, etc.) or alkali aqueous solutions (NaOH, KOH, CaOH, etc.) (Bauer et al., 2013; Muñoz et al., 2015; Awe et al., 2017). This system is also operated like water scrubbing in a countercurrent flow configuration, with a packed bed absorption unit where the biogas and chemical solution come in contact, followed by a chemical regeneration unit equipped with a boiler which heats the chemical agent to 120—150°C to enhance CO_2 desorption (Ryckebosch et al., 2011; Bauer et al., 2013). A premoisture removal unit is needed, and a H_2S removal step is recommended if the biogas H_2S content is more than 300 ppm to improve the life of the chemical agent (Bauer et al., 2013). Amine degradation and corrosion issues are among the important operational problems with this system (Bauer et al., 2013).

4.1.4 Pressure swing adsorption

PSA systems consist of four interconnected parallel vertical units packed with a porous adsorbent with high specific surface area such as activated carbon, activated alumina, silica gel, or zeolite (Fig. 8.2) (Ryckebosch et al., 2011). These adsorbents rely on molecular size exclusion and adsorption affinity mechanisms to selectively separate CO_2 from CH_4. The four interconnected units are operated sequentially in any of the following four phases: adsorption, depressurization/blowdown, desorption/purge, and pressurization (Fig.8.2) (Bauer et al., 2013; Muñoz et al., 2015). During the adsorption phase, the raw biogas is typically between 4 and 10 bars of pressure to increase CO_2 retention

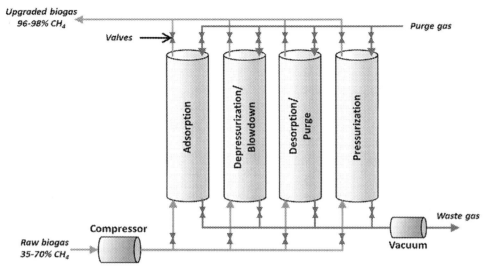

Figure 8.2 Schematic of the pressure swing adsorption biogas upgradation system. *(Adapted from Muñoz, R., Meier, L., Diaz, I., Jeison, D., 2015. A review on the state-of-the-art of physical/chemical and biological technologies for biogas upgrading. Rev. Environ. Sci. Biotechnol. 14 (4), 727–759. https://doi.org/10.1007/s11157-015-9379-1).*

in the pores (Muñoz et al., 2015). When the unit is saturated with CO_2, the raw biogas feed is directed to another, previously regenerated unit. The saturated unit is depressurized countercurrently to the feed step, which results in partial desorption of CO_2 from the adsorbent and releases a CO_2-rich gas from the unit.

H_2S and siloxanes can irreversibly adsorb onto PSA molecular sieves reducing the effectiveness and operational life. Therefore, H_2S and siloxanes, along with moisture, are removed prior to PSA (Bauer et al., 2013). Unlike water, organic solvent, and chemical absorbent system, PSA does not involve costs of pumping liquids and heating for absorbent regeneration, and constituted 21% of biogas upgrading market share in 2015 (Muñoz et al., 2015).

4.1.5 Cryogenic separation

Cryogenic separation involves stepwise reduction of biogas temperature at a constant pressure allowing for selective liquification, solidification, and separation of different biogas contaminants such as water, H_2S, siloxanes, and CO_2. Most cryogenic system configurations involve a multistage alternating gas compression and cooling step. The raw biogas is compressed to more than 10 bars and the temperature is stepwise decreased to about $-25°C$ to first separate water, whereas H_2S, siloxanes, and CO_2 are separated at temperatures of -55 to $-85°C$. The upgraded biogas purity is typically about 97% with CH_4 losses of less than 2%. The technology is not widely used and occupied only 0.4% of the global biogas upgrading market, due to the costs associated with energy required for gas compression and subzero cooling (Bauer et al., 2013; Muñoz et al., 2015).

4.1.6 Membrane separation

Membrane-based biogas upgradation systems rely on selective permeability of polymeric hydrophobic membranes (such as cellulose acetate) to separate biogas constituents (Ryckebosch et al., 2011; Bauer et al., 2013; Muñoz et al., 2015). The membrane systems can be a gas—gas separation system where both sides of the membranes have a gas phase. In the gas—liquid system, the biogas inlet side is a gas phase and the opposite side of the membrane contains a liquid chemical absorbent, which is typically an alkanolamine or alkaline solution (Bauer et al., 2013). The gas—liquid system can be operated at atmospheric pressures, whereas the gas—gas systems are typically operated between 6 and 20 bars (Bauer et al., 2013). The gas—gas systems can be configured as a single pass unit; however, a multiple stage system with internal recirculation of permeates and retentates is recommended for better biogas upgradation (Fig. 8.3) (Bauer et al., 2013; Muñoz et al., 2015). A single pass unit typically produces upgraded biogas with 92%—94% CH_4, whereas a gas—liquid unit or two-stage gas—gas units with recirculation can achieve 98%—99% CH_4 (Bauer et al., 2013). Pretreatment of raw biogas to remove moisture,

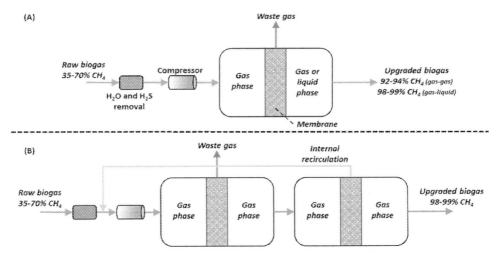

Figure 8.3 Schematic of the membrane separation process showing (A) a single pass gas—gas or gas—liquid system and (B) a multistage gas—gas membrane system. *(Adapted from Muñoz, R., Meier, L., Diaz, I., Jeison, D., 2015. A review on the state-of-the-art of physical/chemical and biological technologies for biogas upgrading. Rev. Environ. Sci. Biotechnol. 14 (4), 727—759. https://doi.org/10.1007/s11157-015-9379-1).*

H$_2$S, volatile organic compounds (VOCs), ammonia (NH$_3$), and siloxanes is often applied to avoid membrane deterioration and clogging (Bauer et al., 2013). The primary operating costs come from membrane replacement (every 5–10 years), biogas compression, and pretreatment costs. Membrane separation is a mature technology constituting ~10% of market share in 2014 (Thrän et al., 2014).

4.1.7 Biological methods

Several biological methods such as H$_2$-assisted CO$_2$ bioconversion, microalgae-based CO$_2$ fixation, enzymatic CO$_2$ removal, fermentative CO$_2$ reduction, and in situ CO$_2$ desorption are being investigated in small-scale and pilot-scale studies (Bauer et al., 2013; Muñoz et al., 2015). Commercial application of biological upgrading methods is currently limited (Thrän et al., 2014).

4.2 H$_2$S removal

Several technologies described for CO$_2$ removal such as water scrubbing, chemical absorption, membrane separation, and cryogenic upgradation can concurrently remove H$_2$S from biogas. However, operational and maintenance issues due to corrosion and chemical scaling of equipment, chemical absorbent contamination, reduction in absorbent life, and membrane clogging impose limits on the raw biogas H$_2$S various technologies can handle (Bauer et al., 2013; Ryckebosch et al., 2011). In most instances, it is recommended to include an independent pretreatment step that specifically removes H$_2$S prior to any CO$_2$ removal step to improve biogas upgradation efficiency and reduce operational and maintenance costs (Bauer et al., 2013).

4.2.1 H$_2$S removal through adsorption

Biogas H$_2$S removal via adsorption involves passing biogas through filter columns containing chemical agents such as iron oxide (Fe$_2$O$_3$), iron hydroxide (Fe(OH)$_3$), or zinc oxide (ZnO) (Muñoz et al., 2015; Ryckebosch et al., 2011). H$_2$S in biogas reacts with metal oxides and hydroxides to form insoluble metal sulfides. Conventionally, metal oxides are immobilized and packed in the column using steel wool (iron sponge) or on wood chips which are inexpensive and provide sufficient surface area (Bauer et al., 2013). Some alternative materials are also commercially available such as Sulphur-Rite, SulfaTreat, SOXSIA, and Meda-G2 that are marketed to improve H$_2$S removal (Muñoz et al., 2015).

H$_2$S adsorption can also be carried out using virgin or impregnated activated carbon. Addition of limited O$_2$ oxidizes H$_2$S to elemental sulfur which is adsorbed onto activated carbon (Ryckebosch et al., 2011). Impregnated activated carbon can provide better removal as it oxidizes H$_2$S at a higher rate. Catalytic impregnation is performed by treating the activated carbon with a nitrogen containing reagent such as urea or ammonia, while regular impregnation is conducted by mixing activated carbon with chemical reagents such as NaOH, KOH, NaHCO$_3$, Na$_2$CO$_3$, KI, or KMnO$_4$ (Ryckebosch et al., 2011). KI and KMnO$_4$ impregnated activated carbons are unique in that they support partial oxidation of H$_2$S without addition of O$_2$, which makes them among the most preferred options (Awe et al., 2017). The adsorption process is commonly conducted using two parallel columns operating in sequential H$_2$S adsorption and adsorbent regeneration modes. The adsorbent is regenerated using heat and depressurization (Awe et al., 2017). The major drawback of this process is that these filters only can be regenerated a few times before significantly losing their adsorption capacity (Abatzoglou and Boivin, 2009).

4.2.2 H$_2$S removal using biological filters

Biological filters employ sulfur oxidizing bacteria (SOB) that use H$_2$S as an electron donor, O$_2$ or NO$_3^-$ as an electron acceptor, and CO$_2$ as a carbon source, to convert H$_2$S in the raw biogas into elemental sulfur (Montebello et al., 2014). This process can be carried out in biotrickling filters packed with suitable surface material (Pall rings or polyurethane foam, etc.) to support microbial growth. The raw biogas is passed through the trickling filters along with a recirculating liquid solution containing nutrients for SOB growth at controlled pH (Fortuny et al., 2011). This technology has been successfully applied in full scale providing 80%—100% H$_2$S removal from biogas containing 5000—10,000 ppm H$_2$S concentration (Muñoz et al., 2015). Biological filters typically have lower operational costs than physical and chemical systems; however, clogging of packing material due to excess microbial growth and elemental sulfur accumulation can be an operational issue (Montebello et al., 2014). SOB biofilms have also been maintained on cloth or other media within anaerobic digester headspace with concurrent addition of controlled, low doses of air. The O$_2$ concentration must be controlled to be near zero to preclude explosive gas mixtures as the system removes H$_2$S and O$_2$ from the biogas (Ramos et al., 2014).

4.3 Siloxane removal

Siloxanes are present in personal care products and some industrial chemicals to impart softening, smoothing, or moistening qualities to chemical products. Therefore, siloxanes can be present in biogas from some municipal and industrial anaerobic digesters when siloxanes are in the digester feed material. However, siloxanes may be absent from biogas in other anaerobic digesters. If siloxanes are in biogas, then they can be effectively removed up to 97%—99% using organic solvent scrubbing (with Selexol absorbent) and cryogenic separation methods (Bauer et al., 2013; Muñoz et al., 2015). However, the high cost associated with absorbent regeneration in organic solvent scrubbing process, and the high investment and energy consumption required in cryogenic technology, has prevented the full-scale application of these technologies for siloxane removal. Adsorption using activated carbon and silica gel currently are the only commercially available technologies applied for biogas siloxane removal (Abatzoglou and Boivin, 2009; Ryckebosch et al., 2011). Activated carbon can provide ~95% siloxane removal; however removal is negatively affected by moisture in the biogas and the regeneration of activated carbon at high temperatures is not economically feasible (Ryckebosch et al., 2011). Silica gel is also sensitive to moisture in the biogas but has shown superior performance compared to activated carbon. Silica gels can remove around 99% of siloxane in the biogas and the adsorbent can be regenerated to 95% of its adsorption capacity by applying heat at 250°C for 20 min (Muñoz et al., 2015).

5. Novel microalgae biogas conditioning technology

Among the carbon capture and utilization technologies, microalgae have obtained growing recognition to remove CO_2 and generate biomass (Rahaman et al., 2011). Microalgal biomass contains approximately 50% carbon by dry weight (Chisti, 2007). While inorganic carbon is a carbon source for microalgae growth, CH_4 is not known to be a major algae carbon source (Alcántara et al., 2015). In this regard, microalgae are suitable for biogas upgrading selectively to remove CO_2 from biogas. Other potential biogas and algae growth media constituents, including sulfur, nitrogen, and phosphorus, may also be removed in algae systems, whereas oxygen is produced. Also, growth of algae can be inhibited by excess CO_2, H_2S and potentially other biogas constituents.

5.1 Carbon dioxide removal

In general, 1.6–2 g CO_2 can produce 1 g of microalgal biomass photosynthetically (Sayre, 2010). Considering the mass of inorganic carbon (i.e., CO_2) in crude biogas flowing into an upgrading system (M_{ICG-in}), the fate of inorganic carbon mainly includes fixation as algae biomass (M_{OCX}), residual carbon dioxide in treated biogas ($M_{ICG-out}$), dissolved inorganic carbon in the liquid (M_{ICL}), and stripped carbon dioxide to the atmosphere via desorption ($M_{ICG-stripping}$) (Meier et al., 2017; Bose et al., 2019). Extracellular organic substances (M_{OCL}) which are secreted metabolites (Liu et al., 2016) as well as precipitated carbonate minerals (M_{ICS}) are other potential inorganic carbon streams in biogas upgrading systems, although they are usually assumed to be negligible (Table 8.5) (Thomas et al., 2004; Ramanan et al., 2010). The biogas CO_2 mass balance equation has been represented as follows (Fig. 8.4):

$$M_{ICG-in} = M_{ICG-out} + M_{ICL} + M_{ICG-stripping} + M_{OCX} + M_{OCL} + M_{ICS} \quad (8.1)$$

5.2 Sulfur

H_2S is soluble in microalgae systems. Total dissolved sulfide is a mixture of H_2S (aq), HS^-, and S^{-2}, and the proportions of species are primarily a function of pH. The high dissolved oxygen concentration and alkaline pH typically encountered in algal

Table 8.5 Carbon mass balance of microalgae-based biogas upgrading system: the flow of carbon dioxide.

Carbon inlet	Inorganic carbon outlet				Organic carbon outlet		References
	CO_2 in treated biogas ($M_{ICG-out}$)	DIC in liquid (M_{ICL})	CO_2 stripping ($M_{ICG-stripping}$)	Carbonation (M_{ICS})	Biomass (M_{OCX})	Secretion (M_{OCL})	
CO_2	11% (light time)	13%	57%	n.c.[a]	19%	n.c	Meier et al. (2017)
	7% (dark time)	30%	63%	n.c.	n.d.[b]	n.c	
CO_2	1%–2%	12%	50%	n.c.	37%	n.c.	Meier et al. (2018)
CO_2 and inorganic carbon in feed liquid	40% (CO_2 concentration 13.5%)	1%	49%	n.c.	9%	1%	Alcántara et al. (2015)
CO_2	~7% (continuous illumination)	58%	5%	n.c.	30%	n.c.	Franco-Morgado et al. (2017)
	~77%	12%	n.c.	7%	n.c.		
	~4% (light/dark cycle)						

[a]*n.c.*, not considered.
[b]*n.d.*, not detected. It means negligible.

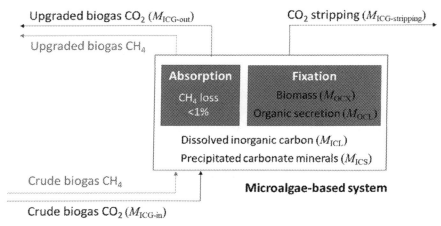

Figure 8.4 Schematic representation of biogas conditioning process with microalgae indicating flow of carbon dioxide removal for mass balance.

ponds lead to the formation of SO_4^{2-} as the major end-product with an excess of O_2 and other S-intermediates (i.e., pentasulfide, elemental sulfur, thiosulfate, or sulfite) of H_2S oxidation, as shown in Table 8.6 (Meier et al., 2018; Rodero et al., 2020c).

5.3 Nitrogen and phosphorus

Macronutrients for microalgae growth, e.g., nitrogen and phosphorus, are mainly assimilated from the liquid. Nitrogen accounts for 8% of microalgae biomass (w/w), based on microalgae molecular formula of $CH_{1.7}O_{0.4}N_{0.15}P_{0.0094}$ (Xie et al., 2013). Both nitrate (Franco-Morgado et al., 2017) and ammonium (Wang et al., 2020b) are possible nitrogen source for microalgae. Moreover, phosphorus accounts for 12% of microalgae biomass (w/w), and is even higher when luxury uptake occurs and phosphorus is stored as polyphosphate inclusions (Powell et al., 2011; Mukherjee et al., 2015). Reactive phosphorus such as orthophosphate and nonreactive phosphorus such as organic beta-glycerol phosphate can provide phosphorus for microalgae (Vrba et al., 2018). These liquid nutrient can be fed in an artificial medium (Franco-Morgado et al., 2017) or may be present in wastewater used to grow algae (Xu et al., 2019; Ran et al., 2021).

5.4 Oxygen production

When carbon dioxide is bioabsorbed by microalgae photosynthetically, half of the O_2 is fixed in the form of organic biomass and the other half is released as O_2. About 1.33 g O_2 are generated for per gram biomass produced (as VSS) (Meier et al., 2017) and contributes high dissolved O_2 concentration of about 9 mg/L in the liquid (Meier et al., 2018). Some O_2 may be consumed through respiration of microalgae or aerobic

Table 8.6 Distribution of sulfur in microalgae-based biogas upgrading system.

Biogas upgrading system	Gas output		Liquid outlet				Solid outlet	References
	Hydrogen sulfide H_2S	Pentasulfide S_5^2	Elemental sulfur S^0	Thiosulfate $S_2O_3^{2-}$	Sulfite SO_3^{2-}	Sulfate SO_4^{2-}	Biomass	
Absorption + photobioreactor	n.d.[a]	n.a.[b]	n.d.	n.d.	n.a.	92%	8%	Meier et al. (2018)
Photobioreactor	n.d.	n.a.	n.a.	n.a.	n.a.	100%	n.a.	Lebrero et al. (2016)
Absorption + photobioreactor	n.d.	n.a.	n.a.	n.a.	n.a.	97%—99%	1%—3%	Franco-Morgado et al. (2017)
Absorption + photobioreactor (IC 1500 mg/L)	3.6% (35°C) 0% (12°C)	n.a.	Considered	Considered	Considered	100% 55%	Considered	Rodero et al. (2018)
Absorption + photobioreactor	2.9%	n.a.	Considered	Considered	Considered	40%	n.a.	Toledo-Cervantes et al. (2016)
Absorption + photobioreactor	0%—0.1%	n.a.	Considered	Considered	Considered	Considered	26% (low IC) 16%—17% (high IC)	Posadas et al. (2017)
Erlenmeyer flasks (Batch)	n.a.	Detected	n.a.	Detected	Detected	Detected	Detected	González-Sánchez and Posten (2017)
Tubular closed photobioreactor (Low O_2/S^{2-} ratio)	n.d.	n.a.	n.a.	98%—100%	n.a.	n.a.	n.a.	Ramírez-Rueda et al. (2020)

[a]n.d., not detected. It means negligible.
[b]n.a., not available.

bacteria (Meier et al., 2017), as well as sulfide oxidation (Franco-Morgado et al., 2017). A fraction of the O_2 can leave the system in the treated biogas or desorb from the photobioreactor to the atmosphere (Meier et al., 2017).

5.5 Microalgae growth inhibition

5.5.1 CO₂ affect

Microalgae generally can withstand exposure to an atmosphere containing high CO_2 concentration of 20%−55% to perform on-site carbon capture and utilization (Bose et al., 2019). However, high CO_2 partial pressure may result in intracellular acidification caused by the activity of carbonic anhydrase which subsequently inhibits microalgal photosynthesis (Nagarajan et al., 2019). For example, *Nannochloropsis* was inhibited above 10% CO_2 (Meier et al., 2015). The response is species specific. Some CO_2-tolerant microalgae like *Chlorella*, *Desmodesmus*, *Scenedesmus*, and *Neochloris* can grow at a CO_2 concentration of 50% and higher (Sergeenko et al., 2000; Park et al., 2020). The selection of suitable microalgal strains is essential for biogas upgrading systems at high CO_2 partial pressure.

5.5.2 H₂S toxicity

H_2S is a trace component (approximately 0.005%−2%) in biogas, but microalgae are susceptible to sulfide toxicity due to damage to the electron transport with the median effective inhibitory concentration of 1.76 mg S/L for *Scenedesmus vacuolatus* (Küster et al., 2005) and observed inhibition at 5 mg S/L for *Chlorella* (Gonzalez-Camejo et al., 2017). When exposed to H_2S gas concentrations higher than 150 ppm, *Chlorella* growth was severely inhibited (Kao et al., 2012). In consideration of sulfide toxicity, H_2S removal via pretreatment usually has been performed prior to algae contact (Kao et al., 2012; Wang et al., 2016; Takabe et al., 2017; Guo et al., 2018), or microalgae are cocultured with SOB (Rodero et al., 2020a). Chemical sulfide oxidation was found to diminish the inhibitory effect of sulfide, and its oxidative product, sulfate, can provide sulfur as a micronutrient to foster microalgal growth (González-Sánchez and Posten, 2017; Ramírez-Rueda et al., 2020). The risk of sulfide toxicity and sulfur management should be taken into account when considering algae biogas upgrading.

5.5.3 Other potential toxicity issues

Other trace constituents including siloxanes (<0.02%) and halogenated volatile organic chemical hydrocarbons (HVOC < 0.6%) are rarely reported to inhibit microalgae, or are not usually present in biogas, respectively. However, free ammonia has inhibited microalgal photosynthesis by affecting the electron transport chain and the O_2-evolution complex as well as dark respiration (Markou et al., 2016). Even though ammonia concentration in biogas is typically low, many wastewaters are rich in ammonium and can exert toxicity if not managed (Andriani et al., 2014; Wang et al., 2020a). In this regard, pH control with maintenance at neutral to acid values or ammonia removal pre-treatment should be considered in high-concentration ammonia nitrogen conditions in order to reduce toxic free ammonia concentration.

The major component of biogas, CH_4, does not exert a toxic effect on microalgae and even may enhance CO_2 fixation efficiency and biomass production in microalgae systems (Meier et al., 2015; Choix et al., 2017). Although one report described a nearly 30% reduction in growth rate when exposing *Chlorella* to a biogas containing 80% CH_4, the toxicity may have been due to other constituents (Kao et al., 2012). Microalgae remove CO_2 in the presence of high levels of CH_4 (Nagarajan et al., 2019). In microalgae biogas upgrading systems, a slight decrease in CH_4 content was observed in some one-step photobioreactors, although no microalgae have been found that utilize CH_4 (Lebrero et al., 2016; Prandini et al., 2016). The CH_4 loss was possibly due to bacterial consumption with oxygen or system leakage.

6. Benefits of algae biogas upgrading

Microalgae are able to capture carbon dioxide using solar energy to support their high growth rates, with a high theoretical photosynthesis efficiency up to 8% (Zelitch, 1971; Meier et al., 2015). Microalgae do not only remove CO_2 from biogas but also transform it into biomass that can be used as feedstock for biofuel or chemical production such as biodiesel, bioethanol, or even more biogas (Bose et al., 2019). Compared to traditional CO_2 removal methods, microalgae biotechnology can be attractive since it achieves carbon dioxide sequestration, utilization, and biomass generation.

Microalgae-based processes are capable of simultaneously removing CO_2, H_2S, and even VOCs in a single stage (Bahr et al., 2014; Posadas et al., 2017). Further, biogas upgrading can be integrated with digestate liquid remediation or other wastewater treatment since microalgae can remove soluble nitrogen and phosphorus from the wastewater (Nagarajan et al., 2019). Therefore, multiple biogas and digestate purification steps may be achievable in a single microalgae system.

Most common biogas upgrading described above requires complex operation and high energy input (Meier et al., 2015). According to Toledo-Cervantes et al. (2017a), microalgae biotechnology requires 26% of the energy of other physical/chemical technologies and has lower environmental impacts in terms of material and water consumption as well as greenhouse gas emissions. Moreover, the operating cost of microalgae biotechnology was estimated to be 15% of other physical/chemical technologies. Microalgae technologies may, however, have higher land requirements and higher capital cost. In addition, algae systems require a light source and warmer temperature which may impact operating and maintenance costs. Overall, full-scale microalgae advantages and disadvantages must still be determined and the appropriateness of microalgae for biogas upgrading will probably depend on site-specific details.

7. Limitations of algae biogas upgrading systems

Microalgae increase the O_2 level in the upgraded biogas even though aerobic bacteria and sulfide oxidation consume some or all of the O_2 (Meier et al., 2015; Prandini et al., 2016). Most natural gas specifications require an O_2 concentration less than 0.01%–3% (Bose et al., 2019; Meier et al., 2019). Therefore, O_2 must be removed from microalgae-treated biogas if it is to be used in natural gas systems.

The CO_2 removal rate from biogas is often limited by low CO_2 mass transfer rates (Bose et al., 2019). When applying biogas for natural gas grid injection and vehicle fuel, the CO_2 concentration must be below 1%–8% (Sun et al., 2015). System optimization is necessary to minimize or eliminate final CO_2 concentration in the upgraded biogas.

Diurnal variability and seasonal fluctuations in operations can affect temperature and light which are critical factors for microalgal growth. Temperature affects CO_2 gas/liquid equilibrium and microalgal metabolism (Prandini et al., 2016). Suitable light irradiance is necessary for microalgal cultivation. Higher light intensity may lead to cell damage and lower light intensity may not be enough to support sufficient photo-autotrophy (Zhu et al., 2016). It's challenging to maintain optimum operating conditions, especially in open systems using sunlight.

8. Potential carbon and nutrient sources for algae

Microalgae can utilize the low CO_2 concentration in air (about 0.04%) and rely on carbon-concentrating mechanisms (Raven et al., 2012), while some microalgal strains also tolerate high CO_2 concentrations in biogas (20%–55%) (Sergeenko et al., 2000; Park et al., 2020; Zhang et al., 2020). The relatively high CO_2 concentration in biogas has significant effects on biomass production and nutrient uptake (Sun et al., 2016). Research is required to define microalgae species as well as system design, and operating conditions for optimum CO_2 fixation rate and biogas upgrading.

Microalgae can treat anaerobic digestate by removing macronutrients, such as nitrogen and phosphorus, as well as micronutrients to support growth. To provide the liquid environment for microalgal biogas upgrading, microalgae are cultured in aqueous media (Toro-Huertas et al., 2019; Rodero et al., 2020b,c). It may be a better choice to use wastewater or treated wastewater to replace media in terms of economic, environmental, and social impacts. The digestate discharged from anaerobic digesters often contains high nitrogen and phosphorus concentrations and may serve as an economical nutrient source for microalgae either alone or diluted with other water (Toledo-Cervantes et al., 2017b; Xu et al., 2019; Arashiro et al., 2020; Ivanova et al., 2020). The combination of digestate treatment with biogas conditioning would help achieve a closed loop system and to realize circular economy (Zhu et al., 2016).

Microalgae can also treat digestate by removing residual organic carbon via a central carbon catabolic pathway like heterotrophic bacteria (Walter et al., 1987). Some microalgae can perform heterotrophy and mixotrophy (i.e., both autotrophy and autotrophy simultaneously), in addition to photosynthetic autotrophy (Kim et al., 2013). Although glucose is the most widely studied carbon source, microalgae can assimilate volatile fatty acids (VFAs) through energy dependent transporters (Hu et al., 2018; Nagarajan et al., 2019). For example, acetate, the principal VFA present in most digestate, can be used as the sole carbon source by microalgae (Vazhappilly and Chen, 1998; Turon et al., 2015a,b; Turon et al., 2016). Therefore, VFA residuals in digestate can be removed from the media and provide carbon for microalgal biomass growth during biogas upgrading.

9. Algal bioreactor configurations and operations

Microalgae-based biogas upgrading system can be classified as either single or two-stage photobioreactor systems (Table 8.7). Current single reactor technologies are mainly comprised of transparent polyethylene bags (Sun et al., 2016) or glass fermenters (Prandini et al., 2016). The CO_2 absorption and bioconversation take place in one space.

Two-stage systems separate CO_2 absorption and other functions, such as algae growth or dissolved inorganic carbon (DIC) air stripping, temporally or spatially. Bose et al. (2019) separated CO_2 adsorption and consumption by microalgae spatially, providing gas/liquid contact in an absorption unit for mass transfer that was connected to a subsequent photobioreactor. The alkaline algal solution was pumped into the absorption column by a continuous recirculation flow to absorb acidic CO_2 and H_2S gasses, whereas microalgal photosynthesis regenerated a high pH liquid (Bose et al., 2019). Bahr et al. (2014) built a 180 L pilot-scale, outdoor high-rate algal pond connected to an external 0.8 L bubble column. The algal pond reactor configuration used by Bahr et al. (2014) can be substituted by other photobioreactor configurations, such as a tubular photobioreactor (Toledo-Cervantes et al., 2018). These two-stage photoreactors were open systems which were easier to construct and operate, although they had the disadvantages of evaporation and CO_2 losses, difficulty maintaining optimum operation conditions, and contamination by invasive microorganisms. The bubble column contactor configuration was described as beneficial to preclude clogging of packed media in other configurations (Toledo-Cervantes et al., 2016). Cocurrent bubble and liquid flow was used to preclude diffuser clogging from sulfur deposition and pipe obstruction caused by biomass accumulation at the column top (Toledo-Cervantes et al., 2017b; Bose et al., 2019). The liquid-to-gas-flow ratio and gas/liquid velocities were modified to achieve CO_2 and H_2S removal as well as O_2 and N_2 stripping to obtain high-quality biomethane (Toledo-Cervantes et al., 2016; Bose et al., 2019; Marín et al., 2020; Rodero et al., 2020c). Other strategies were investigated to

Table 8.7 Performances of microalgae biogas upgrading using different photobioreactors.

Bioreactor type	Bioreactor construction	Upgraded biogas				References
		CH$_4$	CO$_2$	H$_2$S	O$_2$	
Single	Transparent polyethylene bag	90%	10%	<50 ppm	<1%	Sun et al. (2016)
	Bag	80%	n.a.[a]	Chemical absorption pretreatment	n.a.	Wang et al. (2016)
	Glass-made fermenter	64.7%	7.5%	5 ppm	17.8%	Prandini et al. (2016)
Two-stage (Temporally)	Airlift tubular photobioreactor	90%–93% (Using diluted digested swim manure as liquid nutrient) 94%–95% (Cattle manure) 83%–93% (Domestic sewage)	7%–10% 5%–6% 7%–17%	<100 ppm	n.a.	Miyawaki et al. (2021)
Two-stage (Spatially)	HRAP interconnected to an absorption column	95%	3%–5%	n.d.[b]	<0.2%	Bahr et al. (2014)
	HRAP interconnected to an absorption column	97.6%–99.1%	0.2%–1.8%	n.d.	0.05%–0.18%	Franco-Morgado et al. (2018)
	Photo-fermenter interconnected to an absorption column	n.a.	<1.5%	n.d.	2%–3.5%	Meier et al. (2018)

[a]*n.a.*, not available.
[b]*n.d.*, not detected. It means negligible.

optimize the overall system including cell-free liquid recirculation to purify biogas (Toledo-Cervantes et al., 2016), a zero effluent strategy (Franco-Morgado et al., 2018), membrane degasification of the cultivation broth, and biogas scrubbing at high pressure (Ángeles et al., 2020).

9.1 Influence of alkalinity and temperature

Microalgae media alkalinity is a key parameter influencing CO_2/H_2S absorption and final biomethane quality, especially for absorption column photobioreactor systems. Generally, biogas upgrading relies on carbonate/bicarbonate chemistry. Microalgal photosynthesis utilizes bicarbonate. Carbonic anhydrase enzyme catalyzes conversion of bicarbonate to CO_2 with concomitant hydroxide transported out of the cell. This activity leads to an increase in pH and carbonate concentration in the solution, which can scrub biogas to remove CO_2. Employing high alkalinity, the upgraded biomethane can typically comply with most European regulations for natural gas grid injection or vehicle fuel (Rodero et al., 2018; Rodero et al., 2020c). Further, CO_2 absorption regenerates bicarbonate to provide microalgae with sufficient carbon for photosynthesis. This beneficial cycle can help avoid acidification of the scrubbing solution and ensures that the microalgal system continues to operate (Bose et al., 2019).

Temperature plays a role in biomass growth kinetics as well as CO_2 gas/liquid equilibria. Especially at low alkalinity, temperature effects on biomethane quality were significant (Rodero et al., 2018). The solubility of CO_2 is temperature dependent. The solubility increases at lower temperature. In this regard, some have achieved biogas upgrading and CO_2 removal at night with no light since more inorganic carbon can be absorbed in the photobioreactor at lower temperatures during night (Meier et al., 2017; Franco-Morgado et al., 2018; Rodero et al., 2018).

9.2 Influence of light: light wavelength/photoperiod

The light wavelengths and intensity contribute to maximal photosynthetic efficiency and biogas conditioning affects. A light-emitting diode (LED) was investigated as an artificial light source and the optimum monochrome light wavelength was found to be red light (Zhao et al., 2013; Yan et al., 2016a). In other studies, a mixed LED light of red and blue had the best performance for biogas upgrading and nutrient removal (Yan et al., 2014; Zhao et al., 2015; Yan et al., 2016b; Wang et al., 2017; Zhang et al., 2017). It is recommended to employ solar radiation and natural existing day/night photoperiods for full-scale microalgae-based biogas upgrading systems (Meier et al., 2017; Posadas et al., 2017). Although microalgae do not perform photosynthesis without illumination, CO_2 and H_2S removal has been observed throughout the entire light/dark cycle (Prandini et al., 2016; Franco-Morgado et al., 2017; Zhang et al., 2020).

High light intensity is favored for batch culture (Yan et al., 2016a,b). Moderate light intensities are effective for continuous biogas upgrading, for example, about 160 $\mu mol/m^2$ s for *Scenedesmus* (Ouyang et al., 2015) and 350 $\mu mol/m^2$ s for *Chlorella* (Yan and Zheng, 2013). Moreover, low light intensities are typically used in the integration of biogas upgrading with organic wastewater bioremediation in mixotrophic mode (Nagarajan et al., 2019).

9.3 Mono- and cocultivation of microalgae

Microalgae-based technologies, such as mono-cultivation of microalgae or cocultivation with bacteria or fungi, can significantly influence the biogas conditioning and nutrient absorption abilities (Zhang et al., 2020). Monocultivation of microalgae was successfully applied for biogas upgrading. Converti et al. (2009) use *Spirulina platensis*; Kao et al. (2012), Yan and Zheng (2013), and Zhao et al. (2013) used *Chlorella* sp.; and Zhao et al. (2015) use *Scenedesmus oblique*, *Chlorella vulgaris*, and *Neochloris oleoabundans*. Generally, desulfurized biogas and artificial media/sterilized digestate were adopted in these experimental mono- or comicroalgae systems.

Cocultures of microalgae have also been employed for inocula, such as a consortium of *Chlorella vulgaris*, *Stigeoclonium tenue*, *Nitzschia closterium*, and *Navicula amphora* (Marín et al., 2019). In addition, there can be a shift in the microalgae coculture population composition when operational conditions change (Serejo et al., 2015). Franco-Morgado et al. (2018) observed that the predominant microalga species changed from green microalgae to cyanobacteria when the nutrient feeding regime was changed.

Cocultivation of microalgae with other microorganisms has been investigated, especially when using raw biogas and real anaerobic digestate. SOB, nitrifying—denitrifying activated sludge, and other bacteria or fungi have been successfully applied in microalgae-based biogas upgrading systems (Bahr et al., 2014; Serejo et al., 2015; Sun et al., 2016). Based on a mutually beneficial relationship, the cocultivation systems can achieve higher microalgal growth rate as well as increased CO_2 removal (Xu et al., 2017; Zhang et al., 2017; Gao et al., 2018). Additionally, coculture of microalgae with fungi/bacteria is regarded as a way to significantly increase flocculation and settleability while reducing energy consumption for biomass separation in the overall process (Xu et al., 2017).

10. Biogas constituents removed

10.1 CO_2 removal and energy density upgrading

Among the prevalent CO_2 removal methods, microalgae technology stands out with the advantages of cost competitive and environmentally sustainability. In the absorption—fixation two-stage biogas upgrading system, the CO_2 removal rate mainly depends on the operating parameters of the absorption column (Marín et al., 2020; Park et al., 2020; Miyawaki et al., 2021). Under the optimal conditions at a liquid recirculation/biogas ratio of 0.5, the CO_2 removal rate was up to 99.5%, and the residual CO_2 concentration in biomethane was controlled at 0.2% which was well satisfied with the limits of the gas quality standards (USA 2%−3%, EU 1%−8%) (Franco-Morgado et al., 2018). Accordingly, the energy density of biogas is linearly scaled up with improved CH_4 concentration in biogas. The upgraded biogas with 98% CH_4 can yield a volumetric energy density of more than 35 MJ m^{-3} at standard conditions, while the raw biogas with 70% CH_4 has about 25 MJ m^{-3} (Cuéllar and Webber, 2008).

Moreover, the superiority of two-stage system is to separate biogas purification process from microalgal photosynthesis. The photosynthesis converts CO_2 into O_2 and organic biomass carbon. O_2 is generated in the microalgae pond unit separately. Biomethane released from the absorption unit would be not polluted, and high concentration of CH_4 above 99% was feasible. On the other side, the bioconversion of captured carbon in the open algal pond is not complete, and more than a half was stripped to the atmosphere due to high alkalinities, resulting in greenhouse gas emission (Alcántara et al., 2015; Meier et al., 2017; Meier et al., 2018; Rodero et al., 2020c). One-step closed photobioreactors avoid the problem of carbon emission (Prandini et al., 2016), but the high-concentration O_2 (10%−24%) entails a potential explosion which restricted its direct application of natural gas grid injection (Meier et al., 2015). Anyway, this CH_4-O_2 gas can provide ideal mixture for on situ complete combustion and also an ideal functional gas feedstock for biogas-based biopolymer production (Chidambarampadmavathy et al., 2017).

10.2 H₂S removal

H_2S, generated by sulfate reducing bacteria in anaerobic digestion, is an undesired component in raw biogas, which is acidic, corrosive, toxic, and malodorous (Franco-Morgado et al., 2017). Nevertheless, the H_2S removal could achieve a full or near-complete affect in a microalgae treatment system, as shown in Table 8.6. The high removal rate is mediated by the high aqueous solubility of H_2S according to its dimensionless Henry's law constants (C_L/C_G, H_{H2S} of 2.44 vs. H_{CO2} of 0.83 at 25°C) and the rapid oxidation of H_2S in the liquid phase (Rodero et al., 2020c). A series of sulfur intermediates and the end-product sulfate may exist (see section 1.5.2). In the start-up period, sulfide was oxidized gradually by oxic chemical reactions, and microalgae started to grow after the inhibition from sulfide was removed (González-Sánchez and Posten, 2017). After microalgae system was activated, their photosynthetic activity provides sufficient dissolved O_2 for the fast oxidation of sulfide to sulfate (Meier et al., 2018). The presence and activity of SOB made microalgae-based system perform well at conditioning biogas with up to 5000 ppm H_2S (Bahr et al., 2014; Rodero et al., 2018). Moreover, the final product sulfate as a nutritional supplement can be assimilated by microalgae, since sulfur is an essential macronutrient for microalgal growth (González-Sánchez and Posten, 2017). In summary, microalgae-based biogas upgrading system had an excellent H_2S removal ability at the same time as removing CO_2 from biogas, which is dominated by both chemical and biological oxidations. In this context, biogas without desulfurization can be applied in the photosynthetic upgrading system directly.

10.3 Siloxanes removal

Siloxanes, a group of polymeric compounds of Si—O bonds with organic side chains attached to each silicon atom, are considered to be the third most important contaminant. During the anaerobic digestion of waste sludges or in landfills, undecomposed siloxanes are significantly volatilized and transferred to biogas (up to 50 mg Si m^{-3}) (Nagarajan et al., 2019). In contrast, biogas from anaerobic digestion of animal manure typically contains less or no siloxanes. Siloxanes' presence during combustion is detrimental to turbines because they form glassy microcrystalline silica deposits on surfaces which lead to abrasion, malfunctioning spark plugs, overheating of sensitive parts of engines due to coating, and general deterioration of mechanical engine parts (Abatzoglou and Boivin, 2009; Sun et al., 2015). Therefore, the removal of siloxanes is mandatory for many applications to ensure an acceptable and useful lifespan of equipment. Unfortunately, reports of microalgae-based biogas upgrading systems did not include siloxane removal data or the research used pretreated, siloxane-free biogas (Muñoz et al., 2015; Takabe et al., 2017). Research to investigate the biological absorption effect of siloxanes during microalgae-based biogas upgrading is required.

10.4 VOCs removal

Biogas contains about 1% VOCs, which can be grouped into arenes, alkenes, alkanes, carbonyls, alcohols, heterocyclic compounds, chlorinated compounds, and other compounds (Carriero et al., 2018). Some may be corrosive and potentially toxic to humans. Typically, VOC biotreatment can be effectively performed in bacterial or fungal biofilters. For example, a biological trickling filter was used to remove toluene from a gas stream based on the adsorption—biofilm theory (Peishi et al., 2004). Currently, microalgae-based biotechnology is emerging for biogas upgrading to remove multiple impurities. As for microalgal ability to remove VOCs, synergistic algal—bacterial microcosms can achieve substantial biodegradation (more than 85%) of aromatic pollutants salicylate, phenol, and phenanthrene in liquid (Borde et al., 2003). Photosynthetic treatment of diesel exhaust gas achieved almost complete removal of toluene while performing CO_2 removal in a tubular algal—bacterial photobioreactor (Anbalagan et al., 2017). The biogas upgrading system of an absorption column algal pond also removed VOCs (methyl mercaptan, toluene, and hexane) (Franco-Morgado et al., 2018). Removal of VOCs depended on their aqueous solubility, but more research is needed to evaluate the overall VOC removal efficiency of biogas microalgae upgrading processes.

11. Algal reactor products utilization and management

Several operational and byproduct management considerations arise with the use of microalgae systems for biogas upgrading. A primary concern is introduction of O_2 and N_2 gases into the upgraded biogas stream during CO_2 removal. This phenomenon results in dilution of the CH_4 and may also create an explosive mixture if the O_2 concentration is high. These gases can be generated by biological conversion of CO_2 and NH_4^+ and can become enriched in the biogas due to gas stripping. The degree of gas stripping increases due to higher gas turbulence and higher gas to liquid ratios used during the biogas upgrading process, as well as higher ionic strength of the growth media being recirculated (Toledo-Cervantes et al., 2016). Previous studies have shown that high turbulence and/ or gas to liquid ratios can yield O_2 and N_2 concentrations up to 2%—24% (Posadas et al., 2015) and 7% (Toledo-Cervantes et al., 2016), respectively, in the upgraded biogas. Low gas to liquid ratios (less than 1) have shown final O_2 and N_2 concentrations lower than 0.2% and 0.8%, respectively (Rodero et al., 2018).

CH_4 loss during upgrading with microalgae has been observed but is generally low owing in part to the low aqueous solubility of CH_4. Toledo-Cervantes et al. (2016) reported average CH_4 loss as high as 4.9% due to biological oxidation. The specific biological mechanisms contributing to CH_4 consumption in algal systems are currently unclear (Lebrero et al., 2016; Prandini et al., 2016). While CH_4 loss occurs, mitigation steps to prevent CH_4 release to the environment are not thought to be necessary given that the CH_4 loss is attributed to biological consumption and not to atmospheric losses.

The growth medium used for microalgae cultivation and the biogas upgrading process may require one or more treatment processes prior to disposal or reuse. Previous work has indicated that microalgae may not be easily removed via gravity separation, indicating that coagulant addition may be needed in conjunction with gravity separation, or a filtration process may be necessary in order to recover the microalgae and produce an effluent low in suspended solids (Toledo-Cervantes et al., 2016). Additionally, if a high nutrient laden feed source, such as digestate liquid, is used, then the microalgae process may not be capable of reducing nutrient concentrations in the effluent to achieve low final concentrations suitable for discharge to the environment. In these situations, the effluent from the side-stream microalgae process may require further treatment via conventional wastewater treatment processes such as activated sludge.

12. Conclusions and recommendations for future research

Algae can be used to condition biogas by removing CO_2, H_2S, and VOCs. It may be possible to use algae biogas conditioning in combination with O_2 and water vapor removal to produce pipeline quality biomethane. However, microalgae biogas conditioning to date has been predominantly laboratory scale. Commercial, full-scale systems do not presently exist. More research is required to determine the possible technoeconomic benefits of algae biogas conditioning. Hurdles to overcome include the costs of light source, temperature, and algal biomass removal from culture media.

Significant benefits of nutrient and VFA removal from anaerobic digestate can be accrued if biogas upgrading can be economically coupled with digestate treatment. A single process in which algae remove nitrogen, phosphorus, and residual VFAs from digestate while producing pipeline quality biogas may be transformative for the biomethane industry. However, limitations including optimized or economical algae strain selection, algae and solids removal from culture media, inhibition due to digestate ammonia and H_2S, and algae beneficial use scenarios must still be developed if full-scale algae systems are to be employed for these tasks.

References

Abatzoglou, N., Boivin, S., 2009. A review of biogas purification processes. Biofuel. Bioprod. Biorefin. 3 (1), 42–71. https://doi.org/10.1002/bbb.117.

Alcántara, C., García-Encina, P.A., Muñoz, R., 2015. Evaluation of the simultaneous biogas upgrading and treatment of centrates in a high-rate algal pond through C, N and P mass balances. Water Sci. Technol. 72 (1), 150–157. https://doi.org/10.2166/wst.2015.198.

Anbalagan, A., Toledo-Cervantes, A., Posadas, E., Rojo, E.M., Lebrero, R., González-Sánchez, A., Nehrenheim, E., Muñoz, R., 2017. Continuous photosynthetic abatement of CO_2 and volatile organic compounds from exhaust gas coupled to wastewater treatment: evaluation of tubular algal-bacterial photobioreactor. J. CO2 Util. 21, 353–359. https://doi.org/10.1016/j.jcou.2017.07.016.

Andriani, D., Wresta, A., Atmaja, T.D., Saepudin, A., 2014. A review on optimization production and upgrading biogas through CO_2 removal using various techniques. Appl. Biochem. Biotechnol. 172 (4), 1909–1928. https://doi.org/10.1007/s12010-013-0652-x.

Ángeles, R., Rodríguez, Á., Domínguez, C., García, J., Prádanos, P., Muñoz, R., Lebrero, R., 2020. Strategies for N_2 and O_2 removal during biogas upgrading in a pilot algal-bacterial photobioreactor. Algal Res. 48, 101920. https://doi.org/10.1016/j.algal.2020.101920.

Arashiro, L.T., Ferrer, I., Pániker, C.C., Gómez-Pinchetti, J.L., Rousseau, D.P.L., Van Hulle, S.W.H., Garfí, M., 2020. Natural pigments and biogas recovery from microalgae grown in wastewater. ACS Sustain. Chem. Eng. 8 (29), 10691–10701. https://doi.org/10.1021/acssuschemeng.0c01106.

Audrey, 2019. Renewable Natural Gas Quality Specifications in North America, Biogas World. https://www.biogasworld.com/news/renewable-natural-gas-quality-specifications-in-north-america/.

Awe, O.W., Zhao, Y., Nzihou, A., Minh, D.P., Lyczko, N., 2017. A review of biogas utilisation, purification and upgrading technologies. Waste Biomass Valorization 8 (2), 267–283. https://doi.org/10.1007/s12649-016-9826-4.

Bahr, M., Díaz, I., Dominguez, A., González Sánchez, A., Muñoz, R., 2014. Microalgal-biotechnology as a platform for an integral biogas upgrading and nutrient removal from anaerobic effluents. Environ. Sci. Technol. 48 (1), 573–581. https://doi.org/10.1021/es403596m.

Bauer, F., Persson, T., Hulteberg, C., Tamm, D., 2013. Biogas upgrading — technology overview, comparison and perspectives for the future. Biofuel. Bioprod. Biorefin. 7 (5), 499—511. https://doi.org/10.1002/bbb.1423.

Borde, X., Guieysse, B.t., Delgado, O., Muñoz, R., Hatti-Kaul, R., Nugier-Chauvin, C., Patin, H., Mattiasson, B., 2003. Synergistic relationships in algal—bacterial microcosms for the treatment of aromatic pollutants. Bioresour. Technol. 86 (3), 293—300. https://doi.org/10.1016/S0960-8524(02)00074-3.

Bose, A., Lin, R., Rajendran, K., O'Shea, R., Xia, A., Murphy, J.D., 2019. How to optimise photosynthetic biogas upgrading: a perspective on system design and microalgae selection. Biotechnol. Adv. 37 (8), 107444. https://doi.org/10.1016/j.biotechadv.2019.107444.

Carriero, G., Neri, L., Famulari, D., Di Lonardo, S., Piscitelli, D., Manco, A., Esposito, A., Chirico, A., Facini, O., Finardi, S., Tinarelli, G., Prandi, R., Zaldei, A., Vagnoli, C., Toscano, P., Magliulo, V., Ciccioli, P., Baraldi, R., 2018. Composition and emission of VOC from biogas produced by illegally managed waste landfills in Giugliano (Campania, Italy) and potential impact on the local population. Sci. Total Environ. 640—641, 377—386. https://doi.org/10.1016/j.scitotenv.2018.05.318.

Chidambarampadmavathy, K., Karthikeyan, O.P., Huerlimann, R., Maes, G.E., Heimann, K., 2017. Response of mixed methanotrophic consortia to different methane to oxygen ratios. Waste Manag. 61, 220—228. https://doi.org/10.1016/j.wasman.2016.11.007.

Chisti, Y., 2007. Biodiesel from microalgae. Biotechnol. Adv. 25 (3), 294—306. https://doi.org/10.1016/j.biotechadv.2007.02.001.

Choix, F.J., Polster, E., Corona-González, R.I., Snell-Castro, R., Méndez-Acosta, H.O., 2017. Nutrient composition of culture media induces different patterns of CO_2 fixation from biogas and biomass production by the microalga Scenedesmus obliquus U169. Bioproc. Biosyst. Eng. 40 (12), 1733—1742. https://doi.org/10.1007/s00449-017-1828-5.

Converti, A., Oliveira, R.P.S., Torres, B.R., Lodi, A., Zilli, M., 2009. Biogas production and valorization by means of a two-step biological process. Bioresour. Technol. 100 (23), 5771—5776. https://doi.org/10.1016/j.biortech.2009.05.072.

Cuéllar, A.D., Webber, M.E., 2008. Cow power: the energy and emissions benefits of converting manure to biogas. Environ. Res. Lett. 3 (3), 034002. https://doi.org/10.1088/1748-9326/3/3/034002.

Fortuny, M., Gamisans, X., Deshusses, M.A., Lafuente, J., Casas, C., Gabriel, D., 2011. Operational aspects of the desulfurization process of energy gases mimics in biotrickling filters. Water Res. 45 (17), 5665—5674. https://doi.org/10.1016/j.watres.2011.08.029.

Franco-Morgado, M., Alcántara, C., Noyola, A., Muñoz, R., González-Sánchez, A., 2017. A study of photosynthetic biogas upgrading based on a high rate algal pond under alkaline conditions: influence of the illumination regime. Sci. Total Environ. 592, 419—425. https://doi.org/10.1016/j.scitotenv.2017.03.077.

Franco-Morgado, M., Toledo-Cervantes, A., González-Sánchez, A., Lebrero, R., Muñoz, R., 2018. Integral (VOCs, CO_2, mercaptans and H_2S) photosynthetic biogas upgrading using innovative biogas and digestate supply strategies. Chem. Eng. J. 354, 363—369. https://doi.org/10.1016/j.cej.2018.08.026.

Gao, S., Hu, C., Sun, S., Xu, J., Zhao, Y., Zhang, H., 2018. Performance of piggery wastewater treatment and biogas upgrading by three microalgal cultivation technologies under different initial COD concentration. Energy 165, 360—369. https://doi.org/10.1016/j.energy.2018.09.190.

Gonzalez-Camejo, J., Serna-Garcia, R., Viruela, A., Paches, M., Duran, F., Robles, A., Ruano, M.V., Barat, R., Seco, A., 2017. Short and long-term experiments on the effect of sulphide on microalgae cultivation in tertiary sewage treatment. Bioresour. Technol. 244 (Pt 1), 15—22. https://doi.org/10.1016/j.biortech.2017.07.126.

González-Sánchez, A., Posten, C., 2017. Fate of H_2S during the cultivation of Chlorella sp. deployed for biogas upgrading. J. Environ. Manag. 191, 252—257. https://doi.org/10.1016/j.jenvman.2017.01.023.

Guo, P., Zhang, Y., Zhao, Y., 2018. Biocapture of CO_2 by different microalgal-based technologies for biogas upgrading and simultaneous biogas slurry purification under various light intensities and photoperiods. Int. J. Environ. Res. Publ. Health 15 (3), 528. http://europepmc.org/article/PMC/5877073.

Hu, J., Nagarajan, D., Zhang, Q., Chang, J.-S., Lee, D.-J., 2018. Heterotrophic cultivation of microalgae for pigment production: a review. Biotechnol. Adv. 36 (1), 54–67. https://doi.org/10.1016/j.biotechadv.2017.09.009.

Ivanova, J.G., Vasileva, I.A., Kabaivanova, L.V., 2020. Enhancement of algal biomass accumulation using undiluted anaerobic digestate. Int. J. Pharma Med. Bio. Sci. 9 (3).

Kao, C.-Y., Chiu, S.-Y., Huang, T.-T., Dai, L., Wang, G.-H., Tseng, C.-P., Chen, C.-H., Lin, C.-S., 2012. A mutant strain of microalga *Chlorella* sp. for the carbon dioxide capture from biogas. Biomass Bioenergy 36, 132–140. https://doi.org/10.1016/j.biombioe.2011.10.046.

Kim, S., Park, J.-e., Cho, Y.-B., Hwang, S.-J., 2013. Growth rate, organic carbon and nutrient removal rates of *Chlorella sorokiniana* in autotrophic, heterotrophic and mixotrophic conditions. Bioresour. Technol. 144, 8–13. https://doi.org/10.1016/j.biortech.2013.06.068.

Küster, E., Dorusch, F., Altenburger, R., 2005. Effects of hydrogen sulfide to *Vibrio fischeri, Scenedesmus vacuolatus*, and *Daphnia magna*. Environ. Toxicol. Chem. 24 (10), 2621–2629. https://doi.org/10.1897/04-546R.1.

Lebrero, R., Toledo-Cervantes, A., Muñoz, R., del Nery, V., Foresti, E., 2016. Biogas upgrading from vinasse digesters: a comparison between an anoxic biotrickling filter and an algal-bacterial photobioreactor. J. Chem. Technol. Biotechnol. 91 (9), 2488–2495. https://doi.org/10.1002/jctb.4843.

Lide, D.R., 1999. CRC Handbook of Chemistry and Physics, *eightieth ed.* CRC Press, Boca Raton FL.

Liu, L., Pohnert, G., Wei, D., 2016. Extracellular metabolites from industrial microalgae and their biotechnological potential. Mar. Drugs 14 (10), 191. https://doi.org/10.3390/md14100191.

Marín, D., Carmona-Martínez, A.A., Lebrero, R., Muñoz, R., 2020. Influence of the diffuser type and liquid-to-biogas ratio on biogas upgrading performance in an outdoor pilot scale high rate algal pond. Fuel 275, 117999. https://doi.org/10.1016/j.fuel.2020.117999.

Marín, D., Ortíz, A., Díez-Montero, R., Uggetti, E., García, J., Lebrero, R., Muñoz, R., 2019. Influence of liquid-to-biogas ratio and alkalinity on the biogas upgrading performance in a demo scale algal-bacterial photobioreactor. Bioresour. Technol. 280, 112–117. https://doi.org/10.1016/j.biortech.2019.02.029.

Markou, G., Depraetere, O., Muylaert, K., 2016. Effect of ammonia on the photosynthetic activity of *Arthrospira* and *Chlorella*: a study on chlorophyll fluorescence and electron transport. Algal Res. 16, 449–457. https://doi.org/10.1016/j.algal.2016.03.039.

Meier, L., Barros, P., Torres, A., Vilchez, C., Jeison, D., 2017. Photosynthetic biogas upgrading using microalgae: effect of light/dark photoperiod. Renew. Energy 106, 17–23. https://doi.org/10.1016/j.renene.2017.01.009.

Meier, L., Martínez, C., Vílchez, C., Bernard, O., Jeison, D., 2019. Evaluation of the feasibility of photosynthetic biogas upgrading: simulation of a large-scale system. Energy 189, 116313. https://doi.org/10.1016/j.energy.2019.116313.

Meier, L., Pérez, R., Azócar, L., Rivas, M., Jeison, D., 2015. Photosynthetic CO_2 uptake by microalgae: an attractive tool for biogas upgrading. Biomass Bioenergy 73, 102–109. https://doi.org/10.1016/j.biombioe.2014.10.032.

Meier, L., Stará, D., Bartacek, J., Jeison, D., 2018. Removal of H2S by a continuous microalgae-based photosynthetic biogas upgrading process. Process Saf. Environ. Protect. 119, 65–68. https://doi.org/10.1016/j.psep.2018.07.014.

Miyawaki, B., Mariano, A.B., Vargas, J.V.C., Balmant, W., Defrancheschi, A.C., Corrêa, D.O., Santos, B., Selesu, N.F.H., Ordonez, J.C., Kava, V.M., 2021. Microalgae derived biomass and bioenergy production enhancement through biogas purification and wastewater treatment. Renew. Energy 163, 1153–1165. https://doi.org/10.1016/j.renene.2020.09.045.

Montebello, A.M., Mora, M., López, L.R., Bezerra, T., Gamisans, X., Lafuente, J., Baeza, M., Gabriel, D., 2014. Aerobic desulfurization of biogas by acidic biotrickling filtration in a randomly packed reactor. J. Hazard Mater. 280, 200–208. https://doi.org/10.1016/j.jhazmat.2014.07.075.

Mukherjee, C., Chowdhury, R., Ray, K., 2015. Phosphorus recycling from an unexplored source by polyphosphate accumulating microalgae and cyanobacteria-a step to phosphorus security in agriculture. Front. Microbiol. 6. https://doi.org/10.3389/fmicb.2015.01421, 1421-1421.

Muñoz, R., Meier, L., Diaz, I., Jeison, D., 2015. A review on the state-of-the-art of physical/chemical and biological technologies for biogas upgrading. Rev. Environ. Sci. Biotechnol. 14 (4), 727–759. https://doi.org/10.1007/s11157-015-9379-1.

Nagarajan, D., Lee, D.-J., Chang, J.-S., 2019. Integration of anaerobic digestion and microalgal cultivation for digestate bioremediation and biogas upgrading. Bioresour. Technol. 290, 121804. https://doi.org/10.1016/j.biortech.2019.121804.

Ouyang, Y., Zhao, Y., Sun, S., Hu, C., Ping, L., 2015. Effect of light intensity on the capability of different microalgae species for simultaneous biogas upgrading and biogas slurry nutrient reduction. Int. Biodeterior. Biodegrad. 104, 157–163. https://doi.org/10.1016/j.ibiod.2015.05.027.

Park, J., Kumar, G., Bakonyi, P., Peter, J., Nemestóthy, N., Koter, S., Kujawski, W., Bélafi-Bakó, K., Pientka, Z., Muñoz, R., Kim, S.-H., 2020. Comparative evaluation of CO_2 fixation of microalgae strains at various CO_2 aeration conditions. Waste Biomass Valorization. https://doi.org/10.1007/s12649-020-01226-8.

Peishi, S., Xianwan, Y., Ruohua, H., Bing, H., Ping, Y., 2004. A new approach to kinetics of purifying waste gases containing volatile organic compounds (VOC) in low concentration by using the biological method. J. Clean. Prod. 12 (1), 95–100. https://doi.org/10.1016/S0959-6526(02)00195-6.

Petersson, A., Wellinger, A., 2009. Biogas Upgrading Technologies — Developments and Innovations, IEA Bioenergy - Technology Cooperation Programme. https://www.ieabioenergy.com/wp-content/uploads/2009/10/upgrading_rz_low_final.pdf.

Posadas, E., Marín, D., Blanco, S., Lebrero, R., Muñoz, R., 2017. Simultaneous biogas upgrading and centrate treatment in an outdoors pilot scale high rate algal pond. Bioresour. Technol. 232, 133–141. https://doi.org/10.1016/j.biortech.2017.01.071.

Posadas, E., Serejo, M.L., Blanco, S., Pérez, R., García-Encina, P.A., Muñoz, R., 2015. Minimization of biomethane oxygen concentration during biogas upgrading in algal—bacterial photobioreactors. Algal Res. 12, 221–229. https://doi.org/10.1016/j.algal.2015.09.002.

Powell, N., Shilton, A., Pratt, S., Chisti, Y., 2011. Luxury uptake of phosphorus by microalgae in full-scale waste stabilisation ponds. Water Sci. Technol. 63 (4), 704.

Prandini, J.M., da Silva, M.L.B., Mezzari, M.P., Pirolli, M., Michelon, W., Soares, H.M., 2016. Enhancement of nutrient removal from swine wastewater digestate coupled to biogas purification by microalgae Scenedesmus spp. Bioresour. Technol. 202, 67–75. https://doi.org/10.1016/j.biortech.2015.11.082.

Rahaman, M.S.A., Cheng, L.-H., Xu, X.-H., Zhang, L., Chen, H.-L., 2011. A review of carbon dioxide capture and utilization by membrane integrated microalgal cultivation processes. Renew. Sustain. Energy Rev. 15 (8), 4002–4012. https://doi.org/10.1016/j.rser.2011.07.031.

Ramanan, R., Kannan, K., Deshkar, A., Yadav, R., Chakrabarti, T., 2010. Enhanced algal CO2 sequestration through calcite deposition by Chlorella sp. and Spirulina platensis in a mini-raceway pond. Bioresour. Technol. 101 (8), 2616–2622. https://doi.org/10.1016/j.biortech.2009.10.061.

Ramírez-Rueda, A., Velasco, A., González-Sánchez, A., 2020. The effect of chemical sulfide oxidation on the oxygenic activity of an alkaliphilic microalgae consortium deployed for biogas upgrading. Sustainability 12 (16), 6610. https://www.mdpi.com/2071-1050/12/16/6610.

Ramos, I., Pérez, R., Fdz-Polanco, M., 2014. The headspace of microaerobic reactors: sulphide-oxidising population and the impact of cleaning on the efficiency of biogas desulphurisation. Bioresour. Technol. 158, 63–73. https://doi.org/10.1016/j.biortech.2014.02.001.

Ran, C., Zhou, X., Yao, C., Zhang, Y., Kang, W., Liu, X., Herbert, C., Xie, T., 2021. Swine digestate treatment by prior nitrogen-starved Chlorella vulgaris: the effect of over-compensation strategy on microalgal biomass production and nutrient removal. Sci. Total Environ. 768, 144462. https://doi.org/10.1016/j.scitotenv.2020.144462.

Raven, J.A., Giordano, M., Beardall, J., Maberly, S.C., 2012. Algal evolution in relation to atmospheric CO_2: carboxylases, carbon-concentrating mechanisms and carbon oxidation cycles. Phil. Trans. Biol. Sci. 367 (1588), 493–507. https://doi.org/10.1098/rstb.2011.0212.

Rodero, M.d.R., Carvajal, A., Arbib, Z., Lara, E., de Prada, C., Lebrero, R., Muñoz, R., 2020a. Performance evaluation of a control strategy for photosynthetic biogas upgrading in a semi-industrial scale photobioreactor. Bioresour. Technol. 307, 123207. https://doi.org/10.1016/j.biortech.2020.123207.

Rodero, M.d.R., Muñoz, R., Lebrero, R., Verfaillie, A., Blockx, J., Thielemans, W., Muylaert, K., Praveenkumar, R., 2020b. Harvesting microalgal-bacterial biomass from biogas upgrading process and evaluating the impact of flocculants on their growth during repeated recycling of the spent medium. Algal Res. 48, 101915. https://doi.org/10.1016/j.algal.2020.101915.

Rodero, M.d.R., Posadas, E., Toledo-Cervantes, A., Lebrero, R., Muñoz, R., 2018. Influence of alkalinity and temperature on photosynthetic biogas upgrading efficiency in high rate algal ponds. Algal Res. 33, 284–290. https://doi.org/10.1016/j.algal.2018.06.001.

Rodero, M.d.R., Severi, C.A., Rocher-Rivas, R., Quijano, G., Muñoz, R., 2020c. Long-term influence of high alkalinity on the performance of photosynthetic biogas upgrading. Fuel 281, 118804. https://doi.org/10.1016/j.fuel.2020.118804.

Ryckebosch, E., Drouillon, M., Vervaeren, H., 2011. Techniques for transformation of biogas to biomethane. Biomass Bioenergy 35 (5), 1633–1645. https://doi.org/10.1016/j.biombioe.2011.02.033.

SAE J1616, February 1994. Recommended Practice for Compressed Natural Gas Vehicle Fuel, Society of Automotive Engineers J1616_199402.

Sayre, R., 2010. Microalgae: the potential for carbon capture. Bioscience 60 (9), 722–727. https://doi.org/10.1525/bio.2010.60.9.9.

Serejo, M.L., Posadas, E., Boncz, M.A., Blanco, S., García-Encina, P., Muñoz, R., 2015. Influence of biogas flow rate on biomass composition during the optimization of biogas upgrading in microalgal-bacterial processes. Environ. Sci. Technol. 49 (5), 3228–3236. https://doi.org/10.1021/es5056116.

Sergeenko, T.V., Muradyan, E.A., Pronina, N.A., Klyachko-Gurvich, G.L., Tsoglin, L.N., 2000. The effect of extremely high CO_2 concentration on the growth and biochemical composition of microalgae. Russ. J. Plant Physiol. 47 (5), 632–638.

Speece, R.E., 2008. Anaerobic Biotechnology and Odor/corrosion Control for Municipalities and Industries. Arche Press, Nashville, TN.

Sun, Q., Li, H., Yan, J., Liu, L., Yu, Z., Yu, X., 2015. Selection of appropriate biogas upgrading technology-a review of biogas cleaning, upgrading and utilisation. Renew. Sustain. Energy Rev. 51, 521–532. https://doi.org/10.1016/j.rser.2015.06.029.

Sun, S., Ge, Z., Zhao, Y., Hu, C., Zhang, H., Ping, L., 2016. Performance of CO_2 concentrations on nutrient removal and biogas upgrading by integrating microalgal strains cultivation with activated sludge. Energy 97, 229–237. https://doi.org/10.1016/j.energy.2015.12.126.

Takabe, Y., Himeno, S., Okayasu, Y., Minamiyama, M., Komatsu, T., Nanjo, K., Yamasaki, Y., Uematsu, R., 2017. Feasibility of microalgae cultivation system using membrane-separated CO_2 derived from biogas in wastewater treatment plants. Biomass Bioenergy 106, 191–198. https://doi.org/10.1016/j.biombioe.2017.09.004.

Thomas, M.W., Zhang, X., Betsy, A.R., 2004. Expressed sequence tag profiles from calcifying and non-calcifying cultures of *Emiliania huxleyi*. Micropaleontology 50, 145–155. http://www.jstor.org/stable/4097109.

Thrän, D., Billig, E., Persson, T., Svensson, M., Daniel-Gromke, J., Ponitka, J., Michael, S., John, B., 2014. Biomethane Status and Factors Affecting Market Development and Trade. IEA Bioenergy.

Toledo-Cervantes, A., Estrada, J.M., Lebrero, R., Muñoz, R., 2017a. A comparative analysis of biogas upgrading technologies: photosynthetic vs physical/chemical processes. Algal Res. 25, 237–243. https://doi.org/10.1016/j.algal.2017.05.006.

Toledo-Cervantes, A., Madrid-Chirinos, C., Cantera, S., Lebrero, R., Muñoz, R., 2017b. Influence of the gas-liquid flow configuration in the absorption column on photosynthetic biogas upgrading in algal-bacterial photobioreactors. Bioresour. Technol. 225, 336–342. https://doi.org/10.1016/j.biortech.2016.11.087.

Toledo-Cervantes, A., Morales, T., González, Á., Muñoz, R., Lebrero, R., 2018. Long-term photosynthetic CO_2 removal from biogas and flue-gas: exploring the potential of closed photobioreactors for high-value biomass production. Sci. Total Environ. 640–641, 1272–1278. https://doi.org/10.1016/j.scitotenv.2018.05.270.

Toledo-Cervantes, A., Serejo, M.L., Blanco, S., Pérez, R., Lebrero, R., Muñoz, R., 2016. Photosynthetic biogas upgrading to bio-methane: boosting nutrient recovery via biomass productivity control. Algal Res. 17, 46–52. https://doi.org/10.1016/j.algal.2016.04.017.

Toro-Huertas, E.I., Franco-Morgado, M., de los Cobos Vasconcelos, D., González-Sánchez, A., 2019. Photorespiration in an outdoor alkaline open-photobioreactor used for biogas upgrading. Sci. Total Environ. 667, 613–621. https://doi.org/10.1016/j.scitotenv.2019.02.374.

Turon, V., Baroukh, C., Trably, E., Latrille, E., Fouilland, E., Steyer, J.P., 2015a. Use of fermentative metabolites for heterotrophic microalgae growth: yields and kinetics. Bioresour. Technol. 175, 342–349. https://doi.org/10.1016/j.biortech.2014.10.114.

Turon, V., Trably, E., Fouilland, E., Steyer, J.P., 2015b. Growth of *Chlorella sorokiniana* on a mixture of volatile fatty acids: the effects of light and temperature. Bioresour. Technol. 198, 852–860. https://doi.org/10.1016/j.biortech.2015.10.001.

Turon, V., Trably, E., Fouilland, E., Steyer, J.P., 2016. Potentialities of dark fermentation effluents as substrates for microalgae growth: a review. Process Biochem. 51 (11), 1843–1854. https://doi.org/10.1016/j.procbio.2016.03.018.

U.S. Department of Energy Combined Heat and Power Installation Database, 2020. https://doe.icfwebservices.com/downloads/chp.

U.S. EPA, July 2020. An Overview of Renewable Natural Gas from Biogas, U.S. Environmental Protection Agency Combined Heat and Power Partnership Program. https://www.epa.gov/sites/production/files/2020-07/documents/lmop_rng_document.pdf.

U.S. EPA, September 2017. Catalogue of CHP Technologies, U.S. Environmental Protection Agency Combined Heat and Power Partnership Program. https://www.epa.gov/sites/production/files/2015-07/documents/catalog_of_chp_technologies.pdf.

U.S. EPA, October 2011. Opportunities for Combined Heat and Power at Wastewater Treatment Facilities: Market Analysis and Lessons from the Field, U.S. Environmental Protection Agency Combined Heat and Power Partnership Program. https://www.epa.gov/sites/production/files/2015-07/documents/opportunities_for_combined_heat_and_power_at_wastewater_treatment_facilities_market_analysis_and_lessons_from_the_field.pdf.

Vazhappilly, R., Chen, F., 1998. Eicosapentaenoic acid and docosahexaenoic acid production potential of microalgae and their heterotrophic growth. J. Am. Oil Chem. Soc. 75 (3), 393–397. https://doi.org/10.1007/s11746-998-0057-0.

Vrba, J., Macholdová, M., Nedbalová, L., Nedoma, J., Šorf, M., 2018. An experimental insight into extracellular phosphatases - differential induction of cell-specific activity in green algae cultured under various phosphorus conditions. Front. Microbiol. 9. https://doi.org/10.3389/fmicb.2018.00271, 271-271.

Walter, J.K., Rickert, M., Aach, H.G., 1987. The role of glucose on the enzymes involved in the release of mature spores of *Chlorella fusca*. Physiol. Plantarum 71 (2), 219–223.

Wang, Q., Yu, Z., Wei, D., 2020a. High-yield production of biomass, protein and pigments by mixotrophic *Chlorella pyrenoidosa* through the bioconversion of high ammonium in wastewater. Bioresour. Technol. 313, 123499. https://doi.org/10.1016/j.biortech.2020.123499.

Wang, Q., Yu, Z., Wei, D., 2020b. High-yield Production of Biomass, Protein and Pigments by Mixotrophic Chlorella Pyrenoidosa through the Bioconversion of High Ammonium in Wastewater. Bioresource Technology.

Wang, X., Gao, S., Zhang, Y., Zhao, Y., Cao, W., 2017. Performance of different microalgae-based technologies in biogas slurry nutrient removal and biogas upgrading in response to various initial CO_2 concentration and mixed light-emitting diode light wavelength treatments. J. Clean. Prod. 166, 408–416. https://doi.org/10.1016/j.jclepro.2017.08.071.

Wang, Z., Zhao, Y., Ge, Z., Zhang, H., Sun, S., 2016. Selection of microalgae for simultaneous biogas upgrading and biogas slurry nutrient reduction under various photoperiods. J. Chem. Technol. Biotechnol. 91 (7), 1982–1989. https://doi.org/10.1002/jctb.4788.

Xie, T., Liu, J., Du, K., Liang, B., Zhang, Y., 2013. Enhanced biofuel production from high-concentration bioethanol wastewater by a newly isolated heterotrophic microalga, *Chlorella vulgaris* LAM-Q. J. Microbiol. Biotechnol. 23 (10), 1460–1471. https://doi.org/10.4014/jmb.1301.01046.

Xu, J., Wang, X., Sun, S., Zhao, Y., Hu, C., 2017. Effects of influent C/N ratios and treatment technologies on integral biogas upgrading and pollutants removal from synthetic domestic sewage. Sci. Rep. 7 (1), 10897. https://doi.org/10.1038/s41598-017-11207-y.

Xu, Z.-M., Wang, Z., Gao, Q., Wang, L.-L., Chen, L.-L., Li, Q.-G., Jiang, J.-J., Ye, H.-J., Wang, D.-S., Yang, P., 2019. Influence of irrigation with microalgae-treated biogas slurry on agronomic trait, nutritional quality, oxidation resistance, and nitrate and heavy metal residues in Chinese cabbage. J. Environ. Manag. 244, 453—461. https://doi.org/10.1016/j.jenvman.2019.04.058.

Yan, C., Muñoz, R., Zhu, L., Wang, Y., 2016a. The effects of various LED (light emitting diode) lighting strategies on simultaneous biogas upgrading and biogas slurry nutrient reduction by using of microalgae *Chlorella* sp. Energy 106, 554—561. https://doi.org/10.1016/j.energy.2016.03.033.

Yan, C., Zhang, L., Luo, X., Zheng, Z., 2014. Influence of influent methane concentration on biogas upgrading and biogas slurry purification under various LED (light-emitting diode) light wavelengths using *Chlorella* sp. Energy 69, 419—426. https://doi.org/10.1016/j.energy.2014.03.034.

Yan, C., Zheng, Z., 2013. Performance of photoperiod and light intensity on biogas upgrade and biogas effluent nutrient reduction by the microalgae *Chlorella* sp. Bioresour. Technol. 139, 292—299. https://doi.org/10.1016/j.biortech.2013.04.054.

Yan, C., Zhu, L., Wang, Y., 2016b. Photosynthetic CO2 uptake by microalgae for biogas upgrading and simultaneously biogas slurry decontamination by using of microalgae photobioreactor under various light wavelengths, light intensities, and photoperiods. Appl. Energy 178, 9—18. https://doi.org/10.1016/j.apenergy.2016.06.012.

Zelitch, I., 1971. Photosynthesis, Photorespiration and Plant Productivity. Academic Press.

Zhang, W., Zhao, C., Cao, W., Sun, S., Hu, C., Liu, J., Zhao, Y., 2020. Removal of pollutants from biogas slurry and CO_2 capture in biogas by microalgae-based technology: a systematic review. Environ. Sci. Pollut. Control Ser. 27 (23), 28749—28767. https://doi.org/10.1007/s11356-020-09282-2.

Zhang, Y., Bao, K., Wang, J., Zhao, Y., Hu, C., 2017. Performance of mixed LED light wavelengths on nutrient removal and biogas upgrading by different microalgal-based treatment technologies. Energy 130, 392—401. https://doi.org/10.1016/j.energy.2017.04.157.

Zhao, Y., Sun, S., Hu, C., Zhang, H., Xu, J., Ping, L., 2015. Performance of three microalgal strains in biogas slurry purification and biogas upgrade in response to various mixed light-emitting diode light wavelengths. Bioresour. Technol. 187, 338—345. https://doi.org/10.1016/j.biortech.2015.03.130.

Zhao, Y., Wang, J., Zhang, H., Yan, C., Zhang, Y., 2013. Effects of various LED light wavelengths and intensities on microalgae-based simultaneous biogas upgrading and digestate nutrient reduction process. Bioresour. Technol. 136, 461—468. https://doi.org/10.1016/j.biortech.2013.03.051.

Zhu, L., Yan, C., Li, Z., 2016. Microalgal cultivation with biogas slurry for biofuel production. Bioresour. Technol. 220, 629—636. https://doi.org/10.1016/j.biortech.2016.08.111.

CHAPTER 9

Large-scale demonstration of microalgae-based wastewater biorefineries

Zouhayr Arbib, David Marín, Raúl Cano, Carlos Saúco, Maikel Fernandez, Enrique Lara and Frank Rogalla
FCC Aqualia S.A. Innovation and Technology Department, Madrid, Spain

1. Introduction

Although microalgae have been proposed as a suitable platform to produce pharmaceuticals, cosmetics, nutraceuticals, fine chemicals, food and feed, presently industrial applications are limited to only high-value products related to aquaculture and human consumption (Borowitzka, 2013; Spolaore et al., 2006). Most current technologies permit only small amounts of biomass production at a cost that is excessive for use in low-value markets such as biofuel or wastewater treatment (WWT).

Algae biomass production is increasing worldwide and reached 32. 7 Mt [Fresh weight (FW)] in 2016, but in Europe 98% is originating from harvesting wild stocks of macroalgae Araújo et al. (2019) based on data from FAO). Nevertheless, while official statistics on microalgae production for the EU are almost nonexistent, from worldwide estimates of only 20 kt/y 10 years ago (Benemann, 2013), several data now show a European production above 180 t dry weight (DW)/y of microalgae, of which 80% *Chlorella* spp. and *H. pluvialis*, as well as close to 150 t DW/y of *Spirulina*.

To compete in large-scale markets such as energy and commodities, a significant production capacity increase and a major reduction in production cost below 0.50 €/kg (Chisti, 2007) is necessary. To achieve this objective, several biological and engineering issues need to be overcome (Richmond, 2000). One main problem is the lack of real data from commercial production, most of the reported data being based on pilot- or small-scale facilities. The second major problem is the lack of standardization, with each process and application using different technologies, strains, and locations, thus hindering performance comparisons to make robust conclusions possible.

One of the early applications of microalgae, WWT (Oswald et al., 1957), is still considered an attractive solution to assimilate nitrogen (N) and phosphorous (P), as algae produce oxygen through photosynthesis which is used by bacteria to remove organic matter. While heterotrophic bacteria use algae oxygen to degrade organic carbon and

Integrated Wastewater Management and Valorization using Algal Cultures
ISBN 978-0-323-85859-5, https://doi.org/10.1016/B978-0-323-85859-5.00007-5

© 2022 Elsevier Inc.
All rights reserved.

take up nutrients, CO_2 released by bacteria is used by microalgae, closing the cycle, and avoiding one of the main energy expenditures: gas diffusion.

Therefore, very low energy costs can be achieved through this symbiotic bioprocess, driven by algae photosynthesis using shallow ponds highly exposed to sunlight, called High Rate Algae Ponds (HRAPs). In most reported experiences, primary-treated wastewater (WW) after settling is used to feed the HRAP, while biomass is partially separated in secondary clarifiers (García et al., 2000; Park and Craggs, 2010).

With increasing interest in sustainable methods for microalgae production, other waste effluents and harvesting methods have been studied (Pittman et al., 2011; Udom et al., 2013), using WW as a free source of nutrients (N and P). Microalgae—bacteria WWT presents low operational and installation costs; however, its dependence on photosynthesis leads to seasonal variations (Arbib et al., 2017a,b).

The photosynthetic activity of microalgae determines the oxygenation potential and biomass production rates, and light availability is directly correlated to solar radiation. To adapt the performance of HRAPs, the dilution rate is normally regulated according to seasonal variations of the location. This strategy with shorter hydraulic retention times (HRTs) applied during summer, and longer during winter for light limiting conditions and lower temperatures, has been applied for seasonal variations in Europe, California, and New Zealand (Craggs et al., 2012; García et al., 2006).

In order to complete WWT and to generate a biomass which could be potentially processed into bioenergy or biofertilizer, an efficient and low-cost harvesting process must be developed. According to some reports, the resistance of microalgae to settle is due to the small cell size and strong negative surface charges, resulting in low biomass recoveries and effluents which do not achieve the discharge limits required (less than 35 mg/L Total Suspended Solids (TSS) in Europe for domestic WWT, 91/271/CEE (1998)).

Besides, when biomass is used as feedstock for production of biomethane, maximum biomass recovery is required to achieve a positive energy balance. Mature technologies of solids separation used in water and WWT offer a solution for microalgae biomass harvesting, such as Dissolved Air Flotation (DAF), which provides high TSS removal with low energy needs (Bare et al., 1975).

2. Wastewater versus seawater as culture medium

Algal biomass can be a valuable source of biostimulants, biopesticides, feed additives, etc. Current industrial algae production typically focuses on pure cultures and consumes freshwater resources and chemicals, such as fertilizers as a source of N and P, energy for sterilization and CO_2 supply, as well as generating a waste stream. Focusing on open mixed cultures of algae using seawater (abundant, easily available), rather than fresh water, has been proposed as a more sustainable alternative. However, seawater is poor in N and P (with <0.01 mg/L inorganic N or P), and to supply essential nutrients for algal growth, an external source of N and P must thus be added. To be sufficiently productive, the output must be valuable bioproducts and go beyond simple biofertilizers.

Furthermore, the natural symbiosis of bacteria and microalgae, occurring when using WW, eliminates the need for CO_2 supply (i.e., carbonation), which makes up a large part of the cost and energy use of algae production. Using WW (rich in N and P)—alone or mixed with seawater—can therefore be the basis for a zero-waste environmentally sustainable algal biorefinery, using free nutrients to generate high-value products, while providing environmental services of WWT in a way more energetically efficient than conventional technologies based on aeration with electric blowers.

3. Large-scale raceway ponds construction

Depending on the final location, the HRAP construction can vary significantly, since the topography as well as the nature of the terrain will plays a very important role in selecting the design, normally with two main options:
1. Low-cost (Fig. 9.1): Based on earth movement, digging the channels, compacting, and final lining. Earthen raceways with plastic liners cost little and are easy to build. To maintain the walls stable, the slopes need more space in comparison to conventional raceway's ponds with straight walls.

Figure 9.1 Low-cost raceway pond. *(Images owned by Aqualia).*

Figure 9.2 Conventional raceway pond. *(Images owned by Aqualia).*

2. Conventional (Fig. 9.2): Many raceways are made of more expensive reinforced concrete, which is durable and can be shaped in complex ways. Smaller size (<5 m^3) HRAPs can also be built from polyester resin to provide smooth walls, to be mobile and easy to service.

To construct large-scale raceways, these steps must be followed to obtain a better design:
1. Construction of ponds walls and bottom.
2. Construction of (optional) carbonation station and settling sumps.
3. Covering or lining of the ponds.
4. Mixing system.
5. Flow deflector.

3.1 Construction of pond walls and bottom

A variety of wall construction techniques are conceivable, ranging from earthen berms to slip form concrete, depending mainly on economic aspects. As shown in Fig. 9.3, conventional pond walls, both exterior and center walls which divide the two channels, are of identical construction and erected on the ground with hollow blocks filled with concrete, reinforced the top block with bar steel to give the wall adequate strength, then covering the pond walls and bottom with a plastic membrane.

In general, walls need to be about 0.4 m high (at 30 cm water level operation), except in areas where wind waves could require more freeboard. From a hydraulic standpoint, a single loop is preferred, using as basic geometric parameters pond size and length to width (L/W) ratio. The latter is defined as the length of the center divider wall divided by the single channel width. The L/W ratio must take into account the equilibrium between

Figure 9.3 Wall and bottom construction: Erection of the raceway's ponds wall with concrete blocks and reinforcement of the concrete block with bar steel. *(Images owned by Aqualia).*

minimizing wall length, with a low L/W ratio giving the most pond area for the least wall length, and the cost of those elements related to channel width, such as mixing and carbonation stations (Weissman and Goebel, 1987).

3.2 Construction of carbonation station and settling sumps

In some cases, sumps are necessary in order to provide a deepened area for CO_2 addition so that high absorption is achieved and to provide a collection point for draining the pond. A second sump may also provide an area of reduced velocity, where inert solids and settleable organic matter can accumulate to be removed, typically located on the reverse side of the channel of the carbonation sump, as shown in Fig. 9.4. The carbonation system usually consists of a pit with a vertical deflector, where the gas with CO_2 is bubbled from the bottom, against the direction of the flow. The deflector represents an abrupt obstacle for the horizontal flow in the HRAP reactor, causing an important head

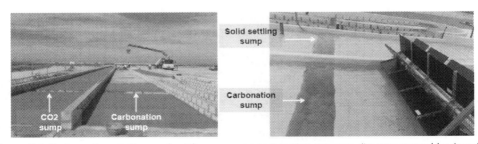

Figure 9.4 Carbonation station and settling sumps construction process. *(Images owned by Aqualia).*

loss. To reduce these losses significantly, recent improvements of carbonation include eliminating the partition wall and relocating the gas injection device (Patent EP 2712917 A1).

3.3 Covering or lining of the ponds

HRAP costs are particularly sensitive to the lining material used to prevent loss of water and nutrients, which must be fixed very carefully to the ground to avoid displacement by winds, as illustrated in Fig. 9.5. Bubble formation due to gas or water accumulation below the liner is another problem frequently encountered, where large trapped gas can result in anaerobic fermentation and gas build-up under the liner can produce seriously damage the lining.

3.4 Mixing system

Conventional raceways have two main problems: one is the high amount of power consumed during agitation, caused by hydraulic losses in the carbonation sumps and at the curves. The second problem is low productivity due to the settling of biomass. Both problems can be solved by the use of an innovative Low Energy Algae Reactor (LEAR). Intensive research in HRAP optimization has defined two different possibilities for improvement. First, by drastically reducing head losses, especially in the bends and sump, and second, by replacing the typical paddle wheel (low hydraulic efficiency) for recirculation with a submergible impeller installed in a venturi flow booster.

3.4.1 Conventional HRAP with paddlewheel

In conventional raceways, the cultivation broth is recirculated with a velocity between 0.2 and 0.3 cm/s by 8-blade paddlewheels, which sit in a depression (sump) on the pond bottom to reduce the back flow, as shown in Fig. 9.6. Also, eccentrically placed curved walls are assembled at the end furthest away from the paddle wheel. This creates curved zones of accelerating flow, followed by a flow expansion zone after the directional changes.

Figure 9.5 Process of lining. *(Images owned by Aqualia).*

Figure 9.6 Conventional raceway pond mixed by a paddlewheel. *(Images owned by Aqualia)*.

3.4.2 Low energy algae reactor

Microalgae cultures in open ponds have relatively low energy requirements, yet the classical raceway design has not been optimized for industrial production. Hydraulic inefficiencies related to curves and mixing result in energy dissipation, formation of backflows, and biomass settling into the reactor. A complete modification of the raceway was proposed, including the change of the mixing device and shape of curves, resulting in a new configuration named LEAR patented as "Open reactor for the cultivation of microalgae" (EP 2875724 B1). In this reactor, the conventional paddle wheel is replaced by an acceleration channel with a propeller, as shown in Fig. 9.7. The flat loops of the conventional raceway are modified into a narrower and deeper curve with a gradient of deep and width from the channel (at the beginning of the loop) to the apex point of the curve.

The new design was simulated by CFD and head loss was calculated resulting in a reduction of 25%. These CFD simulations allowed to compare two HRAPs with a surface of 500 m^2 each: one with the new design and another with the classical configuration (paddle wheel mixing and flat curves), as illustrated in Fig. 9.8. Measurements of the head loss and the Particle Tracking Velocity (PTV) were applied in both ponds to

Figure 9.7 Low energy algae reactor. *(LEAR-images owned by Aqualia)*.

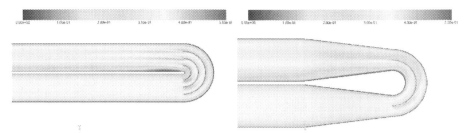

Figure 9.8 Simulations in CFD of the conventional raceway design (left) and LEAR reactor.

Table 9.1 Power consumption of HRAP with conventional paddlewheels and LEAR.

500 m² HRAP	Power (W)	Power (W/m²)
Conventional paddlewheel	434.13	0.8683
LEAR system	47.95	0.0959
Total energy savings	**89%**	**89%**

estimate the energy savings of the system, as summarized in Table 9.1 (Davila and Lara, 2018). The observed reduction of almost 90% of mixing energy allows to achieve a positive energy balance in microalgae cultures.

3.5 Flow deflectors

The curved flow deflectors accelerate flow and minimize eddy formation, reducing associated stagnant zones downstream of the bend that cause solids deposition. The deflectors can be built in the same manner as the straight walls, although it is better to select materials easily adaptable to the required curvature such as PVC or fiberglass, as shown in Fig. 9.9.

Figure 9.9 Construction of flow deflectors. *(Images owned by Aqualia).*

4. Microalgae consortium

Selection of organisms is carried out naturally in the raceways, as diverse species of microalgae and photosynthetic bacteria are present in most municipal waters, and the ponds simply provide favorable conditions (light and nutrients) for these to grow. Fig. 9.10 shows typical microalgae species developing in WW, but environmental temperature and solar radiation govern their growth and productivity, and therefore the nutrient removal capacity of the system (Acién et al., 2016).

One advantage of using WW as a substrate for algae cultivation is the natural selection of bacteria—algae consortia growing in symbiosis. The organic matter present in the WW provides a substrate for heterotrophic bacteria. The CO_2 resulting from heterotrophic growth is used by the algae, while the algae, in turn, provide O_2 for heterotrophic growth. This naturally occurring symbiosis eliminates the supply of CO_2 to the raceways, avoiding one of the biggest costs of HRAP construction and operation.

5. Harvesting process

Concerning harvesting, one of the major challenges in microalgae-based WWT is the recovery of the biomass from treated WW. To avoid light limitation, biomass concentration in raceways should not exceed 1 g TSS/L. To prevent this overgrowth of microalgae, an efficient and low cost harvesting process is required to separate the biomass from the culture broth.

Currently, at commercial plants for high-value products, harvesting is performed in a single-step by centrifugation or tangential microfiltration. Depending on species, cell density, and culture conditions, algae harvesting is estimated to contribute 20%—30% of production cost (Chini Zittelli et al., 2006; Gudin and Thepenier, 1986; Molina Grima et al., 2003), while the biomass is concentrated from 0.1 to 2.0 g/L in the culture

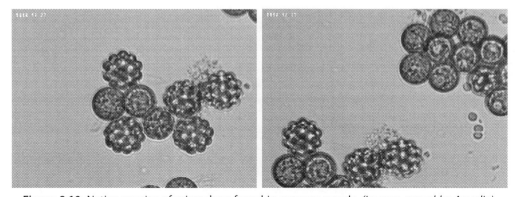

Figure 9.10 Native species of microalgae found in raceway ponds. *(Images owned by Aqualia).*

to a slurry up to 100 times denser (10–150 g/L). These energy-intensive methods to concentrate the microalgae biomass (between 2–8 kWh/m^3 Danquah et al. (2009)) are too expensive for WWT where large volumes of culture need to be processed, and final product value often does not reach 1 €/kg.

5.1 Optimal strategy to efficiently harvest microalgae biomass

Only a two-step harvesting and dewatering strategy is feasible when the goal is WWT. Sedimentation by gravity is the most common and cost-effective method of algal biomass removal in WWT (Nurdogan and Oswald, 1996). However, typical algal settlers have relatively long retention times (1–2 days) and remove only 50%–80% of the biomass (Park et al., 2011). These removal rates are not enough to meet the discharge limits for TSS (a maximum of 35 mg TSS/L, or 90% removal, according to Directive 91/271/CEE). According to the results of FP 7 Allgas project (www.all-gas.eu), DAF was demonstrated for microalgae harvesting with an energy consumption as low as 0.03 kWh/m^3, while concentrating the biomass 100 times. For the second step of biomass concentration, a conventional dewatering system can be applied, such as a decanter centrifuge, widely used in WWTPs.

5.2 Effluent conditioning for enhanced harvesting

Several methods are available to modify the properties of microalgae biomass and enhance the harvesting yield (Gerardo et al., 2015). Pretreatment can neutralize the microalgae negative charges (coagulation) and increase particle size (flocculation), before using separation methods such as flotation or sedimentation. Different combinations and doses of coagulants and flocculants must be tested to obtain a suitable floc for subsequent separation (by gravity settlers or dissolved air flotation DAF), and achieve high solids removal rates and efficient recovery of more than 90% of the biomass.

5.3 Preconcentration

In a first step of preconcentration, DAF is used widely in water and WWT for the separation of suspended solids and sludge thickening. This effective and robust concentration system is based on flotation by addition of a saturated air–water mixture to release micro bubbles into the flocculated water. As shown in Fig. 9.11, the micro bubbles adhere to or enmesh in the composite particles, which then rise to the surface due to their reduced density. Skimming removes the particles floated to the surface, which can reach concentrations of up to 50 g/L. When necessary to achieve higher solids, a thickening or dewatering stage (centrifugation, filtration, solar drying) is added.

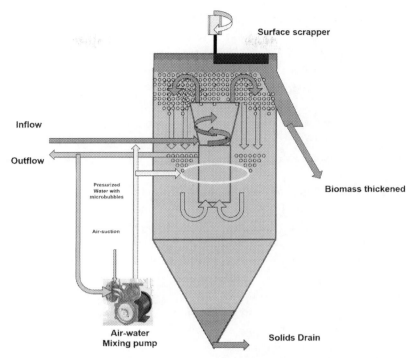

Figure 9.11 Main features of dissolved air flotation (DAFAST system).

5.4 Filtration or centrifugation

In a second step, the solids removed from the first step will be treated in screw press filter or in a decanter centrifuge, with the main objective of achieving a final concentration of at least 15 wt%. Fig. 9.12 shows a diagram of an optimal harvesting process.

6. Potential WWT using microalgae at large scale

WWT is a niche market where microalgae can contribute a commercial solution, as first proposed in the 1960s (Oswald and Golueke, 1968) but today being redesigned for resource recovery (Morales-Amaral et al., 2015; Muñoz et al., 2015; Olguín, 2012;

Figure 9.12 Two-step harvesting strategy.

Park et al., 2011). Microalgae production from WW can recover nutrients and produce up to 1 kg/m^3 of biomass at a much lower energy consumption and cost (Acién et al., 2016) than conventional WWT.

However, despite the benefits of microalgae processes versus conventional WWT, they have not yet been extended to commercial scale, other than large facultative ponds without specific algae harvesting and use. To recover algae biomass from WW, only a few pilot and demonstration plants are presently operating worldwide. One drawback of this technology is the larger surface area involved, meaning that an increase in the efficiency of the current processes is required to reduce both the costs and optimize the energy balance.

One of largest demonstration facilities of WWT with microalgae was developed in Chiclana (Spain), part of the FP7 ALL-GAS project funded by the EU. Sited on a conventional WWT plant capable of treating WW from 60,000 population equivalent (PE) of Chiclana, a low-cost solution consists of reducing pretreatment to grit and grease removal and screening. Furthermore, no additional nutrients/fertilizers nor CO_2 were supplied, as organic and inorganic carbon inherently present in the WW deliver the sole carbon source for the biomass production.

As shown in Fig. 9.13, step by step upscaling was achieved, until a demonstration plant of 2 ha was built with a treatment capacity around 2000 m^3/d. An average yearly biomass production capacity of more than 90 t/ha y was observed, close to the theoretical

Figure 9.13 Microalgae cultivation and harvesting at El Torno WWTP. *(Chiclana/Spain-images owned by Aqualia)*.

values for autotrophic growth (Arbib et al., 2016). The algal–bacterial biomass productivity varied between 53 g VS/m^2 d in summer and 16 g VS/m^2 d in winter. The HRAP developed in this project reduced the land required to about 2 m^2/PE by operating the process at an average HRT of 2 days in summer. This elevated solids productivity was achieved with "mixotrophic" cultures consisting of a symbiosis of microalgae and bacteria, using only screened WW without previous settling.

The biomass is capable of efficiently taking up nutrients from WW to produce clean water complying with national and European regulations. As algae biomass contains around 9% nitrogen (Taylor et al., 1988), a productivity of 25 g VS/m^2 d at an HRT of 3 d would allow to take up around 22 mg N/L by assimilation, although more nitrogen was missing in the balances. This additional nitrogen removal is probably due to nitrification/denitrification because of the varying oxygen content (aerobic during the day, anoxic at night) and stripping, as the pH also can reach high values during the day.

In the ALL-GAS case study, removal efficiencies of up to 80% total nitrogen and 90% of total phosphorus are achieved, at an energy consumption of less than 0.2 kWh/m^3. The objective of the EU demonstration was energy production, as the biomass was used to produce biogas in conventional anaerobic digesters. After adequate upgrading, a vehicle grade compressed biomethane production of more than 13 t/ha per year is obtained (Arbib et al., 2016), resulting in an Energy Return of Investment (EROI) of around 4 (Maga, 2017).

Another demo plant was operated in El Toyo WWT Plant in Almeria (Spain), which consists of a dual system with HRAP (Saúco et al., 2021) followed by constructed wetlands (CWs). The HRAP has an area of 3000 m^2 and is able to treat up to 250 m^3/d of WW (1500 PE). As shown in Fig. 9.14, the reactor was designed with a compact configuration to fit it in the available surface. An overflow drain ensured a constant water height of 0.3 m. Two alternative mixing systems were installed and compared: a conventional paddlewheel and a LEAR propeller.

Because of the concentrated WW in Almeria (see Table 9.2 below), the apparent algal–bacterial biomass productivity was much higher than in Chiclana and varied from a minimum of 23 g TSS/m^2 d in January to 137 g TSS/m^2 d in July, directly related to variations in temperature and solar radiation. The lowest values of temperature and solar radiation were recorded in the month of January, 13.1°C and 97 kWh/m^2, while the highest values were recorded in the month of July, 24.5°C and 242 kWh/m^2(Saúco et al., 2021).

From the HRAP, the cultivation broth was pumped to a DAF to separate algae biomass from treated WW. After adding coagulant and flocculant, the effluent is mixed with air microbubbles, to ascend the flocculated solids to the top of the DAF reactor. After separating the algae from the liquid phase with a scraper, the clarified water was stored in a tank for reuse. One option was to feed the effluent obtained after harvesting into four CWs, for polishing to further nitrify and reduce turbidity, TSS, organic matter, and

Figure 9.14 Panoramic view of demo plant at El Toyo WWTP. *(Almería - images owned by Aqualia).*

pathogens. Each CW occupies an area of 50 m^2, with a total depth of 78 cm filled with gravel (15 cm) and sand (63 cm, grain size of 0—4 mm), acting as the filter material.

6.1 Wastewater performance

The demo plant at El Toyo WWTP was studied by monitoring the removal of chemical oxygen demand (COD), TSS, total nitrogen (TN), total phosphorus (TP), turbidity, and pathogens in each treatment step, comparing conventional extended aeration and tertiary treatment (Sand filters + Chlorination + Ozonation) with the algae system (DAF outlet and CW). It is important to highlight that the HRAP treated a highly concentrated WW, with organic load and solids after pretreatment with screens reaching values higher than 1 g/L of COD and 0.5 g/L of TSS. These values were twice as high than conventional municipal WWT, typically with COD and TSS concentration values between 250 and 800 mg COD/L and 120—400 mg TSS/L (Metcalf, 2003). Nevertheless, COD and TSS removal achieved stable performance, with efficiencies over 90% throughout the year.

Table 9.2 summarizes treatment performance and shows high concentration values of both global TN and TP in the pretreated water, 88 ± 14 and 14 ± 3 mg/L, respectively. Despite these high influent values, HRAP could remove more than 50 mg TN/L, with the CW eliminating another 10 mg TN/L to reach a final effluent of 21 ± 12 mg TN/L

Large-scale demonstration of microalgae-based wastewater biorefineries   229

Table 9.2 Comparison of effluent quality from conventional WWT with HRAP + CW demo.

| Parameter (units) | Sample point | | | Removal efficiency (%) | | |
	Pretreated water	Effluent[a] or DAF	Tertiary treatment[a] or CW	Biological	Polishing	Total
Conventional plant[a]						
COD (mg/L)	1147 ± 623	36 ± 22	26 ± 13	97	29	98
TN (mg/L)	70 ± 14	7 ± 10	5 ± 10	90	25	92
TP (mg/L)	15 ± 4	5 ± 3	5 ± 2	68	5	70
TSS (mg/L)	568 ± 449	10 ± 16	2 ± 3	98	79	100
Turbidity (NTU)	656 ± 525	7 ± 15	2 ± 2	99	66	100
Coliforms (CFU/ 100 mL)	1.74E + 07 ± 8.85E + 06	b	7.70E + 04 ± 5.30E + 04	—	—	100
E. coli (CFU/ 100 mL)	3.60E + 06 ± 2.50E + 06	b	7.90E + 04 ± 7.10E + 03	—	—	98
HRAP demo plant						
COD (mg/L)	1142 ± 363	73 ± 29	39 ± 11	94	47	97
TN (mg/L)	88 ± 14	32 ± 16	21 ± 12	63	36	77
TP (mg/L)	14 ± 3	2 ± 2	1 ± 1	84	36	90
TSS (mg/L)	472 ± 243	24 ± 26	0 ± 0	95	99	100
Turbidity (NTU)	656 ± 525	15 ± 6	0 ± 0	98	97	100
Coliforms (CFU/ 100 mL)	1.74E + 07 ± 8.85E + 06	8.87E + 04 ± 9.38E + 04	2646 ± 3758	99	97	100
E. coli (CFU/ 100 mL)	3.60E + 07 ± 2.50E+06	2.61E + 04 ± 4.05E+04	54 ± 168	99	100	100

[a]Data from a conventional WWTP.
[b]Not determined.

and removals >75% TN. With the chemical addition in DAF, total phosphorus removal achieved >85% in HRAP, and final effluent quality of 1 mg TP/L was reached with the CW.

6.2 Energy consumption and process sustainability

In order to obtain a global assessment of energy efficiency and process sustainability, both technical and economic aspects have to be analyzed. Most research based results on small pilot plants or prototypes which, when scaling-up, provide a wrong estimate with

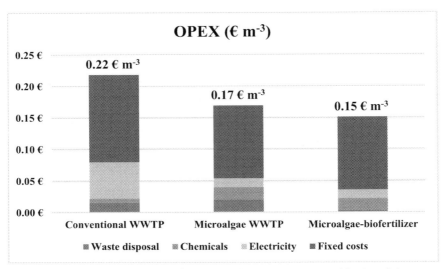

Figure 9.15 Comparison of operational cost. *(Figure owned by Aqualia).*

considerable deviations from industrial plants. In this study, however, the uncertainty is minimized thanks to the demonstration scale of both HRAP and DAF units.

Commonly, operational costs are divided in four parts: fixed cost, chemicals, electricity, and waste disposal, as shown in Fig. 9.15 for a comparison of conventional WWT with two algae options. Conventional WWT with primary and secondary treatment to remove organic matter and nutrients has a global operational cost of 0.26 €/m^3, where 0.07 €/m^3 are energy costs, 0.06 €/m^3 staff costs, and 0.13 €/m^3 other costs (Hernández-Sancho et al., 2015; Rodriguez-Garcia et al., 2011). For this study, the cost of conventional WWT such as activated sludge was determined at 0.22 €/m^3, slightly higher than the microalgae-based technology developed (0.17 €/m^3), with the microalgae-based process being 23% cheaper overall (Acién et al., 2017).

While the chemical requirements (flocculants) to recover large amounts of biomass and achieve phosphorus removal are almost three times higher in HRAP than conventional WWT, 0.020 €/m^3 compared to 0.007 €/m^3, that cost component is less than 10% of the overall expense. With regard to energy consumption, the microalgae-based process requires only a quarter of the energy of that of conventional WWT, 0.014 €/m^3 compared to 0.058 €/m^3 (Acién et al., 2017).

If the algae are used to produce biofertilizers, the cost of biomass disposal is avoided, while returns from the sale of this product are obtained. Without accounting for income, the overall WWT cost decreases to 0.15 €/m^3, about one third lower than conventional WWT, mainly thanks to minimizing electricity consumption for aeration by a blower and avoiding the management of excess sludge.

6.3 From commodities to higher value products

While the cost of conventional WWT is higher than for an HRAP, additional benefits can be generated from the use of the biomass compared to sludge. By converting solids to biofuel, an average algae productivity of 30 g VS/m^2 d can yield biomethane close to 0.2 kg CH$_4$/kg VS. If the ponds are fed at an HRT of 3 d, each m^3 treated can thus generate up to 0.6 kWh of thermal energy. The sale of this biofuel for vehicles can therefore partially compensate the WWT cost with algae, especially if a premium for green energy is factored such as the biomethane feed-in tariff in France, Italy, or UK, which between €60 MWh^{-1} and €120 MWh^{-1} multiplies the CNG bulk price three to five times.

But even higher value can be expected as the excess biomass contains up to 40% of proteins and 30% of lipids, in addition to carbohydrates, and can be used to produce biofertilizers by enzymatic hydrolysis of the thickened solids (Concentration ca. 10–20 wt%). Microalgae biomass has been demonstrated to be rich in essential amino acids, which are especially useful as growth stimulants in agriculture (Romero García et al., 2012), and can be extracted with a hydrolysis efficiency above 50% for *Scenedesmus almeriensis* grown in wastewater (Navarro-Lopez et al., 2020).

These amino acids can be used in agriculture, with typical commercial products containing 4–5 wt% of amino acids (://fliphtml5.com/pbac/yabi/basic or ://fitosanitariosmavesa.com/algafert-5L/), so that each m^3 of wastewater could generate up to 1.5 L of solution. Depending on their purity and characteristics, commercial solutions reach bulk market prices between 10 and 15 €/L, with a potential revenue above 10 €/m^3, not counting the cost of production and distribution.

Europe is currently the biggest market for biostimulants with close to 10 M ha of area treated (Liebig et al., 2020), and the new Fertilizing Products Regulation (FPR) (EU) 2019/1009 including biostimulants for the first time. The Global Biostimulant Market was estimated above USD 2.5 billion in 2019 with expected yearly growth above 10% (Ferreira et al., 2021). Other high value products such as biopesticides can be extracted by mild processes, with a market potential even larger than biostimulants and at higher growth to replace the usage of synthetic chemicals.

7. Conclusion

While the hype on biodiesel from algae is overblown, as there is no easy way to extract lipids from algae with a positive energy balance, HRAP can be optimized to provide sustainable WWT to tertiary standards while producing excess biomass. With an energy efficient harvesting and thickening using DAF, followed by anaerobic digestion and biogas upgrading, this algae biomass can be converted to vehicle grade biomethane.

232 Integrated Wastewater Management and Valorization using Algal Cultures

With an EROI of up to 4, and avoiding the electricity consumption of aeration, WWT with microalgae can become a productive activity for smaller communities in favorable climates, as the costs of the infrastructure and operation are rapidly amortized by the benefits generated. The financial returns can even be an order of magnitude higher than biofuel, if the biomass is used for hydrolysis to produce amino acids that have positive effects in agriculture as biostimulants.

References

Acién, F.G., Gómez-Serrano, C., Morales-Amaral, M.M., Fernández-Sevilla, J.M., Molina-Grima, E., 2016. Wastewater treatment using microalgae: how realistic a contribution might it be to significant urban wastewater treatment? Appl. Microbiol. Biotechnol. 100, 9013—9022. https://doi.org/10.1007/s00253-016-7835-7.

Acién, F.G., Molina, E., Fernández-Sevilla, J.M., Barbosa, M., Gouveia, L., Sepúlveda, C., Bazaes, J., Arbib, Z., 2017. Economics of microalgae production. Microalgae-Based Biofuels and Bioproducts From Feedstock Cultivation to End-Products, pp. 485—503. https://doi.org/10.1016/B978-0-08-101023-5.00020-0.

Araújo, R., Lusser, M., Sanchez Lopez, J., Avraamides, M., 2019. Brief on Algae Biomass Production. Luxemb. Publ. Off. Eur. Union.

Arbib, Z., Cano, R., Fernández, M., de Godos, I., 2016. All-gas Project: From Wastewater to Bioenergy. Proc. IWA LET.

Arbib, Z., de Godos Crespo, I., Corona, E.L., Rogalla, F., 2017a. Understanding the biological activity of high rate algae ponds through the calculation of oxygen balances. Appl. Microbiol. Biotechnol. 101, 5189—5198. https://doi.org/10.1007/s00253-017-8235-3.

Arbib, Z., de Godos, I., Ruiz, J., Perales, J.A., 2017b. Optimization of pilot high rate algal ponds for simultaneous nutrient removal and lipids production. Sci. Total Environ. 589, 66—72. https://doi.org/10.1016/j.scitotenv.2017.02.206.

Bare, W.F.R., Jones, N.B., Middlebrooks, E.J., 1975. Algae removal using dissolved air flotation. J. Water Pollut. Control Fed. 47, 153—169.

Benemann, J., 2013. Microalgae for biofuels and animal feeds. Energies 6, 5869—5886. https://doi.org/10.3390/en6115869.

Borowitzka, M.A., 2013. High-value products from microalgae-their development and commercialisation. J. Appl. Phycol. 25, 743—756. https://doi.org/10.1007/s10811-013-9983-9.

Chini Zittelli, G., Rodolfi, L., Biondi, N., Tredici, M.R., 2006. Productivity and photosynthetic efficiency of outdoor cultures of Tetraselmis suecica in annular columns. Aquaculture 261, 932—943. https://doi.org/10.1016/j.aquaculture.2006.08.011.

Chisti, Y., 2007. Biodiesel from microalgae. Biotechnol. Adv. 25, 294—306. https://doi.org/10.1016/j.biotechadv.2007.02.001.

Craggs, R., Sutherland, D., Campbell, H., 2012. Hectare-scale demonstration of high rate algal ponds for enhanced wastewater treatment and biofuel production. J. Appl. Phycol. 24, 329—337. https://doi.org/10.1007/s10811-012-9810-8.

Danquah, M.K., Ang, L., Uduman, N., Moheimani, N., Forde, G.M., 2009. Dewatering of microalgal culture for biodiesel production: exploring polymer flocculation and tangential flow filtration. J. Chem. Technol. Biotechnol. 84, 1078—1083. https://doi.org/10.1002/jctb.2137.

Davila, J., Lara, E., 2018. Large raceway design for waste water treatment BY algae cultivation. In: Algae Tech Conference (Sept. 2018, Munich).

Ferreira, A., Melkonyan, L., Carapinha, S., Ribeiro, B., Figueiredo, D., Avetisova, G., Gouveia, L., 2021. Biostimulant and biopesticide potential of microalgae growing in piggery wastewater. Environ. Adv. 4, 100062. https://doi.org/10.1016/j.envadv.2021.100062.

García, J., Mujeriego, R., Hernández-Mariné, M., 2000. High rate algal pond operating strategies for urban wastewater nitrogen removal. J. Appl. Phycol. 12, 331–339. https://doi.org/10.1023/A:1008146421368.

García, J., Green, B.F., Lundquist, T., Mujeriego, R., Hernández-Mariné, M., Oswald, W.J., 2006. Long term diurnal variations in contaminant removal in high rate ponds treating urban waste-water. Bioresour. Technol. 97, 1709–1715. https://doi.org/10.1016/j.biortech.2005.07.019.

Gerardo, M.L., Van Den Hende, S., Vervaeren, H., Coward, T., Skill, S.C., 2015. Harvesting of microalgae within a biorefinery approach: a review of the developments and case studies from pilot-plants. Algal Res. 11, 248–262. https://doi.org/10.1016/j.algal.2015.06.019.

Gudin, C., Thepenier, C., 1986. Bioconversion of solar energy into organic chemicals by microalgae. Adv. Biotechnol. Process. 6, 73–110.

Hernández-Sancho, F., Lamizana-Diallo, B., Mateo-Sagasta, M., Qadir, M., 2015. Economic Valuation of Wastewater: The Cost of Action and the Cost of No Action. United Nations Environment Programme.

Liebig, N., Salaun, M., Monnier, C., 2020. Development of Standards and Guidance Documents for Biostimulants Approval under European Fertilizer Regulation (EU) 2019/1009. Eurofins Agroscience Serv, pp. 2019–2021.

Maga, D., 2017. Life cycle assessment of biomethane produced from microalgae grown in municipal waste water. Biomass Convers. Biorefinery 7, 1–10. https://doi.org/10.1007/s13399-016-0208-8.

Metcalf, E., 2003. Wastewater Engineering and Reuse. Mc GrawHill.

Molina Grima, E., Belarbi, E.-H., Acién Fernández, F.G., Robles Medina, A., Chisti, Y., 2003. Recovery of microalgal biomass and metabolites: process options and economics. Biotechnol. Adv. 20, 491–515. https://doi.org/10.1016/s0734-9750(02)00050-2.

Morales-Amaral, M. del M., Gómez-Serrano, C., Acién, F.G., Fernández-Sevilla, J.M., Molina-Grima, E., 2015. Production of microalgae using centrate from anaerobic digestion as the nutrient source. Algal Res. 9, 297–305. https://doi.org/10.1016/j.algal.2015.03.018.

Muñoz, R., Meier, L., Diaz, I., Jeison, D., 2015. A review on the state-of-the-art of physical/chemical and biological technologies for biogas upgrading. Environ. Sci. Biotechnol. 14, 727–759. https://doi.org/10.1007/s11157-015-9379-1.

Navarro-López, E., Ruíz-Nieto, A., Ferreira, A., Acién, F.G., Gouveia, L., 2020. Biostimulant potential of Scenedesmus obliquus grown in Brewery wastewater. Molecules 25, 664. https://doi.org/10.3390/molecules25030664.

Nurdogan, Y., Oswald, W.J., 1996. Tube settling of high-rate pond algae. Water Sci. Technol. 33, 229–241. https://doi.org/10.1016/0273-1223(96)00358-7.

Olguín, E.J., 2012. Dual purpose microalgae—bacteria-based systems that treat wastewater and produce biodiesel and chemical products within a Biorefinery. Biotechnol. Adv. 30, 1031–1046. https://doi.org/10.1016/j.biotechadv.2012.05.001.

Oswald, W.J., Golueke, C.G., 1968. Large scale production of microalgae. MIT Press Cambridge, pp. 271–305.

Oswald, W.J., Gotaas, H.B., Golueke, C.G., Kellen, W.R., Gloyna, E.F., Hermann, E.R., 1957. Algae in waste treatment [with discussion]. Sewage Ind. Waste. 29, 437–457.

Park, J.B.K., Craggs, R.J., 2010. Wastewater treatment and algal production in high rate algal ponds with carbon dioxide addition. Water Sci. Technol. 633–639. https://doi.org/10.2166/wst.2010.951.

Park, J.B.K., Craggs, R.J., Shilton, A.N., 2011. Bioresource Technology Wastewater treatment high rate algal ponds for biofuel production. Bioresour. Technol. 102, 35–42. https://doi.org/10.1016/j.biortech.2010.06.158.

Pittman, J.K., Dean, A.P., Osundeko, O., 2011. The potential of sustainable algal biofuel production using wastewater resources. Bioresour. Technol. 102, 17–25. https://doi.org/10.1016/j.biortech.2010.06.035.

Richmond, A., 2000. Microalgal biotechnology at the turn of the millennium: a personal view. J. Appl. Phycol. 12, 441–451. https://doi.org/10.1023/a:1008123131307.

Rodriguez-Garcia, G., Molinos-Senante, M., Hospido, A., Hernández-Sancho, F., Moreira, M.T., Feijoo, G., 2011. Environmental and economic profile of six typologies of wastewater treatment plants. Water Res. 45, 5997–6010. https://doi.org/10.1016/j.watres.2011.08.053.

Romero García, J.M., Acién Fernández, F.G., Fernández Sevilla, J.M., 2012. Development of a process for the production of l-amino-acids concentrates from microalgae by enzymatic hydrolysis. Bioresour. Technol. 112, 164–170. https://doi.org/10.1016/j.biortech.2012.02.094.

Saúco, C., Cano, R., Marín, D., Lara, E., Rogalla, F., Arbib, Z., 2021. Hybrid wastewater treatment system based in a combination of high rate algae pond and vertical constructed wetland system at large scale. J. Water Process Eng. 43.

Spolaore, P., Joannis-Cassan, C., Duran, E., Isambert, A., 2006. Commercial applications of microalgae. J. Biosci. Bioeng. 101, 87–96. https://doi.org/10.1263/jbb.101.87.

Taylor, R.W., Sistani, K.R., Floyd, M., 1988. Algal biomass production, N uptake and N2 fixation in synthetic medium. Biomass 15 (4), 249–257. https://doi.org/10.1016/0144-4565(88)90060-1.

Udom, I., Zaribaf, B.H., Halfhide, T., Gillie, B., Dalrymple, O., Zhang, Q., Ergas, S.J., 2013. Harvesting microalgae grown on wastewater. Bioresour. Technol. 139, 101–106. https://doi.org/10.1016/j.biortech.2013.04.002.

Weissman, J.C., Goebel, R.P., 1987. Microbial products. U.S. Department of Energy Contract No. DE-AC02-83CH10093. In: Design and Analysis of Microalgal Open pond Systems for the Purpose of Producing Fuels.

CHAPTER 10

Cultivation of microalgae on agricultural wastewater for recycling energy, water, and fertilizer nutrients

Lijun Wang and Bo Zhang
Department of Natural Resources and Environmental Design, North Carolina Agricultural and Technical State University, Greensboro, NC, United States

1. Introduction

Agriculture is the largest user of freshwater worldwide, representing about 70% of total water use. Climate change reduces the available water resources and increases the water demand for agricultural production (Knox et al., 2012). It has become essential to secure adequate water and increase the sustainability of water use for agricultural production. Agricultural production generates large amounts of residues, solid wastes, and wastewater. Agricultural wastewater, particularly animal wastewater from farms, is a major contributor to the eutrophication of surface water due to the release of nitrogen and phosphate into the water (Cronk, 1996).

The cultivation of microalgae on wastewater can produce algal biomass as a feedstock for the production of biofuels, fertilizers, animal and fish feed, and other bio-based chemicals, while remediating the wastewater. The cultivation of microalgae on wastewater provides a promising approach to reduce the cost and increase the sustainability of microalgal production. A life cycle analysis study indicated that algae production may have negative environmental effects if commercial fertilizers are used (Clarens et al., 2010). Wastewater such as swine wastewater and anaerobic digestion (AD) effluent contains high levels of nutrients such as N and P, which are essential for algal growth. Microalgae can metabolize nutrients of N and P in wastewater even at very low levels. Research showed that a locally collected algae *polyculture* grown on municipal wastewater removed over 99% of both the ammonium and phosphate (Woertz et al., 2009). This chapter reviewed the feasibility, challenges, and enhancement strategies to grow microalgae on various agricultural wastewater for recycling energy, water, and fertilizer nutrients.

2. Microalgae for agricultural wastewater treatment
2.1 Recovery of resources from agricultural wastewater using microalgae

Agricultural solid wastes and wastewater have caused environmental pollution, public health issues, and loss of valuable resources. Sustainable and intensive agricultural

Integrated Wastewater Management and Valorization using Algal Cultures
ISBN 978-0-323-85859-5, https://doi.org/10.1016/B978-0-323-85859-5.00006-3

© 2022 Elsevier Inc.
All rights reserved.

production demands sustainable management of agricultural residues and wastes. The circular economy in which wastes are preferentially avoided, reduced, reused and valorized, or alternatively fully recycled has recently promoted to address the global problems of resources depletion and climate change by policymakers around the world. Innovative conversion technologies for the recycling and valorization of agricultural wastes are crucial in the circular economy and the transition to sustainable agriculture (Donner et al., 2021).

Wastewater is generally treated to remove nutrients in order to limit the impact of its discharge on the receiving environment. Several physical and chemical water treatment technologies such as adsorption, oxidation, and high-pressure membrane filtration have been proved to be effective for the removal of N, P, and K fertilizers and organic micropollutants (OMPs) in wastewater, but they are expensive. By contrast, biofiltration systems are robust and simple to construct and have low energy requirements. Natural systems such as riverbank filtration can provide significant removal of OMPs (Reungoat et al., 2011). However, existing biological treatment processes, such as activated sludge systems, are not very effective in removing nutrients such as N and P from the wastewater (Mantzavinos and Kalogerakis, 2005). Constructed wetlands have been studied to treat animal wastewater from dairy and swine operations. These wetlands have been found to be effective in sequestering N but less effective reducing P (Stone et al., 2004). Sequestration efficiencies of constructed wetlands decrease with lower air temperatures and higher volumes of rain (Poach et al., 2004).

The cultivation of microalgae on wastewater is an economic and effective approach to produce algal biomass and meanwhile remediate the wastewater. Microalgae are 10−50 times more efficient in capturing solar energy than terrestrial plants. As microalgae grow in an aquatic environment, they do not compete with terrestrial food crops. Wastewater such as swine wastewater and AD effluent contains high levels of nutrients such as N and P, which are essential for algal growth. Microalgae can metabolize nutrients of N and P in wastewater at very low levels (Amini et al., 2016b).

2.2 Microalgae grown on swine wastewater

Each hog produces 4−8 liters of wastewater each day (García et al., 2017). Swine wastewater typically contains 2000−30,000 mg/L BOD, 200−2055 mg/L total nitrogen, 110−1650 mg/L NH_4−N, and 100−620 mg/L total phosphorus (Cheng et al., 2019). The selection of robust microalgal strains that can consistently grow in swine wastewater for varied environmental conditions is important for better production of biomass and wastewater treatment (Salama et al., 2017). Several microalgae species including *Chlorella* sp., *Scenedesmus* sp., *Neochloris* sp., and *Coelastrella* sp. have been studied for the treatment of the swine wastewater under various cultivation conditions (Cheng et al., 2019).

Hasan et al. (2014) compared the growth of *Scenedesmus dimorphus*, *Neochloris oleoabundans*, *Chlorella vulgaris*, and *Chlamydomonas reinhardtii* on swine wastewater (Hasan et al., 2014). It was discovered that *C. vulgaris* and *C. reinhardtii* could grow on the swine wastewater, but *S. dimorphus* and *N. oleoabundans* were unable to grow on the swine wastewater. The removal efficiency of nutrients from the wastewater was a function of the microalgal growth. The optimal culture conditions for *C. vulgaris* and *C. reinhardtii* were 600 $\mu mol/m^2 s$ and 25°C, and 300 $\mu mol/m^2 s$ and 20°C, respectively. The final biomass yield and lipid content were 1.25 g/L and 15.2% of the total cell dry weight for *C. vulgaris,* and 0.73 g/L and 21.7% for *C. reinhardtii*, respectively. Amini et al. (2016a,b) found that the productivity of *C. vulgaris* in 1 liter of swine wastewater with 102 mg N and 76 mg P was 0.160 g per day, compared to 0.191 g per day in 1 liter of Bold's medium with 100 mg N and 53 mg P (Amini et al., 2016b). Wang et al. (2017) investigated the growth of a carbohydrate-rich microalga *Neochloris aquatica* CL-M1 in swine wastewater. Their results indicated that the highest COD and NH_3-N removal efficiencies were 81.7% and 96.2% that was achieved at 150 $\mu mol/m^2 s$ light intensity, 25°C and N/P ratio of 1.5/1. The highest biomass concentration and carbohydrate content were 6.10 g/L and 50.46%, respectively, that were obtained at N/P ratio of 5/1 (Wang et al., 2017).

Several studies have been done on the selection of native microalgal strains for the treatment of swine wastewater. Chiang et al. (2018) isolated a novel genus of microalgae from swine wastewater, which was designated as *Chlorellaceae* sp. P5. This strain can grow at a large range of temperatures from 20 to 40°C in a pH range of 4—7 with the salinity as high as 3% NaCl. Lab-scale cultivation yielded 1.66 g/L d of biomass under 15.19 g/L glucose and 8.26 mM ammonium bicarbonate. The lipid from *Chlorellaceae* sp. P5 that had a low degree of unsaturated fatty acids is a desirable feedstock for biodiesel production. The cultivation of this strain on swine wastewater could remove ammonia and phosphate over 97.3 mg/L of NH_3-N and 20.3 mg/L of PO_4^{3-} in the sterilized and non-sterilized swine wastewater, respectively (Chiang et al., 2018). Zhang et al. (2014) discovered a native green microalgal species of *Chlamydomonas debaryana* AT24 from a local lagoon. The biomass yields of *C. debaryana AT24* cultivated on swine wastewater under various conditions were between 0.6 and 1.62 g/L with a median value of 1.11 g/L. The microalgal growth was intrinsically enhanced by increasing mass transfer rates and providing sufficient light intensity of 300 $\mu mol/m^2 s$. The growth of *C. debaryana* reduced most nutritional contents of the wastewater except iron. Each gram of *C. debaryana* biomass was able to utilize $1.3-1.6 \times 10^3$ mg COD (chemical oxygen demand), 55—90 ppm ammonia, and 48—89 ppm phosphorous. The lipid content of *C. debaryana* was $19.9 \pm 4.3\%$ of cell dry weight. The transesterified microalgal oil mostly consisted of 14 kinds of fatty acids, ranging from C5 to C22, which can be refined into renewable jet fuel or used as sources of omega-3 and omega-6 fatty acids (Zhang et al., 2014). Zhang et al. (2016) further analyzed the effects of the growth conditions including temperatures of 8–25°C, light intensity of 150–900 μmol/m²s, and light duration of 6–24 h on the biomass yields of *C. debaryana* in swine wastewater. The results showed that the factors of temperature, light duration, the interaction of light intensity– light duration, and the quadratic effect of temperature were statistically significant in terms of algal biomass yield. When evaluating different scenarios for the sustainable production of algal biomass and biofuels in North Carolina, U.S.A., the predictions using the regression quadratic model showed that (a) Growing *C. debaryana* in a 10-acre pond on swine wastewater under local weather conditions would yield algal biomass of 113 tons each year; (b) If all swine wastewater generated in North Carolina was treated with algae, it will require 137–485 acres of ponds, yielding biomass of 5048–10,468 tons each year and algal oil (total lipid) of 1010–2094 tons each year. Annually, hundreds of tons of nitrogen and phosphorus could be removed from swine wastewater. The required land area is mainly dependent on the growth rate of algal species (Zhang et al., 2016).

2.3 Microalgae grown on aquaculture wastewater

Aquaculture wastewater contains large amounts of N and P. It was reported that 78 kg N and 9.5 kg P could be released in the aquaculture wastewater for each ton of fish produced using the fish meal with 7.2% N and 0.9% P (Ackefors and Enell, 1994). There is a large variation in the concentrations of N and P fertilizer nutrients in the wastewater of various aquaculture species. The NH_4–N, NO_3–N, and PO_4–P concentrations in the wastewater for various aquaculture species including Nile Tilapia, American eel, Catfish, giant gourami, shrimp, and polyculture were found to be in the ranges of 0–443 mg/L, 1.17–152.8 mg/L, and 2.5–16.1 mg/L, respectively (Nie et al., 2020).

There are synergic benefits between the cultivation of microalgae on aquaculture wastewater and fish farming. Microalgae can be cultivated in the aquaculture effluent or directly in a fishpond to remove the excess nutrients in the water. The algae can be used as fish feed. The algal productivity and the removal efficiency nutrients are affected by the microalgal species, cultivating conditions, the concentration of nutrients, and pretreatment of wastewater (Nie et al., 2020; Tejido-Nuñez et al., 2019). Studies showed that both *C. vulgaris* and *Tetradesmus obliquus* were able to grow in aquaculture wastewater and effectively remove nutrients in it. However, *C. vulgaris* showed a higher growth rate and nutrient removal efficiency than *T. obliquus* in sterile wastewater, while *T. obliquus* had better performance than *C. vulgaris* in nonsterile wastewater because of its higher tolerance to grazing protozoa (Tejido-Nuñez et al., 2019). The further study showed that cocultivation of *C. vulgaris* and *T. obliquus* could achieve better algal biomass productivity and nutrient removal efficiency and was more reliable and robust for a long cultivation period than monocultures. The productivity of coculture in aquaculture wastewater was 13.3 g/m² d and final dry algal biomass concentration after 29 days was 11.1 g/L (Tejido-Nuñez et al., 2020).

Although microalgae provide economic and ecological benefits to the aquaculture industry, their synergic benefits face several technical problems including the bacterial bloom with algae culture, the alkaline environment caused by algal growth, the nutrient deficiency in aquaculture effluent, and the seasonal fluctuation in algal productivity and nutrients removal efficiency (Lu et al., 2020).

2.4 Microalgae grown on the effluent of anaerobic digestion

The effluent of AD is considered as a high strength wastewater, which contains significant amounts of ammonia and phosphorus. However, due to the opaqueness and highly variable chemical composition of the AD effluent, it is an unfavorable nutrient supplement for algae cultivation. Studies have been conducted to discover new algal strains that could grow in the AD effluent. A new unicellular green microalgae species of *Desmodesmus* sp. EJ9-6 could remove 100% NH_4-N (68.691 mg/L), TP (4.565 mg/L) and PO_4-P (4.053 mg/L), and 75.50% TN (84.236 mg/L) in the medium with 10% AD effluent. The highest algal biomass production was 0.412 g/L after 14 d cultivation (Ji et al., 2014b). Morales-Amaral et al. (2015) studied the biomass production of freshwater microalgae of *Muriellopsis* sp. and *Pseudokirchneriella subcapitata* using the AD centrate as an only nutrient source. Their results demonstrated that with 40%–50% of centrate in the culture medium, the biomass productivity reached 1.13 and 1.02 g/L·day for *Muriellopsis* sp. and *P. subcapitata*, respectively. In addition, the removal efficiencies of nitrogen and phosphorous contained in the culture medium exceeded 90%, with the COD at the outlet being lower than 100 mg/L. It was also found that the microalgae productivity decreased using more than 50% centrate due to toxicity. *Muriellopsis* sp. showed higher tolerance to high ammonium concentrations (Morales-Amaral et al., 2015).

Bohutskyi et al. (2015) evaluated 18 mixed microalgae cultures for their ability to grow and remove nutrients from unsterilized primary or secondary wastewater effluents as well as wastewater supplemented with the AD effluent. Most of the tested species were unable to grow efficiently, which may be due to the lack of certain genetic traits important for robust growth in the unsterilized wastewater (Bohutskyi et al., 2015). Singh et al. (2011) evaluated the productivities and nutrient removal efficiencies of mixotrophic algae (*Chlorella minutissima*, *Chlorella sorokiniana*, and *Scenedesmus bijuga*) in the effluent of poultry litter AD. The maximum biomass productivity was found to be 76 mg/L d in the medium with 6% (v/v) of the effluent. The protein and carbohydrates of the algal biomass were 39% w/w and 22%, while the lipid content was lower than 10%, making it suitable to be used as an animal feed supplement (Singh et al., 2011). Nwoba et al. (2016) investigated the growth and ammonium nitrogen removal rate of semicontinuous mixed microalgae culture in paddle wheel-driven raceway pond and helical tubular closed photobioreactor for treating sand-filtered, undiluted AD piggery effluent under outdoor climatic conditions. Similar average ammonium nitrogen removal rates were observed for both bioreactors. The average volumetric biomass productivity of microalgae grown in the helical tubular bioreactor was 25.03 ± 0.24 mg/L day which was 2.1 times higher than that of the raceway pond. The consortium could be maintained in semicontinuous culture for more than 3 months without changes in the algal composition, and *Chlorella* sp. was the most dominant species. In summary, the selected mixotrophic microalgae could show sustainable growth against variations in the AD effluent composition, thus proving to be a suitable candidate for large scale wastewater treatment and production of renewable biomass (Nwoba et al., 2016).

3. Microalgae for feeds, fuels, and fertilizers

3.1 Microalgae biofuels

Microalgae are viewed as the source of third-generation biofuels (John et al., 2011). The major advantages offered by microalgae are as follows: (1) no competition with conventional agricultural plants for land, and utilization of different water sources (seawater, brackish water, and wastewater), (2) high productivity, (3) nondependence on agricultural input (fertilizer, pesticides, etc.), and (4) high lipid and carbohydrates, and low lignin content for the production of biofuels of biodiesel and bioethanol (Chen et al., 2015).

3.1.1 Biodiesel

Biodiesel is mono alkyl esters of long-chain fatty acids that can be produced from the triglycerides in vegetable oils and animal fats (Abdo et al., 2016; Chen et al., 2012; Issariyakul and Dalai, 2014). Microalgae have been considered as one of the most promising feedstocks for the production of biodiesel due to their high growth rate and high lipid content (Maity et al., 2014; Nigam and Singh, 2011; Rahman et al., 2015). Different methods have been explored to produce biodiesel from microalgae. The popular approach involves the extraction of algal oil from dried algae followed by the transesterification of the algal oil to biodiesel using an alcohol in the presence of a catalyst (Chisti, 2007; Farobie and Matsumura, 2015). However, the drying of wet microalgae is an energy-intensive process (Macías-Sánchez et al., 2007; Santana et al., 2012). Alternatively, direct extraction of lipids from wet algae through catalytic hydrolysis (Sathish and Sims, 2012; Takisawa et al., 2013a) and noncatalytic lipid hydrolysis and esterification under supercritical or subcritical conditions (Levine et al., 2010; Takisawa et al., 2013a,b) were explored to avoid drying wet algae. Organic solvents can assist the thermal treatment of wet microalgae for increased quality and yields of biodiesel and bio–oil through the extraction of algal lipids, in situ transesterification of lipids, and liquefaction (Zhang et al., 2017). Methanol that is readily available at a low price has been widely used in supercritical transesterification of algal lipids (Demirbas, 2009; Saka and Kusdiana, 2001). However, the toxic properties of methanol and its production from petroleum-based resources restrict the development of byproducts from residual biomass as livestock feed for cattle and aqua culture (Caporgno et al., 2016; Reddy et al., 2014). The use of ethanol that can be produced exclusively from carbohydrate rich renewable sources can make the process more sustainable and renewable (Joshi et al., 2010; Knothe, 2005; Reddy et al., 2014).

3.1.2 Bioethanol

Bioethanol is a clean, safe, and bio-based energy, which is commonly regarded as one of the primary candidates to replace gasoline (Morales et al., 2015). As some microalgal species have higher carbohydrate and low lignin content, they are excellent substrates for bioethanol production (Harun et al., 2010; John et al., 2011; Suali and Sarbatly, 2012). The general process for ethanol production from microalgal biomass involves strain selection and cultivation, pretreatment with chemicals and/or enzymes, and ethanol fermentation process. Some early studies about the production of ethanol from microalgae have been summarized in Chen et al. (2015). Recently, Sanchez Rizza et al. (2017) performed a preliminary analysis of 17 microalgal strains and selected a strain of SP2-3 that could accumulate carbohydrates up to 70 wt% of its biomass and another strain of *Desmodesmus* sp. FG for their high fermentable sugars productivity. By the optimization of microalgae culture conditions, acid hydrolysis, and fermentation with *Saccharomyces cerevisiae*, the ethanol yields were up to 0.24 g/g dry biomass and an ethanol concentration in the fermentation broth was 24 g/L which was 87.4% of the maximum theoretical value (Sanchez Rizza et al., 2017). Another study showed that a consortium of *S. cerevisiae* and *Pichia stipitis* was able to efficiently ferment a mixture of hexoses and pentoses produced from the hydrolysis of microalgae. Salinity significantly decreased the ethanol yield and increased the fermentation time. The studies also showed that the simultaneous saccharification and fermentation (SSF) was identified as a superior process for bioethanol production from microalgae (Kim et al., 2017).

Coproduction of biodiesel from algal lipid and bioethanol from algal carbohydrates has been considered as an economic method for the production of biofuels from microalgae (Jones and Mayfield, 2012; Wang et al., 2014). It was demonstrated that the fermentation of the carbohydrates in wet *Chlorella* sp. and *Nannochloropsis* sp. into ethanol followed by ethanol-assisted liquefaction could increase biodiesel yield (Rahman et al., 2019a,b). The process simulation and preliminary economic assessment also demonstrated the techno-economic feasibility of the proposed combined fermentation and ethanol-assisted liquefaction process at the industrial scale (Rahman et al., 2019a,b).

3.1.3 Methane

Methane (CH_4) is the primary component of natural gas and biogas. Biogas is typically produced through AD of organic compounds. As the water content of wet microalgae is usually as high as 90%—95%, AD is an attractive method that can convert wet microalgal biomass into biogas. However, there are three main issues on AD of algae: (1) the biodegradability of algae can be low depending on both the biochemical composition and the nature of the cell wall; (2) the high protein content ranging in 30%—75% of cell dry weight results in the release of ammonia, which are toxic to microbes; and (3) the presence of sodium for marine species can also affect the AD performance. Microalgae typically yield less methane than wastewater sludge ($\sim$0.3 vs. 0.4 L CH_4 per gram of volatile solids) (Salerno et al., 2009). It was reported that 50% of *C. vulgaris* biomass without pretreatment could not be anaerobically digested at an organic loading rate of 1 g COD/L and hydraulic retention times (HRT) of 16—28 days (Ras et al., 2011). Various pretreatment technologies including ozonation (Cardeña et al., 2017), electrolysis and ultrasonication (Kumar et al., 2017), enzymes (Córdova et al., 2019; Kendir Çakmak and Ugurlu, 2020), and dilute acids (Park et al., 2020; Rincón-Pérez et al., 2020) have been studied to enhance the anaerobic digestibility of microalgae. The results showed that the methane yield from AD of microalgae pretreated with different methods could be increased by 6%—190%.

The unbalanced nutrients of microalgae sludge due to its low C/N ratio were also considered as a significant limitation factor to the AD process. Anerobic codigestion of microalgae and lignocellulosic biomass or its derives with high C/N ratio can balance the C/N ratio of microalgae-containing feedstock in AD. The addition of 50% (volatile solid based) of wastepaper in algal sludge increased the methane production rate to 1170 mL/L day, as compared to 573 mL/L day AD of algal sludge alone when both were operated at 4 g VS/L day, 35°C, and 10 days of HRT. The maximum methane production rate was 1607 mL/L day that was achieved at a 5 g VS/L day loading rate with 60% (VS based) of paper. The results suggested that an optimum C/N ratio for codigestion of algal sludge and wastepaper was in the range of 20—25 (Yen and Brune, 2007). Another study showed that anaerobic codigestion of primary sludge and microalgae (*Scenedesmus* and *Chlorella*) in a lab-scale semicontinuous anaerobic membrane bioreactor at 35°C at a solid retention time of 100 days achieved 73% biodegradability (Serna-García et al., 2020). Codigestion of microalgae, sludge, and fat oil and grease (FOG) improved the AD kinetics by up to 67% and increased the methane yield by 25%—42%, compared to AD of microalgae only (Solé-Bundó et al., 2020). The synergic benefits of codigestion of microalgae with other wastes depend on the mass ratio of microalgae to the other wastes (Du et al., 2019).

Gonzalez-Fernandez et al. (2018) studied the biochemical methane potential of *C. sorokiniana* and *Scenedesmus* sp. using different microbial inocula (Gonzalez-Fernandez et al., 2018). Their results showed that sludge samples adapted to digest microalgae exhibited a concomitant increase in methane yield together with increasing digestion temperatures. The methane yields were 63.4 ± 1.5, 79.2 ± 3.1, and 108.2 ± 1.9 mL CH_4 g/COD for psychrophilic, mesophilic, and thermophilic digestions, respectively. The relative abundance of *Firmicutes*, particularly *Clostridia*, and *Proteobacteria* together with an important abundance of hydrogenotrophic methanogens was highlighted in this inoculum. They concluded that tailored anaerobic microbiome could help avoiding pretreatments devoted to methane yield enhancement. Klassen et al. (2017) reported a similar conclusion. In their work, AD of nitrogen limited microalgal biomass was characterized as a stable process with low levels of inhibitory substances, which resulted in extraordinary high methane productivity [750 ± 15 and 462 ± 9 mLN/g volatile solids (VS) day^{-1}, respectively], corresponding to biomass-to-methane energy conversion efficiency of up to 84% (Klassen et al., 2017). The microbial community structure within this highly efficient digester revealed a clear predominance of the phyla *Bacteroidetes* and the family Methanosaetaceae among the Bacteria and Archaea, respectively. The fermentation of replete nitrogen biomass was demonstrated to be less productive and failed completely due to acidogenesis that was caused through high ammonia/ammonium concentrations. Another study showed that AD of algal residues after lipid extraction is a favorable process that can recover more energy than the energy from the cell lipids. The lipid content of algae has a wide range of 5%—70% of cell dry weight, and most algae do not contain over 25% of lipid. When the cell lipid content does not exceed 40%, AD of the whole biomass appears to be the optimal strategy on an energy balance basis, for the energetic recovery of cell biomass (Sialve et al., 2009). When evaluating the potential of algae as feedstock for methane production from a process technical and economic point of view, production of mixed algae culture in raceway ponds on nonagricultural sites, such as landfills, was identified as a preferred approach (Zamalloa et al., 2011).

In summary, ammonia toxicity and recalcitrant cell walls are the common reasons of the low biogas yield. Ammonia toxicity might be counteracted by codigesting microalgae with high-carbon organic wastes. Carbon-rich organic wastes that are available near major wastewater pond systems include primary and secondary municipal sludge, sorted municipal organic solid waste, waste FOGs, food industry waste, waste paper, and various agricultural residues. Adapting the digester microbial community to microalgae digestion may also improve the yield (Salerno et al., 2009).

3.1.4 Hydrogen

The increasing requirement for carbon emission reduction makes H_2 more attractive as a fuel as its combustion yields H_2O only. Oxygen evolving photosynthesis (also called oxygenic photosynthesis) is the most advanced biological H_2 production approach, which uses solar energy to split water into protons (H^+), electrons (e^-), and O_2, and then recombines the derived H^+ and e^- by either hydrogenase or nitrogenase enzymes to produce H_2 (Oey et al., 2016). Microalgae including cyanobacteria can perform the indirect process, in which solar energy is first converted into chemical energy in the form of carbohydrates and then used as substrates for H_2 production. The indirect process has been extensively reviewed by Antal et al. (2009), Mathews and Wang (2009), and Rathore and Singh (2013). Microalgae can also produce hydrogen via direct photolysis, which involves the funneling of e^- derived from the light-driven water splitting reaction of photosystem II directly to a H_2-producing hydrogenase (Antal et al., 2009). *C. reinhardtii* is one of the best studied microalgae species with respect to H_2 production processes, while several other algal species such as *Chlorella, Scenedesmus*, and *Tetraselmis* have also been reported to produce H_2 at lower levels. Direct photolysis by microalgae has been comprehensively reviewed in the literature (Oey et al., 2016).

3.2 Microalgae fertilizers

Algal biomass has significantly high NPK contents and better bioavailability than traditional organic fertilizers such as manure but much lower than chemical fertilizer. It was reported that *C. minutissima* grown on wastewater had 5.87% N, 1.15% P, and 0.28% K on a dry weight basis (Khan et al., 2019). Therefore, microalgae provide a biological means of recovering and concentrating NPK nutrients in wastewater. The general biology of phosphorus and metabolism of microalgae from nutrient-rich waste streams was reviewed in literature (Solovchenko et al., 2016). Life cycle analysis of microalgae based biofertilizer showed that microalgal biofertilizer presented positive environmental impacts in all categories investigated. The most significant environmental benefits of microalgal biofertilizer are in the climate change and terrestrial ecotoxicity categories. It was concluded that microalgae have a significant potential for recycling nutrients from wastewater, while dewatering and drying steps need further improvement for the production of microalgae-based biofertilizers (Castro et al., 2020).

Recently, research on the biofertilizer value of microalgae has been conducted on various crops. The microalgal liquid fertilizer at three different doses was applied to the four crop plants, namely *Cucumis sativus, Solanum lycopersicum, Capsicum annuum*, and *Vigna radiata* at various concentrations (0%—100%) of the algae liquid fertilizer (ALF) prepared by mixing 500 mg of dried algal powder of *Chorococcum* sp. with 200 mg of distilled water (Deepika and MubarakAli, 2020). The application of the microalgal fertilizer for the crop plants was a lead substitute for the commercially available chemical fertilizers. A low concentration (20% this ALF) showed the maximum growth in all four crop plants. The growth parameters such as root and shoot length, number of leaves, and number of lateral roots were also recorded maximum at 20% ALF. Dineshkumar et al. (2019) assessed the effect of the microalgal fertilizer on the growth and yield of maize (*Zea mays* L.) which were raised in soil supplemented with *C. vulgaris* and *Spirulina platensis* along with dairy manure for 75 days under the greenhouse condition (Dineshkumar et al., 2019). The microalgal treatment increased the growth performance at the early stage of growth and improved the yield characteristics, in addition to increase the seed germination. They further studied the efficacy of microalgae as biofertilizer of onion plants (Dineshkumar et al., 2020). The maximum micro/macro nutrient availability, the growth parameters viz., plant height, leaf numbers/plant leaves weight/plant, fresh weight/plant, and dry weight were found to be higher in plants treated with dairy manure and *S. platensis* followed by dairy manure and *C. vulgaris* treated plants. Coppens et al. (2016) compared the tomato growth using microalgal and commercial organic fertilizer treatments. The microalgal fertilizers improved the fruit quality through an increase in sugar and carotenoid content, although a lower tomato yield was obtained (Coppens et al., 2016).

3.3 Microalgae-based feeds

Microalgae feeds are traditionally used as a fish feed in aquaculture for the production of various fishes such as larvae, zooplankton, finfish, and juvenile shell-fish (Hemaiswarya et al., 2011). The most frequently used species are *Chlorella, Tetraselmis, Isochrysis, Pavlova, Phaeodactylum, Chaetoceros, Nannochloropsis, Skeletonema*, and *Thalassiosira*. Combination of different algal species provides better balanced nutrition and improves animal growth better than a diet composed of only one algal species (Spolaore et al., 2006). Microalgae-based ingredients produced for aquafeeds could have competitive market advantages over terrestrial crops in terms of input costs, lower aerial footprint, and potential for wastewater remediation and carbon credits from the conversion of greenhouse gases. Microalgae-based products are also considered to have a tremendous potential as "next-generation" feed ingredients for sustainable salmonid aquaculture (e.g., salmon, trout, charr), though few have yet been successfully commercialized (Tibbetts, 2018). Kiron et al. (2012) examined two marine algal species MAP3 and MAP8 for their suitability as fishmeal protein substitutes in feeds of three prominent farmed species including Atlantic salmon, common carp, and whiteleg shrimp (Kiron et al., 2012). After 9— 12 weeks feeding period, the growth and feed performance of all species did not reveal any significant difference between those offered the microalgae-based feed and those offered the control feed. The lipid content in common carp fed higher level of MAP3 was significantly lower than that of the fish fed the control feed. The protein content in the shrimp fed the higher level of MAP8 was significantly lower than that of shrimp on the control feed. The three species could accept the algal meals in their feeds at the tested levels, though there were some noticeable effects on body composition at higher levels of algae (Kiron et al., 2012).

Studies also indicated that microalgal strains of *Schizochytrium* and *Crypthecodinium* as source ingredients for essential n-3 LC-PUFA and *Haematococcus* that effectively accumulates natural-source astaxanthin are promising high-value replacements for conventional fish oils and synthetic astaxanthin, respectively (Sprague et al., 2016). However, due to the fragmented and inconsistent information on the microalgal biochemical composition, inconsistent nutrient characterization analytics, variable digestibility related to recalcitrant cell walls, and general scarcity of adequate nutritional investigations, more research is required to further evaluate the salmonid species-specific safety and efficacy of many microalgae-based products including their effects on growth performance, nutrient utilization, fish health, and product quality (Tibbetts, 2018).

4. Microalgae cultivation systems

4.1 Open ponds

Open pond systems such as natural ponds, raceway ponds, and circular ponds are widely used commercially to produce microalgal biomass. The advantages of open pond systems are low construction and operational costs. Open ponds are popular for treatment of a huge amount of wastewater that closed systems cannot handle. However, the open pond systems face several technical barriers including contamination with other microorganisms, low biomass productivity, high water loss due to evaporation, low CO_2 supply, large land area requirement, and high sensitivity to local weather conditions, and difficulty in harvesting (Nie et al., 2020). Fig. 10.1A shows a lab-scale open raceway pond (ORP). Mixing in an ORP is achieved by using one or two paddlewheels to circulate water through the system. An integrated growth kinetics and computational fluid dynamics (CFD) model can be used for the design, optimization, and operating of ORPs

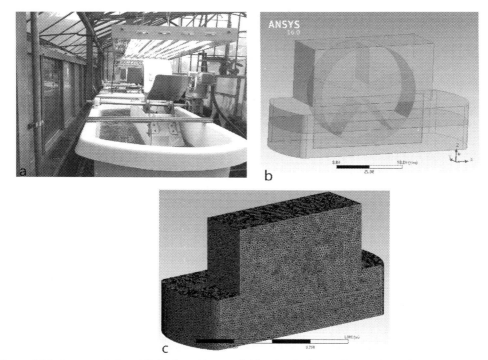

Figure 10.1 A computational fluid dynamics model for the design and operation of an open raceway pond for microalgal cultivation (A) an open raceway pond, (B) CFD geometric design, and (C) mesh for numerical analysis. *(Produced with permission from Amini, H., Hashemisohi, A., Wang, L., Shahbazi, A., Bikdash, M., Kc, D., et al., 2016a. Numerical and experimental investigation of hydrodynamics and light transfer in open raceway ponds at various algal cell concentrations and medium depths. Chem. Eng. Sci. 156, 11–23. https://doi.org/10.1016/j.ces.2016.09.003).*

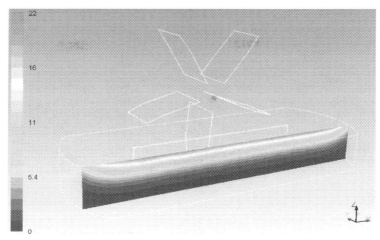

Figure 10.2 Contours of the predicted light intensity distribution (in W/m^2) in a lab-scale ORP channel containing *Chlorella vulgaris* at a concentration of 0.1 g/L. *(Produced with permission from Amini, H., Hashemisohi, A., Wang, L., Shahbazi, A., Bikdash, M., Kc, D., et al., 2016a. Numerical and experimental investigation of hydrodynamics and light transfer in open raceway ponds at various algal cell concentrations and medium depths. Chem. Eng. Sci. 156, 11–23. https://doi.org/10.1016/j.ces.2016.09.003).*

(Amini et al., 2016a, 2018). Fig. 10.1B and C show the geometric design and mesh of numeric analysis of the ORP for the CFD model. As seen from Fig. 10.2, the CFD simulation shows that light intensity sharply drops with several centimeters from the medium surface. The simulations showed that medium depth and harvesting cell density have significant effects on the productivity of the ORP. The CFD model was further used to analyze the effects of different harvesting strategies on the algal productivity in ORPs. The predicted average algal productivity was 10.5 g/m^2 day for the 3-week cultivation in the ORP with a 0.2 m culture depth and the maximum cell density of 0.2 g/L that was maintained by harvesting 50% of the algae when the cell density reached 0.2 g/L. This was 43.8% higher than the average algal productivity of 7.3 g/m^2 day for the 3-week cultivation under the same condition without harvesting that the final cell density reached 0.48 g/L. The average algal productivity decreased with the increase of harvesting cell density (Amini et al., 2018).

4.2 Photobioreactors

Photobioreactors are the closed cultivation systems. The most commonly used types of photobioreactors include tubular, flat plate, column, soft-frame, and hybrid as shown in Fig. 10.3 (Vo et al., 2019). Soft frame photobioreactors are foldable, movable, and flexible compared to other types. Various photobioreactors for culturing microalgae were reviewed in literature (Vo et al., 2019). Photobioreactors provide better control of pH and temperature, lower contamination and water evaporation rates, and higher

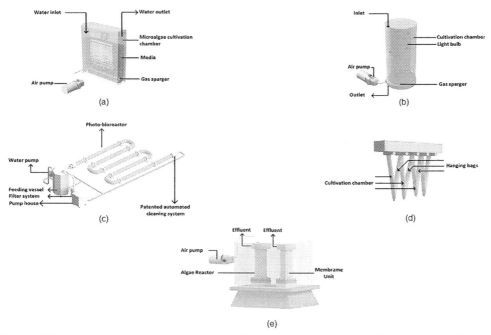

Figure 10.3 Common types of photobioreactors (A) flat plate, (B) column, (C) tubular, (D) soft-frame, and (E) hybrid. *(Produced with permission from Vo, H.N.P., Ngo, H.H., Guo, W., Nguyen, T. M. H., Liu, Y., Liu, Y., et al., 2019. A critical review on designs and applications of microalgae-based photobioreactors for pollutants treatment. Sci. Total Environ. 651, 1549–1568. https://doi.org/10.1016/j.scitotenv.2018.09.282).*

cell densities and productivity than open ponds. However, the constructions costs are higher. It was reported that the average cost of a 500 m^2 photobioreactor was $30 and the pay-back period was from 9 to 13 years (Vo et al., 2019). Closed photobioreactors cannot handle a large amount of wastewater at an industrial scale operation. As microalgae are suspended in the culture medium in photobioreactors, they still face technical and economic challenge in harvesting as well (Nie et al., 2020).

4.3 Attached microalgae cultivation

Suspended microalgae cultivation systems face several technical and economic challenges such as high water demand, low biomass productivity, and high energy and cost inputs for harvesting. Attached microalgal cultivation systems are studied to resolve these problems of suspended cultivation. The microalgae attached on a support medium can be directly harvested via scrapping. Fig. 10.4 shows a rotating algal biofilm reactor, which can simultaneously grow and dewater microalgae to achieve high biomass concentration (Bernstein et al., 2014). It was reported that attached cultivation in a fixed bed bioreactor

Figure 10.4 A rotating algal biofilm reactor at (A) a field scale and (B) a lab scale. *(Produced with permission from Bernstein, H.C., Kesaano, M., Moll, K., Smith, T., Gerlach, R., Carlson, R.P., 2014. Direct measurement and characterization of active photosynthesis zones inside wastewater remediating and biofuel producing microalgal biofilms. Bioresour. Technol. 156, 206−215. https://doi.org/10.1016/j.biortech.2014.01.001).*

could reduce the energy consumption to 4.71 MJ to produce 1 kg of microalgae biomass compared to 9.18 MJ for 1 kg microalgal biomass produced by suspended cultivation in an open pond (Ozkan et al., 2012). Another benefit of attached cultivation is that the adjacent environment of attached microalgal cells has a higher nutrient concentration than the culture medium due to the adsorption of nutrients on the support (Zhuang et al., 2018). It was reported that the total nitrogen and phosphorus in the attached microalgal cells were 2.1 and 15.5 times, respectively, higher than the suspended culture medium (Zhuang et al., 2018).

Various support materials such as polyurethane, polyethylene, polystyrene, and glass wool have been studied for growing microalgae in either fixed bed or fluidized bed photobioreactors (Rosli et al., 2020). The type and physiochemical properties of support materials have significant effects on the growth of different microalgae species. It was reported that the productivity of *C. vulgaris* attached on a filter membrane achieved 14 g/m^{2}d (Rosli et al., 2020). The physiochemical properties of support materials are critical to form microalgae film on the support materials. It was found that *Pseudochlorococcum* sp. grew better on a hydrophilic surface than hydrophobic surface (Ji et al., 2014a). Attached cultivation of microalgae is still at the early stage of development. More studies are needed to address the technical challenges of inefficient light penetration, limited nutrient absorption capability, and proper support materials. Attached cultivation systems have been studied to grow microalgae on various types of wastewater due to their low harvesting cost and higher nutrients removal efficiency. However, there is a large variation of nutrients removal rates and microalgae growth rates due to the difference of wastewater sources, cultivation conditions, and harvesting times (Rosli et al., 2020).

Figure 10.5 Photograph of a floating photo bioreactor for microalgae cultivation. *(Produced with permission from Zhu, C., Xi, Y., Zhai, X., Wang, J., Kong, F., Chi, Z., 2021. Pilot outdoor cultivation of an extreme alkalihalophilic Trebouxiophyte in a floating photobioreactor using bicarbonate as carbon source. J. Clean. Prod. 283, 124648. https://doi.org/10.1016/j.jclepro.2020.124648).*

4.4 Floating photobioreactors

Floating photobioreactors as shown in Fig. 10.5 were initially designed as a low cost and closed system to grow microalgae in ocean (Zhu et al., 2021). As the temperature of the algal culture medium can be controlled by the ocean water and the waves can provide mixing, the operating cost of a floating photobioreactor is low (Zhu et al., 2018). Research showed that cultivation of *I. zhangjiangensis* in a floating photobioreactor that was made from inflatable membrane on the ocean resulted in productivity at the same level as that obtained for a flat-panel photobioreactor at a large scale (Zhu et al., 2019).

5. Challenges of resource recovery using microalgae cultivation in wastewater

During the photosynthesis, microalgae use the light as an energy source and consume CO_2 as a carbon source and fertilizer nutrients in the culture medium to grow biomass. The culture conditions have significant effects on the microalgal growth. There are optimum light intensity, CO_2, pH, temperature, and nutrients for microalgae growth. However, it should be noted that environmental and operational factors have interactive effects on the microalgae growth rather than the effects exerted by the individual parameters. Therefore, it is essential to study the interactive effects to optimize the environmental conditions for the growth of specific algal strains.

5.1 CO$_2$ and fertilizer nutrient effects

Microalgae growth needs carbon, nitrogen, and phosphorus. Microalgae have been used to sequestrate industrial and atmospheric CO$_2$ to mitigate climate change. Research showed that *Chlorella* sp. achieved the highest biomass productivity of 0.335 g/L d and a CO$_2$ fixing rate of 0.35 g CO$_2$/L d in a suspended culture at an optimum CO$_2$ concentration of 5% (Ryu et al., 2009). Another study showed that the optimal growth of the marine algae of *Nannochloropsis salina* occurred at 6% CO$_2$ and excess CO$_2$ resulted in medium acidification and growth inhibition (Chen et al., 2020). Flue gases are an inexpensive CO$_2$ source. However, flue gases usually contain a large amount of O$_2$. It was found that the CO$_2$ fixation capacity and lipid productivity of *Nannochloropsis salina* were improved by 4.8- and 4.4-fold, respectively, if the O$_2$ was removed from the inlet flue gas (Chen et al., 2020).

Nitrogen and phosphorus compounds are the main fertilizer nutrients to support the microalgal growth. Research showed that the nitrogen sources and C/N and N/P ratios have significant impacts on the nutrient removal efficiency (Shen et al., 2019; Xu et al., 2017). The optimum C/N and N/P ratios for microalgae productivity and removal efficiency of nutrients in wastewater vary from 5 to 30 depending on the microalgae strains and growth conditions. The optimal pH for microalgae cultivation is usually around neural. It was reported that the N removal efficiency using *C. vulgaris* at unregulated pH from 7.6 to 8.1 was 78%, compared to 86% for regulated pH at 7 through CO$_2$ incision (Liang et al., 2013). The pH values below 4 and above 9 are considered extreme conditions. A pH above 9 leads to the increase in CO$_3^{2-}$ and HCO$_3^-$ and reduces the availability of CO$_2$ (Nie et al., 2020). The pH value of wastewater is affected by dissolved CO$_2$ and NH$_3$ contents. One of the major technical challenges to grow microalgae on wastewater is the large variation of the contents of nutrients in the wastewater. The variation of nutrients in the wastewater makes it difficult to standardize the cultivation of microalgae (Nie et al., 2020). Another challenge is the presence of microalgal growth inhibitors in wastewater. Research showed that excessive Zn(II) in swine wastewater inhibited microalgal growth (Li et al., 2020). Gradual increases in the CO$_2$ concentration in the culture medium, dilution of wastewater, and supplement of deficient nutrients may be necessary to obtain an optimum C/N/P ratio and pH value in the wastewater for the improvement of the microalgae growth. However, the C/N/P ratio of specific wastewater may not be easily adjusted. In this case, a microalgae strain can be trained with a set of culture media from it C/N/P to the C/N/P ratio of the wastewater to be evolutionarily adopted to the wastewater. Alternatively, coculture of various algal strains could lead to better outcomes (Nie et al., 2020).

5.2 Environmental effects

Light intensity and temperature are two main environmental factors that have significant effects on the growth of microalgae. Microalgae growth rate increases with light intensity until reaching the light saturation point. The saturation point of light intensity depends on the microalgal strains. The photosynthetic efficiency of microalgae may range from 1% to 5% depending on light duration and intensity, and other factors such as pH value and temperature. The removal efficiency of nutrients such as N and P directly depends on the photosynthetic efficiency. The light supply of a microalgae cultivation system is affected by the configuration of cultivation systems and the cultivation period. The microalgae growth rate decreases with the prolonged cultivation period due to the reduced light supply capacity at increased cell density. The optimal temperature for microalgae is around $25°C$, which usually range from 20 to $30°C$ depending on microalgal strains (Rosli et al., 2020). It was reported that the specific growth rate of *C. vulgaris* in an open pond did not have a change when the temperature increased from 25 to $30°C$. However, the increase of temperature to $35°C$ decreased the specific growth rate by 17% and the temperature at $38°C$ caused abrupt cell death (Converti et al., 2009). Therefore, it is important to control the temperature of culture medium, particularly in summer seasons.

5.3 Biological effects

The effect of biological factors on the microalgae growth must be considered for cultivation of microalgae in open ponds. Many grazers can present in open ponds, which feed on microalgal biomass. It was reported that the removal efficiencies of N and P by *C. vulgaris* decreased by 21% and 2% when protozoan significantly presented in wastewater (Tejido-Nuñez et al., 2019). Research also showed that *C. vulgaris* with a cell size of 5 microns in diameter was more prone to herbivorous protozoa than *T. obliquus* at 5—12 microns (Tejido-Nuñez et al., 2019). Therefore, the use of extreme environmental conditions and the selection of highly resistant algal strains may be considered to minimize the presence of grazers in open algal ponds (Nie et al., 2020).

The effects of bacteria in wastewater on the microalgal growth may be inhibitory or beneficial. Microalgae growth and nutrients removal rate can be increased by the synergism with suitable bacterial species. Those bacteria provide CO_2 to enhance the growth of microalgae, while the microalgae provide O_2 and organic compounds to support the bacteria (Luo et al., 2019). However, bacteria may also inhibit the microalgal growth by aggregating microalgae and thus decreasing light distribution in the culture (Arcila and Buitrón, 2017). The metabolites secreted by some bacteria may also inhibit the algal growth (Cole, 1982). Therefore, it is necessary to study the interactions between microalgae and bacteria in wastewater, which is vital for proper management of a microalgae-based wastewater treatment system. The algal growth conditions were found to have significant influence on the bacterial and zooplanktonic community changes in the wastewater (Ferro et al., 2020). Selective growth of bacteria that are readily available in wastewater can create some forms of bacteria—microalgae interactions to enhance the bioremediation process due to the symbiotic relationship between microalgae and bacteria. It was reported that cultivation of microalgae in unsterilized palm oil effluent achieved higher COD removal than that in sterilized effluent due to microalgae—bacteria consortia interaction. However, the microalgae growth rate in unsterilized effluent is affected by the inoculum size of microalgae due to availability of nutrients for growth and self-shading effect (Mohd Udaiyappan et al., 2020).

6. Enhancement of microalgae cultivation in wastewater

6.1 Enhancement of microalgae cultivation with microbial fuel cells

The accumulation of the oxygen that is produced during algal photosynthesis becomes inhibitory to the algal growth. Microalgae cultivation can be integrated with microbial fuel cells (MFCs). An MFC is a bioelectrochemical device for generating electricity from wastewater which usually consists of an anode chamber, a cathode chamber, and a separation membrane as shown in Fig. 10.6 (Jaiswal et al., 2020). The cathode chamber of an MFC can be used to grow microalgae by removing the oxygen generated from the microalgal photosynthesis (Reaction 10.1) in the culture medium via the oxygen reduction reaction (Reaction 10.2). Meanwhile, the CO_2 generated by anaerobic bacteria in

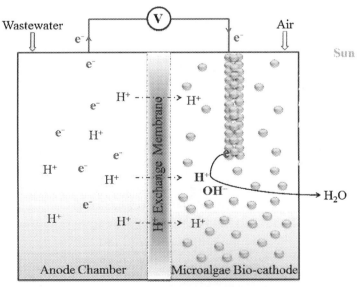

Figure 10.6 A typical microalgae-MFC system. *(Produced with permission from Jaiswal, K.K., Kumar, V., Vlaskin, M.S., Sharma, N., Rautela, I., Nanda, M., 2020. Microalgae fuel cell for wastewater treatment: recent advances and challenges. J. Water Process Eng. 38, 101549. https://doi.org/10.1016/j.jwpe.2020.101549).*

the anode chamber of the MFC via the AD reaction (Reaction 10.3) can be circulated to the cathode chamber to enhance the photosynthesis of microalgae (Khandelwal et al., 2020; Neethu et al., 2018).

$$6CO_2 + 12H_2O \rightarrow C_6H_{12}O_6 + 6O_2 \tag{10.1}$$

$$nO_2 + 4nH^+ + 4ne^- \rightarrow 2nH_2O \tag{10.2}$$

$$(CH_2O)_n + nH_2O \rightarrow nCO_2 + 4nH^+ + 4ne^- \tag{10.3}$$

A microalgae-based MFC is an eco-friendly and sustainable approach to produce renewable energy products from wastewater. However, the microalgae-based MFC technology faces several technical and economic challenges including low power density, low power generation rate, and high capital costs. Therefore, economic electrodes, efficient catalysts for the electrodes, availability of the electron acceptor, and proton exchange membranes are needed to develop low-cost large-scale MFCs (Elshobary et al., 2021).

6.2 Nanotechnology for the enhancement of microalgae cultivation and harvesting

Trace metals play a key role in the growth rate and composition of microalgae and their effectiveness on the algal photosynthesis depends on their concentrations in a culture medium and their antagonistic effect with other environmental factors. Iron is a vital regulatory element in gene expression and metabolism of algae. Research showed that both concentrations and sources of iron compounds affected the lipid content and biomass yield of microalgae. Polat et al. (2020) found that highest saturated fatty acid and highest biomass yield of *Auxenochlorella protothecoides* were obtained at 1.15 mM ferric chloride and 1.08 mM ferrous sulfate, respectively (Polat et al., 2020). Pádrová et al. (2015) found that a trace amount of nanoscale Zero Iron (nZVI) particles at 5.1 mg/L could strongly enhance lipid accumulation, increased the content of polyunsaturated fatty acids, but decreased the content of saturated and monounsaturated fatty acids except palmitoleic acid in both green algae and eustigmatophycean algae (Pádrová et al., 2015). It was also reported that the growth of microalgae was even favored by the nano particles in comparison with their bulk analogues (Kadar et al., 2012). However, it was reported that the increase of iron concentration could increase the growth rate and lipid content of microalgae, but too high iron concentration could have negative impacts on algal biomass and lipid production (Cao et al., 2014).

Harvesting microalgae is a major economic bottleneck due to the small cell size and low biomass density. Conventional harvesting technologies of microalgae include centrifugation, flocculation by organic/inorganic coagulants, precipitation by pH increment, filtration and flotation (Kim et al., 2013). However, all those methods have some technical and economic barriers for large scale of applications. Centrifugation is an energy-intensive separation. Sedimentation needs a large footprint and has low recovery rate. Flotation requires flocculants. Filtration has a high cost and low flux, and faces the fouling issue (Nie et al., 2020). In recent years, magnetophoresis has emerged as an energy-efficient method for harvesting microalgae by tagging microalgae cells on magnetic particles and separating them from the culture medium by external magnetic field (Prochazkova et al., 2013). Fe_3O_4 nanoparticles (NPs) are commonly used due to its effect of specific surface area, superparamagnetism and biocompatibility (Hu et al., 2013). However, tagging magnetic Fe_3O_4 NPs to the negatively charged algal cells requires a specific pH range at which zeta potential of magnetic nanoparticles shows positively charged surface. As a result, cationic materials such as polyethylenimine and polyamido-amine are normally coated onto Fe_3O_4 NPs (Ge et al., 2015; Hu et al., 2013; Wang et al., 2016).

6.3 Enhancement of microalgae cultivation with artificial intelligence

Artificial intelligence (AI) has been used to predict, monitor, and control microalgae cultivation. Several machine learning algorithms such as k-Nearest Neighbor (k-NN), Random Forest, and Neural Networks (ANN) have been used to classify microalgae by predicting microalgae properties and analyzing the population dynamics of microalgae cultures using image recognition. Random forest is a learning method for classification by creating a multitude of decision trees during data training and outputting classes of individual trees. Reimann et al. (2020) found that the random forest was a suitable method for classifying the living and dead microalgae population and analyzing population dynamics of microalgae cultures (Reimann et al., 2020). The k-NN algorithm which is a nonparametric classification method is used to process the images with different features that are determined by their Euclidean distance and presented in an array. The column and row of the array represented the number of features to be studied and the distance of samples. A sample is classified by the weighted average value of its k nearest neighbors. Yew et al. (2020) used k-NN algorithm to process images of microalgae samples for rapidly determining the microalgae concentration, nitrogen concentration, and pH values without the need of tedious analytical processes. They found that the k-value at 4 provided the smallest normalized Root-Mean-Square-Error (RMSE) between the predicted and measured values (Yew et al., 2020). The ANN is a learning method based on a collection of connected artificial neurons and the output of each neuron is computed by some nonlinear function of the sum of its inputs. Franco et al. (2019) found that the ANN could be used to classify the biological compositions of mixed algal cultures by using the spectral signatures of various microalgae species. A three-layer ANN was used to predict the concentration of polyculture microalgae in an ORP as a function of eight input parameters of initial algal concentration, harvesting period, HRT, addition of sodium acetate, solar radiation, temperature, pH, and nitrate ion concentration with a high prediction accuracy of $R^2 = 0.93$ (Supriyanto et al., 2019). Ansari et al. (2021) used the ANN to predict the microalgae yield as a function of six input parameters of temperature, pH, dissolved oxygen, electrical conductivity, NO_3^-, and PO_4^{3-} in wastewater with a prediction accuracy of $R^2 = 0.983$. Their sensitivity analysis using the ANN model showed that the relative importance order of the environmental factor for the algal yield was $NO_3^- > PO_4^{3-} > pH$ and dissolve oxygen $>$ temperature $> $ electric conductivity (Ansari et al., 2021). Micronutrients play a key role in microalgae culture. Liyanaarachchi et al. (2020) used an ANN model to predict the *C. vulgaris* growth yield as a function of 23 growth parameters including the growth conditions, and the concentrations of various macronutrients and micronutrients in the culture medium. They found that the CO_2 supply, nitrogen macronutrient, and copper micronutrient were the most import parameters in the parameter categories of growth conditions, macronutrients, and micronutrients that affected the algal growth (Liyanaarachchi et al., 2020).

7. Conclusions

Agricultural production generates large amounts of solid wastes and wastewater. The cultivation of microalgae on wastewater can produce algal biomass as a feedstock for production of biofuels, fertilizers, animal and fish feed, and other bio-based chemicals, meanwhile remediate the wastewater, which provides a promising approach to reduce the cost of microalgae. Microalgae growth needs carbon, nitrogen, and phosphorus at an optimum ratio. Flue gases and wastewater are an inexpensive and abundant C, and N and P sources, respectively. However, flue gases usually contain a large amount of O_2 and impurities which inhibit the growth of microalgae. The C/N/P ratio of a specific wastewater may not be easily adjusted. Therefore, pretreatment of the flue gases and wastewater may be necessary for their use as feedstocks to cultivate microalgae. The specific microalgae strain can be adapted to the wastewater. Coculture of mixed algal strains or microalgae—bacteria consortium could lead to better performance. As environmental and operational factors have interactive effects on the microalgae growth, it is essential to study the interactive effects to optimize the algal growth conditions. Many grazers and microorganisms may present in the wastewater, which may be minimized by the extreme environmental conditions. The selection of highly resistant algal strains that are suitable to grow in wastewater in open ponds may also be considered. The effects of bacteria in wastewater on the microalgae may be inhibitory due to the aggregation of microalgae and toxic secretes caused by bacteria or beneficial due to the exchange of CO_2 and O_2 between microalgae and aerobic bacteria. Therefore, it is necessary to study the interactions between microalgae and bacteria in wastewater, which is vital for proper management of a microalgae-based wastewater treatment system. New attached and floating cultivation systems and new technologies including MFCs, nanotechnology, and artificial intelligence have been studied to enhance microalgae cultivation in wastewater, which have shown promising results toward the large-scale cultivation of microalgae for resource recovery and wastewater treatment.

Acknowledgment

This publication was made possible by the grant from the U.S. Department of Agriculture-National Institute of Food and Agriculture (USDA-NIFA, award number NC.X 345-5-22-130-1) and National Science Foundation (NSF, award number 1736173). Its contents are solely the responsibility of the authors and do not necessarily represent the official views of the USDA-NIFA and NSF.

References

Abdo, S.M., Abo El-Enin, S.A., El-Khatib, K.M., El-Galad, M.I., Wahba, S.Z., El Diwani, G., Ali, G.H., 2016. Preliminary economic assessment of biofuel production from microalgae. Renew. Sustain. Energy Rev. 55, 1147–1153. https://doi.org/10.1016/j.rser.2015.10.119.

Ackefors, H., Enell, M., 1994. The release of nutrients and organic matter from aquaculture systems in Nordic countries. J. Appl. Ichthyol. 10 (4), 225–241. https://doi.org/10.1111/j.1439-0426.1994.tb00163.x.

Amini, H., Hashemisohi, A., Wang, L., Shahbazi, A., Bikdash, M., Kc, D., Yuan, W., 2016a. Numerical and experimental investigation of hydrodynamics and light transfer in open raceway ponds at various algal cell concentrations and medium depths. Chem. Eng. Sci. 156, 11–23. https://doi.org/10.1016/j.ces.2016.09.003.

Amini, H., Wang, L., Hashemisohi, A., Shahbazi, A., Bikdash, M., Kc, D., Yuan, W., 2018. An integrated growth kinetics and computational fluid dynamics model for the analysis of algal productivity in open raceway ponds. Comput. Electron. Agric. 145, 363–372. https://doi.org/10.1016/j.compag.2018.01.010.

Amini, H., Wang, L., Shahbazi, A., 2016b. Effects of harvesting cell density, medium depth and environmental factors on biomass and lipid productivities of Chlorella vulgaris grown in swine wastewater. Chem. Eng. Sci. 152, 403–412. https://doi.org/10.1016/j.ces.2016.06.025.

Ansari, F.A., Nasr, M., Rawat, I., Bux, F., 2021. Artificial neural network and techno-economic estimation with algae-based tertiary wastewater treatment. J. Water Process Eng. 40, 101761. https://doi.org/10.1016/j.jwpe.2020.101761.

Antal, T.K., Volgusheva, A.A., Kukarskih, G.P., Krendeleva, T.E., Rubin, A.B., 2009. Relationships between H2 photoproduction and different electron transport pathways in sulfur-deprived Chlamydomonas reinhardtii. Int. J. Hydrog. Energy 34 (22), 9087–9094. https://doi.org/10.1016/j.ijhydene.2009.09.011.

Arcila, J.S., Buitrón, G., 2017. Influence of solar irradiance levels on the formation of microalgae-bacteria aggregates for municipal wastewater treatment. Algal Res. 27, 190–197. https://doi.org/10.1016/j.algal.2017.09.011.

Bernstein, H.C., Kesaano, M., Moll, K., Smith, T., Gerlach, R., Carlson, R.P., et al., 2014. Direct measurement and characterization of active photosynthesis zones inside wastewater remediating and biofuel producing microalgal biofilms. Bioresour. Technol. 156, 206–215. https://doi.org/10.1016/j.biortech.2014.01.001.

Bohutskyi, P., Liu, K., Nasr, L.K., Byers, N., Rosenberg, J.N., Oyler, G.A., et al., 2015. Bioprospecting of microalgae for integrated biomass production and phytoremediation of unsterilized wastewater and anaerobic digestion centrate. Appl. Microbiol. Biotechnol. 99 (14), 6139–6154. https://doi.org/10.1007/s00253-015-6603-4.

Cao, J., Yuan, H., Li, B., Yang, J., 2014. Significance evaluation of the effects of environmental factors on the lipid accumulation of Chlorella minutissima UTEX 2341 under low-nutrition heterotrophic condition. Bioresour. Technol. 152, 177–184. https://doi.org/10.1016/j.biortech.2013.10.084.

Caporgno, M.P., Pruvost, J., Legrand, J., Lepine, O., Tazerout, M., Bengoa, C., 2016. Hydrothermal liquefaction of Nannochloropsis oceanica in different solvents. Bioresour. Technol. 214, 404–410. https://doi.org/10.1016/j.biortech.2016.04.123.

Cardeña, R., Moreno, G., Bakonyi, P., Buitrón, G., 2017. Enhancement of methane production from various microalgae cultures via novel ozonation pretreatment. Chem. Eng. J. 307, 948–954. https://doi.org/10.1016/j.cej.2016.09.016.

Castro, J.d.S., Calijuri, M.L., Ferreira, J., Assemany, P.P., Ribeiro, V.J., 2020. Microalgae based biofertilizer: a life cycle approach. Sci. Total Environ. 724, 138138. https://doi.org/10.1016/j.scitotenv.2020.138138.

Chen, J., Bai, J., Li, H., Chang, C., Fang, S., 2015. Prospects for bioethanol production from macroalgae. Trends Renew. Energy 1 (3), 185–197. https://doi.org/10.17737/tre.2015.1.3.0016.

Chen, L., Liu, T., Zhang, W., Chen, X., Wang, J., 2012. Biodiesel production from algae oil high in free fatty acids by two-step catalytic conversion. Bioresour. Technol. 111, 208–214. https://doi.org/10.1016/j.biortech.2012.02.033.

Chen, Y., Xu, C., Vaidyanathan, S., 2020. Influence of gas management on biochemical conversion of CO2 by microalgae for biofuel production. Appl. Energy 261, 114420. https://doi.org/10.1016/j.apenergy.2019.114420.

Cheng, D.L., Ngo, H.H., Guo, W.S., Chang, S.W., Nguyen, D.D., Kumar, S.M., 2019. Microalgae biomass from swine wastewater and its conversion to bioenergy. Bioresour. Technol. 275, 109–122. https://doi.org/10.1016/j.biortech.2018.12.019.

Chiang, Y.-L., Chen, Y.-P., Yu, M.-C., Hsieh, S.-Y., Hwang, I.-E., Liu, Y.-J., et al., 2018. Biomass and lipid production of a novel microalga, Chlorellaceae sp. P5, through heterotrophic and swine wastewater cultivation. J. Renew. Sustain. Energy 10 (3), 033102.

Chisti, Y., 2007. Biodiesel from microalgae. Biotechnol. Adv. 25 (3), 294–306. https://doi.org/10.1016/j.biotechadv.2007.02.001.

Clarens, A.F., Resurreccion, E.P., White, M.A., Colosi, L.M., 2010. Environmental life cycle comparison of algae to other bioenergy feedstocks. Environ. Sci. Technol. 44 (5), 1813–1819. https://doi.org/10.1021/es902838n.

Cole, J.J., 1982. Interactions between bacteria and algae in aquatic ecosystems. Annu. Rev. Ecol. Syst. 13 (1), 291–314. https://doi.org/10.1146/annurev.es.13.110182.001451.

Converti, A., Casazza, A.A., Ortiz, E.Y., Perego, P., Del Borghi, M., 2009. Effect of temperature and nitrogen concentration on the growth and lipid content of Nannochloropsis oculata and Chlorella vulgaris for biodiesel production. Chem. Eng. Process 48 (6), 1146–1151. https://doi.org/10.1016/j.cep.2009.03.006.

Coppens, J., Grunert, O., Van Den Hende, S., Vanhoutte, I., Boon, N., Haesaert, G., De Gelder, L., 2016. The use of microalgae as a high-value organic slow-release fertilizer results in tomatoes with increased carotenoid and sugar levels. J. Appl. Phycol. 28 (4), 2367–2377. https://doi.org/10.1007/s10811-015-0775-2.

Córdova, O., Passos, F., Chamy, R., 2019. Enzymatic pretreatment of microalgae: cell wall disruption, biomass solubilisation and methane yield increase. Appl. Biochem. Biotechnol. 189 (3), 787–797. https://doi.org/10.1007/s12010-019-03044-8.

Cronk, J.K., 1996. Constructed wetlands to treat wastewater from dairy and swine operations: a review. Agric. Ecosyst. Environ. 58 (2), 97–114. https://doi.org/10.1016/0167-8809(96)01024-9.

Deepika, P., MubarakAli, D., 2020. Production and assessment of microalgal liquid fertilizer for the enhanced growth of four crop plants. Biocatal. Agric. Biotechnol. 28, 101701. https://doi.org/10.1016/j.bcab.2020.101701.

Demirbas, A., 2009. Biodiesel from waste cooking oil via base-catalytic and supercritical methanol transesterification. Energy Convers. Manag. 50 (4), 923–927. https://doi.org/10.1016/j.enconman.2008.12.023.

Dineshkumar, R., Subramanian, J., Arumugam, A., Ahamed Rasheeq, A., Sampathkumar, P., 2020. Exploring the microalgae biofertilizer effect on onion cultivation by field experiment. Waste Biomass Valorization 11 (1), 77—87. https://doi.org/10.1007/s12649-018-0466-8.

Dineshkumar, R., Subramanian, J., Gopalsamy, J., Jayasingam, P., Arumugam, A., Kannadasan, S., Sampathkumar, P., 2019. The impact of using microalgae as biofertilizer in maize (Zea mays L.). Waste Biomass Valorization 10 (5), 1101—1110. https://doi.org/10.1007/s12649-017-0123-7.

Donner, M., Verniquet, A., Broeze, J., Kayser, K., De Vries, H., 2021. Critical success and risk factors for circular business models valorising agricultural waste and by-products. Resour. Conserv. Recycl. 165, 105236. https://doi.org/10.1016/j.resconrec.2020.105236.

Du, X., Tao, Y., Li, H., Liu, Y., Feng, K., 2019. Synergistic methane production from the anaerobic co-digestion of Spirulina platensis with food waste and sewage sludge at high solid concentrations. Renew. Energy 142, 55—61. https://doi.org/10.1016/j.renene.2019.04.062.

Elshobary, M.E., Zabed, H.M., Yun, J., Zhang, G., Qi, X., 2021. Recent insights into microalgae-assisted microbial fuel cells for generating sustainable bioelectricity. Int. J. Hydrog. Energy 46 (4), 3135—3159. https://doi.org/10.1016/j.ijhydene.2020.06.251.

Farobie, O., Matsumura, Y., 2015. A comparative study of biodiesel production using methanol, ethanol, and tert-butyl methyl ether (MTBE) under supercritical conditions. Bioresour. Technol. 191, 306—311. https://doi.org/10.1016/j.biortech.2015.04.102.

Ferro, L., Hu, Y.O.O., Gentili, F.G., Andersson, A.F., Funk, C., 2020. DNA metabarcoding reveals microbial community dynamics in a microalgae-based municipal wastewater treatment open photobioreactor. Algal Res. 51, 102043. https://doi.org/10.1016/j.algal.2020.102043.

Franco, B.M., Navas, L.M., Gomez, C., Sepulveda, C., Acien, F.G., 2019. Monoalgal and mixed algal cultures discrimination by using an artificial neural network. Algal Res. 38, 101419. https://doi.org/10.1016/j.algal.2019.101419.

García, D., Posadas, E., Grajeda, C., Blanco, S., Martínez-Páramo, S., Acién, G., et al., 2017. Comparative evaluation of piggery wastewater treatment in algal-bacterial photobioreactors under indoor and outdoor conditions. Bioresour. Technol. 245, 483—490. https://doi.org/10.1016/j.biortech.2017.08.135.

Ge, S., Agbakpe, M., Wu, Z., Kuang, L., Zhang, W., Wang, X., 2015. Influences of surface coating, UV irradiation and magnetic field on the algae removal using magnetite nanoparticles. Environ. Sci. Technol. 49 (2), 1190—1196. https://doi.org/10.1021/es5049573.

Gonzalez-Fernandez, C., Barreiro-Vescovo, S., de Godos, I., Fernandez, M., Zouhayr, A., Ballesteros, M., 2018. Biochemical methane potential of microalgae biomass using different microbial inocula. Biotechnol. Biofuels 11 (1), 184. https://doi.org/10.1186/s13068-018-1188-7.

Harun, R., Singh, M., Forde, G.M., Danquah, M.K., 2010. Bioprocess engineering of microalgae to produce a variety of consumer products. Renew. Sustain. Energy Rev. 14 (3), 1037—1047. https://doi.org/10.1016/j.rser.2009.11.004.

Hasan, R., Zhang, B., Wang, L., Shahbazi, A., 2014. Bioremediation of swine wastewater and biofuel potential by using Chlorella vulgaris, Chlamydomonas reinhardtii, and Chlamydomonas debaryana. J. Petrol Environ. Biotechnol. 5 (3), 175—180. https://doi.org/10.13140/2.1.3348.4168.

Hemaiswarya, S., Raja, R., Ravi Kumar, R., Ganesan, V., Anbazhagan, C., 2011. Microalgae: a sustainable feed source for aquaculture. World J. Microbiol. Biotechnol. 27 (8), 1737—1746. https://doi.org/10.1007/s11274-010-0632-z.

Hu, Y.-R., Wang, F., Wang, S.-K., Liu, C.-Z., Guo, C., 2013. Efficient harvesting of marine microalgae Nannochloropsis maritima using magnetic nanoparticles. Bioresour. Technol. 138, 387—390. https://doi.org/10.1016/j.biortech.2013.04.016.

Issariyakul, T., Dalai, A.K., 2014. Biodiesel from vegetable oils. Renew. Sustain. Energy Rev. 31, 446—471. https://doi.org/10.1016/j.rser.2013.11.001.

Jaiswal, K.K., Kumar, V., Vlaskin, M.S., Sharma, N., Rautela, I., Nanda, M., et al., 2020. Microalgae fuel cell for wastewater treatment: recent advances and challenges. J. Water Process Eng. 38, 101549. https://doi.org/10.1016/j.jwpe.2020.101549.

Ji, B., Zhang, W., Zhang, N., Wang, J., Lutzu, G.A., Liu, T., 2014a. Biofilm cultivation of the oleaginous microalgae Pseudochlorococcum sp. Bioproc. Biosyst. Eng. 37 (7), 1369—1375. https://doi.org/10.1007/s00449-013-1109-x.

Ji, F., Liu, Y., Hao, R., Li, G., Zhou, Y., Dong, R., 2014b. Biomass production and nutrients removal by a new microalgae strain Desmodesmus sp. in anaerobic digestion wastewater. Bioresour. Technol. 161, 200—207. https://doi.org/10.1016/j.biortech.2014.03.034.

John, R.P., Anisha, G.S., Nampoothiri, K.M., Pandey, A., 2011. Micro and macroalgal biomass: a renewable source for bioethanol. Bioresour. Technol. 102 (1), 186–193. https://doi.org/10.1016/j.biortech.2010.06.139.

Jones, C.S., Mayfield, S.P., 2012. Algae biofuels: versatility for the future of bioenergy. Curr. Opin. Biotechnol. 23 (3), 346–351. https://doi.org/10.1016/j.copbio.2011.10.013.

Joshi, H., Moser, B.R., Toler, J., Walker, T., 2010. Preparation and fuel properties of mixtures of soybean oil methyl and ethyl esters. Biomass Bioenerg. 34 (1), 14–20. https://doi.org/10.1016/j.biombioe.2009.09.006.

Kadar, E., Rooks, P., Lakey, C., White, D.A., 2012. The effect of engineered iron nanoparticles on growth and metabolic status of marine microalgae cultures. Sci. Total Environ. 439, 8–17. https://doi.org/10.1016/j.scitotenv.2012.09.010.

Kendir Çakmak, E., Ugurlu, A., 2020. Enhanced biogas production of red microalgae via enzymatic pretreatment and preliminary economic assessment. Algal Res. 50, 101979. https://doi.org/10.1016/j.algal.2020.101979.

Khan, S.A., Sharma, G.K., Malla, F.A., Kumar, A., Rashmi, Gupta, N., 2019. Microalgae based biofertilizers: a biorefinery approach to phycoremediate wastewater and harvest biodiesel and manure. J. Clean. Prod. 211, 1412–1419. https://doi.org/10.1016/j.jclepro.2018.11.281.

Khandelwal, A., Chhabra, M., Yadav, P., 2020. Performance evaluation of algae assisted microbial fuel cell under outdoor conditions. Bioresour. Technol. 310, 123418. https://doi.org/10.1016/j.biortech.2020.123418.

Kim, H.M., Oh, C.H., Bae, H.-J., 2017. Comparison of red microalgae (Porphyridium cruentum) culture conditions for bioethanol production. Bioresour. Technol. 233, 44–50. https://doi.org/10.1016/j.biortech.2017.02.040.

Kim, J., Yoo, G., Lee, H., Lim, J., Kim, K., Kim, C.W., et al., 2013. Methods of downstream processing for the production of biodiesel from microalgae. Biotechnol. Adv. 31 (6), 862–876. https://doi.org/10.1016/j.biotechadv.2013.04.006.

Kiron, V., Phromkunthong, W., Huntley, M., Archibald, I., De Scheemaker, G., 2012. Marine microalgae from biorefinery as a potential feed protein source for Atlantic salmon, common carp and whiteleg shrimp. Aquacult. Nutr. 18 (5), 521–531. https://doi.org/10.1111/j.1365-2095.2011.00923.x.

Klassen, V., Blifernez-Klassen, O., Wibberg, D., Winkler, A., Kalinowski, J., Posten, C., Kruse, O., 2017. Highly efficient methane generation from untreated microalgae biomass. Biotechnol. Biofuels 10 (1), 186. https://doi.org/10.1186/s13068-017-0871-4.

Knothe, G., 2005. Dependence of biodiesel fuel properties on the structure of fatty acid alkyl esters. Fuel Process. Technol. 86 (10), 1059–1070. https://doi.org/10.1016/j.fuproc.2004.11.002.

Knox, J.W., Kay, M.G., Weatherhead, E.K., 2012. Water regulation, crop production, and agricultural water management—understanding farmer perspectives on irrigation efficiency. Agric. Water Manag. 108, 3–8. https://doi.org/10.1016/j.agwat.2011.06.007.

Kumar, G., Sivagurunathan, P., Zhen, G., Kobayashi, T., Kim, S.-H., Xu, K., 2017. Combined pretreatment of electrolysis and ultra-sonication towards enhancing solubilization and methane production from mixed microalgae biomass. Bioresour. Technol. 245, 196–200. https://doi.org/10.1016/j.biortech.2017.08.154.

Levine, R.B., Pinnarat, T., Savage, P.E., 2010. Biodiesel production from wet algal biomass through in situ lipid hydrolysis and supercritical transesterification. Energy Fuels 24 (9), 5235–5243. https://doi.org/10.1021/ef1008314.

Li, X., Yang, C., Zeng, G., Wu, S., Lin, Y., Zhou, Q., et al., 2020. Nutrient removal from swine wastewater with growing microalgae at various zinc concentrations. Algal Res. 46, 101804. https://doi.org/10.1016/j.algal.2020.101804.

Liang, Z., Liu, Y., Ge, F., Xu, Y., Tao, N., Peng, F., Wong, M., 2013. Efficiency assessment and pH effect in removing nitrogen and phosphorus by algae-bacteria combined system of Chlorella vulgaris and Bacillus licheniformis. Chemosphere 92 (10), 1383–1389. https://doi.org/10.1016/j.chemosphere.2013.05.014.

Liyanaarachchi, V.C., Nishshanka, G.K.S.H., Nimarshana, P.H.V., Ariyadasa, T.U., Attalage, R.A., 2020. Development of an artificial neural network model to simulate the growth of microalga Chlorella vulgaris incorporating the effect of micronutrients. J. Biotechnol. 312, 44–55. https://doi.org/10.1016/j.jbiotec.2020.02.010.

Lu, Q., Yang, L., Deng, X., 2020. Critical thoughts on the application of microalgae in aquaculture industry. Aquaculture 528, 735538. https://doi.org/10.1016/j.aquaculture.2020.735538.

Luo, L.-z., Lin, X.-a., Zeng, F.-j., Wang, M., Luo, S., Peng, L., Tian, G.-m., 2019. Using co-occurrence network to explore the effects of bio-augmentation on the microalgae-based wastewater treatment process. Biochem. Eng. J. 141, 10—18. https://doi.org/10.1016/j.bej.2018.10.001.

Macías-Sánchez, M.D., Mantell, C., Rodríguez, M., Martínez de la Ossa, E., Lubián, L.M., Montero, O., 2007. Supercritical fluid extraction of carotenoids and chlorophyll a from Synechococcus sp. J. Supercrit. Fluids 39 (3), 323—329. https://doi.org/10.1016/j.supflu.2006.03.008.

Maity, J.P., Bundschuh, J., Chen, C.-Y., Bhattacharya, P., 2014. Microalgae for third generation biofuel production, mitigation of greenhouse gas emissions and wastewater treatment: present and future perspectives — a mini review. Energy 78, 104—113. https://doi.org/10.1016/j.energy.2014.04.003.

Mantzavinos, D., Kalogerakis, N., 2005. Treatment of olive mill effluents: Part I. Organic matter degradation by chemical and biological processes—an overview. Environ. Int. 31 (2), 289—295. https://doi.org/10.1016/j.envint.2004.10.005.

Mathews, J., Wang, G., 2009. Metabolic pathway engineering for enhanced biohydrogen production. Int. J. Hydrog. Energy 34 (17), 7404—7416. https://doi.org/10.1016/j.ijhydene.2009.05.078.

Mohd Udaiyappan, A.F., Hasan, H.A., Takriff, M.S., Abdullah, S.R.S., Maeda, T., Mustapha, N.A., et al., 2020. Microalgae-bacteria interaction in palm oil mill effluent treatment. J. Water Process Eng. 35, 101203. https://doi.org/10.1016/j.jwpe.2020.101203.

Morales-Amaral, M.d.M., Gómez-Serrano, C., Acién, F.G., Fernández-Sevilla, J.M., Molina-Grima, E., 2015. Production of microalgae using centrate from anaerobic digestion as the nutrient source. Algal Res. 9, 297—305. https://doi.org/10.1016/j.algal.2015.03.018.

Morales, M., Quintero, J., Conejeros, R., Aroca, G., 2015. Life cycle assessment of lignocellulosic bioethanol: environmental impacts and energy balance. Renew. Sustain. Energy Rev. 42, 1349—1361. https://doi.org/10.1016/j.rser.2014.10.097.

Neethu, B., Bhowmick, G.D., Ghangrekar, M.M., 2018. Enhancement of bioelectricity generation and algal productivity in microbial carbon-capture cell using low cost coconut shell as membrane separator. Biochem. Eng. J. 133, 205—213. https://doi.org/10.1016/j.bej.2018.02.014.

Nie, X., Mubashar, M., Zhang, S., Qin, Y., Zhang, X., 2020. Current progress, challenges and perspectives in microalgae-based nutrient removal for aquaculture waste: a comprehensive review. J. Clean. Prod. 277, 124209. https://doi.org/10.1016/j.jclepro.2020.124209.

Nigam, P.S., Singh, A., 2011. Production of liquid biofuels from renewable resources. Prog. Energy Combust. Sci. 37 (1), 52—68. https://doi.org/10.1016/j.pecs.2010.01.003.

Nwoba, E.G., Ayre, J.M., Moheimani, N.R., Ubi, B.E., Ogbonna, J.C., 2016. Growth comparison of microalgae in tubular photobioreactor and open pond for treating anaerobic digestion piggery effluent. Algal Res. 17, 268—276. https://doi.org/10.1016/j.algal.2016.05.022.

Oey, M., Sawyer, A.L., Ross, I.L., Hankamer, B., 2016. Challenges and opportunities for hydrogen production from microalgae. Plant Biotechnol. J. 14 (7), 1487—1499. https://doi.org/10.1111/pbi.12516.

Ozkan, A., Kinney, K., Katz, L., Berberoglu, H., 2012. Reduction of water and energy requirement of algae cultivation using an algae biofilm photobioreactor. Bioresour. Technol. 114, 542—548. https://doi.org/10.1016/j.biortech.2012.03.055.

Pádrová, K., Lukavský, J., Nedbalová, L., Čejková, A., Cajthaml, T., Sigler, K., et al., 2015. Trace concentrations of iron nanoparticles cause overproduction of biomass and lipids during cultivation of cyanobacteria and microalgae. J. Appl. Phycol. 27 (4), 1443—1451. https://doi.org/10.1007/s10811-014-0477-1.

Park, J., Naresh Kumar, A., Cayetano, R.D.A., Kim, S.-H., 2020. Assessment of Chlorella sp. as a potential feedstock for biological methane production. Bioresour. Technol. 305, 123075. https://doi.org/10.1016/j.biortech.2020.123075.

Poach, M.E., Hunt, P.G., Reddy, G.B., Stone, K.C., Johnson, M.H., Grubbs, A., 2004. Swine wastewater treatment by marsh—pond—marsh constructed wetlands under varying nitrogen loads. Ecol. Eng. 23 (3), 165—175. https://doi.org/10.1016/j.ecoleng.2004.09.001.

Polat, E., Yüksel, E., Altınbas, M., 2020. Effect of different iron sources on sustainable microalgae-based biodiesel production using Auxenochlorella protothecoides. Renew. Energy 162, 1970—1978. https://doi.org/10.1016/j.renene.2020.09.030.

Prochazkova, G., Safarik, I., Branyik, T., 2013. Harvesting microalgae with microwave synthesized magnetic microparticles. Bioresour. Technol. 130, 472–477. https://doi.org/10.1016/j.biortech.2012.12.060.

Rahman, Q.M., Wang, L., Zhang, B., Xiu, S., Shahbazi, A., 2015. Green biorefinery of fresh cattail for microalgal culture and ethanol production. Bioresour. Technol. 185, 436–440. https://doi.org/10.1016/j.biortech.2015.03.013.

Rahman, Q.M., Zhang, B., Wang, L., Joseph, G., Shahbazi, A., 2019a. A combined fermentation and ethanol-assisted liquefaction process to produce biofuel from Nannochloropsis sp. Fuel 238, 159–165. https://doi.org/10.1016/j.fuel.2018.10.116.

Rahman, Q.M., Zhang, B., Wang, L., Shahbazi, A., 2019b. A combined pretreatment, fermentation and ethanol-assisted liquefaction process for production of biofuel from Chlorella sp. Fuel 257, 116026. https://doi.org/10.1016/j.fuel.2019.116026.

Ras, M., Lardon, L., Bruno, S., Bernet, N., Steyer, J.-P., 2011. Experimental study on a coupled process of production and anaerobic digestion of Chlorella vulgaris. Bioresour. Technol. 102 (1), 200–206. https://doi.org/10.1016/j.biortech.2010.06.146.

Rathore, D., Singh, A., 2013. Biohydrogen production from microalgae. In: Gupta, V.K., Tuohy, M.G. (Eds.), Biofuel Technologies: Recent Developments. Springer Berlin Heidelberg, Berlin, Heidelberg, pp. 317–333.

Reddy, H.K., Muppaneni, T., Patil, P.D., Ponnusamy, S., Cooke, P., Schaub, T., Deng, S., 2014. Direct conversion of wet algae to crude biodiesel under supercritical ethanol conditions. Fuel 115, 720–726. https://doi.org/10.1016/j.fuel.2013.07.090.

Reimann, R., Zeng, B., Jakopec, M., Burdukiewicz, M., Petrick, I., Schierack, P., Rödiger, S., 2020. Classification of dead and living microalgae Chlorella vulgaris by bioimage informatics and machine learning. Algal Res. 48, 101908. https://doi.org/10.1016/j.algal.2020.101908.

Reungoat, J., Escher, B.I., Macova, M., Keller, J., 2011. Biofiltration of wastewater treatment plant effluent: effective removal of pharmaceuticals and personal care products and reduction of toxicity. Water Res. 45 (9), 2751–2762. https://doi.org/10.1016/j.watres.2011.02.013.

Rincón-Pérez, J., Razo-Flores, E., Morales, M., Alatriste-Mondragón, F., Celis, L.B., 2020. Improving the biodegradability of Scenedesmus obtusiusculus by thermochemical pretreatment to produce hydrogen and methane. BioEnergy Res. 13 (2), 477–486. https://doi.org/10.1007/s12155-019-10067-w.

Rosli, S.S., Amalina Kadir, W.N., Wong, C.Y., Han, F.Y., Lim, J.W., Lam, M.K., et al., 2020. Insight review of attached microalgae growth focusing on support material packed in photobioreactor for sustainable biodiesel production and wastewater bioremediation. Renew. Sustain. Energy Rev. 134, 110306. https://doi.org/10.1016/j.rser.2020.110306.

Ryu, H.J., Oh, K.K., Kim, Y.S., 2009. Optimization of the influential factors for the improvement of CO2 utilization efficiency and CO2 mass transfer rate. J. Ind. Eng. Chem. 15 (4), 471–475. https://doi.org/10.1016/j.jiec.2008.12.012.

Saka, S., Kusdiana, D., 2001. Biodiesel fuel from rapeseed oil as prepared in supercritical methanol. Fuel 80 (2), 225–231. https://doi.org/10.1016/S0016-2361(00)00083-1.

Salama, E.-S., Kurade, M.B., Abou-Shanab, R.A.I., El-Dalatony, M.M., Yang, I.-S., Min, B., Jeon, B.-H., 2017. Recent progress in microalgal biomass production coupled with wastewater treatment for biofuel generation. Renew. Sustain. Energy Rev. 79, 1189–1211. https://doi.org/10.1016/j.rser.2017.05.091.

Salerno, M., Nurdogan, Y., Lundquist, T.J., 2009. Biogas Production from Algae Biomass Harvested at Wastewater Treatment Ponds. ASABE Paper Number: Bio098023.

Sanchez Rizza, L., Sanz Smachetti, M.E., Do Nascimento, M., Salerno, G.L., Curatti, L., 2017. Bioprospecting for native microalgae as an alternative source of sugars for the production of bioethanol. Algal Res. 22, 140–147. https://doi.org/10.1016/j.algal.2016.12.021.

Santana, A., Jesus, S., Larrayoz, M.A., Filho, R.M., 2012. Supercritical carbon dioxide extraction of algal lipids for the biodiesel production. Procedia Eng. 42, 1755–1761. https://doi.org/10.1016/j.proeng.2012.07.569.

Sathish, A., Sims, R.C., 2012. Biodiesel from mixed culture algae via a wet lipid extraction procedure. Bioresour. Technol. 118, 643–647. https://doi.org/10.1016/j.biortech.2012.05.118.

Serna-García, R., Zamorano-López, N., Seco, A., Bouzas, A., 2020. Co-digestion of harvested microalgae and primary sludge in a mesophilic anaerobic membrane bioreactor (AnMBR): methane potential and microbial diversity. Bioresour. Technol. 298, 122521. https://doi.org/10.1016/j.biortech.2019.122521.

Shen, Y., Yu, T., Xie, Y., Chen, J., Ho, S.-H., Wang, Y., Huang, F., 2019. Attached culture of Chlamydomonas sp. JSC4 for biofilm production and TN/TP/Cu(II) removal. Biochem. Eng. J. 141, 1—9. https://doi.org/10.1016/j.bej.2018.09.017.

Sialve, B., Bernet, N., Bernard, O., 2009. Anaerobic digestion of microalgae as a necessary step to make microalgal biodiesel sustainable. Biotechnol. Adv. 27 (4), 409—416. https://doi.org/10.1016/j.biotechadv.2009.03.001.

Singh, M., Reynolds, D.L., Das, K.C., 2011. Microalgal system for treatment of effluent from poultry litter anaerobic digestion. Bioresour. Technol. 102 (23), 10841—10848. https://doi.org/10.1016/j.biortech.2011.09.037.

Solé-Bundó, M., Garfí, M., Ferrer, I., 2020. Pretreatment and co-digestion of microalgae, sludge and fat oil and grease (FOG) from microalgae-based wastewater treatment plants. Bioresour. Technol. 298, 122563. https://doi.org/10.1016/j.biortech.2019.122563.

Solovchenko, A., Verschoor, A.M., Jablonowski, N.D., Nedbal, L., 2016. Phosphorus from wastewater to crops: an alternative path involving microalgae. Biotechnol. Adv. 34 (5), 550—564. https://doi.org/10.1016/j.biotechadv.2016.01.002.

Spolaore, P., Joannis-Cassan, C., Duran, E., Isambert, A., 2006. Commercial applications of microalgae. J. Biosci. Bioeng. 101 (2), 87—96. https://doi.org/10.1263/jbb.101.87.

Sprague, M., Dick, J.R., Tocher, D.R., 2016. Impact of sustainable feeds on omega-3 long-chain fatty acid levels in farmed Atlantic salmon, 2006—2015. Sci. Rep. 6 (1), 21892. https://doi.org/10.1038/srep21892.

Stone, K.C., Poach, M.E., Hunt, P.G., Reddy, G.B., 2004. Marsh-pond-marsh constructed wetland design analysis for swine lagoon wastewater treatment. Ecol. Eng. 23 (2), 127—133. https://doi.org/10.1016/j.ecoleng.2004.07.008.

Suali, E., Sarbatly, R., 2012. Conversion of microalgae to biofuel. Renew. Sustain. Energy Rev. 16 (6), 4316—4342. https://doi.org/10.1016/j.rser.2012.03.047.

Supriyanto, Noguchi, R., Ahamed, T., Rani, D.S., Sakurai, K., Nasution, M.A., et al., 2019. Artificial neural networks model for estimating growth of polyculture microalgae in an open raceway pond. Biosyst. Eng. 177, 122—129. https://doi.org/10.1016/j.biosystemseng.2018.10.002.

Takisawa, K., Kanemoto, K., Kartikawati, M., Kitamura, Y., 2013a. Simultaneous hydrolysis-esterification of wet microalgal lipid using acid. Bioresour. Technol. 149, 16—21. https://doi.org/10.1016/j.biortech.2013.09.031.

Takisawa, K., Kanemoto, K., Miyazaki, T., Kitamura, Y., 2013b. Hydrolysis for direct esterification of lipids from wet microalgae. Bioresour. Technol. 144, 38—43. https://doi.org/10.1016/j.biortech.2013.06.008.

Tejido-Nuñez, Y., Aymerich, E., Sancho, L., Refardt, D., 2019. Treatment of aquaculture effluent with Chlorella vulgaris and Tetradesmus obliquus: the effect of pretreatment on microalgae growth and nutrient removal efficiency. Ecol. Eng. 136, 1—9. https://doi.org/10.1016/j.ecoleng.2019.05.021.

Tejido-Nuñez, Y., Aymerich, E., Sancho, L., Refardt, D., 2020. Co-cultivation of microalgae in aquaculture water: interactions, growth and nutrient removal efficiency at laboratory- and pilot-scale. Algal Res. 49, 101940. https://doi.org/10.1016/j.algal.2020.101940.

Tibbetts, S.M., 2018. The potential for 'next-generation', microalgae-based feed ingredients for salmonid aquaculture in context of the blue revolution. In: Microalgal Biotechnology. IntechOpen, pp. 151—175.

Vo, H.N.P., Ngo, H.H., Guo, W., Nguyen, T.M.H., Liu, Y., Liu, Y., et al., 2019. A critical review on designs and applications of microalgae-based photobioreactors for pollutants treatment. Sci. Total Environ. 651, 1549—1568. https://doi.org/10.1016/j.scitotenv.2018.09.282.

Wang, H., Ji, C., Bi, S., Zhou, P., Chen, L., Liu, T., 2014. Joint production of biodiesel and bioethanol from filamentous oleaginous microalgae Tribonema sp. Bioresour. Technol. 172, 169—173. https://doi.org/10.1016/j.biortech.2014.09.032.

Wang, T., Yang, W.-L., Hong, Y., Hou, Y.-L., 2016. Magnetic nanoparticles grafted with amino-riched dendrimer as magnetic flocculant for efficient harvesting of oleaginous microalgae. Chem. Eng. J. 297, 304—314. https://doi.org/10.1016/j.cej.2016.03.038.

Wang, Y., Ho, S.-H., Cheng, C.-L., Nagarajan, D., Guo, W.-Q., Lin, C., et al., 2017. Nutrients and COD removal of swine wastewater with an isolated microalgal strain Neochloris aquatica CL-M1 accumulating high carbohydrate content used for biobutanol production. Bioresour. Technol. 242, 7—14. https://doi.org/10.1016/j.biortech.2017.03.122.

Woertz, I., Feffer, A., Lundquist, T., Nelson, Y., 2009. Algae grown on dairy and municipal wastewater for simultaneous nutrient removal and lipid production for biofuel feedstock. J. Environ. Eng. 135 (11), 1115–1122. https://doi.org/10.1061/(ASCE)EE.1943-7870.0000129.

Xu, J., Tang, Y., Yan, X., Zhou, J., 2017. Experimental and theoretical methods on determination of stress-dependent natural frequency of marine sedimentary clay. Mar. Georesour. Geotechnol. 35 (8), 1168–1180. https://doi.org/10.1080/1064119X.2017.1302527.

Yen, H.-W., Brune, D.E., 2007. Anaerobic co-digestion of algal sludge and waste paper to produce methane. Bioresour. Technol. 98 (1), 130–134. https://doi.org/10.1016/j.biortech.2005.11.010.

Yew, G.Y., Puah, B.K., Chew, K.W., Teng, S.Y., Show, P.L., Nguyen, T.H.P., 2020. Chlorella vulgaris FSP-E cultivation in waste molasses: photo-to-property estimation by artificial intelligence. Chem. Eng. J. 402, 126230. https://doi.org/10.1016/j.cej.2020.126230.

Zamalloa, C., Vulsteke, E., Albrecht, J., Verstraete, W., 2011. The techno-economic potential of renewable energy through the anaerobic digestion of microalgae. Biosour. Technol. 102 (2), 1149–1158. https://doi.org/10.1016/j.biortech.2010.09.017.

Zhang, B., Wang, L., Hasan, R., Shahbazi, A., 2014. Characterization of a native algae species chlamydomonas debaryana: strain selection, bioremediation ability, and lipid characterization. BioResources 9 (4), 6130–6140.

Zhang, B., Wang, L., Li, R., Rahman, Q.M., Shahbazi, A., 2017. Catalytic conversion of Chlamydomonas to hydrocarbons via the ethanol-assisted liquefaction and hydrotreating processes. Energy Fuels 31 (11), 12223–12231. https://doi.org/10.1021/acs.energyfuels.7b02080.

Zhang, B., Wang, L., Riddicka, B.A., Li, R., Able, J.R., Boakye-Boaten, N.A., Shahbazi, A.J.S., 2016. Sustainable production of algal biomass and biofuels using swine wastewater in North Carolina, US. Sustainability 8 (5), 477.

Zhu, C., Han, D., Li, Y., Zhai, X., Chi, Z., Zhao, Y., Cai, H., 2019. Cultivation of aquaculture feed Isochrysis zhangjiangensis in low-cost wave driven floating photobioreactor without aeration device. Bioresour. Technol. 293, 122018. https://doi.org/10.1016/j.biortech.2019.122018.

Zhu, C., Xi, Y., Zhai, X., Wang, J., Kong, F., Chi, Z., 2021. Pilot outdoor cultivation of an extreme alkalihalophilic Trebouxiophyte in a floating photobioreactor using bicarbonate as carbon source. J. Clean. Prod. 283, 124648. https://doi.org/10.1016/j.jclepro.2020.124648.

Zhu, C., Zhu, H., Cheng, L., Chi, Z., 2018. Bicarbonate-based carbon capture and algal production system on ocean with floating inflatable-membrane photobioreactor. J. Appl. Phycol. 30 (2), 875–885. https://doi.org/10.1007/s10811-017-1285-1.

Zhuang, L.-L., Azimi, Y., Yu, D., Wu, Y.-H., Hu, H.-Y., 2018. Effects of nitrogen and phosphorus concentrations on the growth of microalgae Scenedesmus. LX1 in suspended-solid phase photobioreactors (ssPBR). Biomass Bioenergy 109, 47–53. https://doi.org/10.1016/j.biombioe.2017.12.017.

Index

Note: 'Page numbers followed by "*f*" indicate figures and "*t*" indicate tables.'

A

Abiotic removal, 134—135
Acid mine drainage (AMD), 36—37
Acinetobacter pittii, 32—33
Acrylamides, 20—21
Activated carbon adsorption process, 131
Activated Sludge Model No. 3 (ASM), 39—40
Activated sludge process (AS process), 51—52
Active immobilized systems, 20—21
Adsorption, H_2S removal through, 192—193
Aeration process, 16
Agricultural wastewater treatment
  enhancement of microalgae cultivation in
    wastewater, 253—256
    with artificial intelligence, 255—256
    with microbial fuel cells, 253—254
    nanotechnology for, 254—255
  microalgae cultivation systems, 246—250
  microalgae for, 235—240
    challenges of resource recovery using,
      250—253
    feeds, fuels, and fertilizers, 240—245
    recovery of resources from, 235—236
  microalgae growth
    on aquaculture wastewater, 238—239
    on effluent of anaerobic digestion, 239—240
    on swine wastewater, 236—238
Agriculture, 235
Air lift photobioreactor, 19
Air stripping, 132
Algae, 14
  biogas upgrading
    benefits of, 199
    limitations of, 200
  in biological treatment of wastewater, 14—27
  biomass, 89—90
    production, 215
  cultivation, 92
  examples of wastewater treatment by
    algae—bacteria consortia, 32—38
  mechanism of symbiosis between bacteria and,
    27—32
    carbon—oxygen recycle, 29—30

    growth stimulation, 30—31
    toxicity reduction, 31—32
  ponds, 88
  process design, and modeling aspects of
    algae—bacteria based wastewater treatment,
    38—42
    dynamic algae—bacteria models for wastewater
      treatment, 39—42
Algae liquid fertilizer (ALF), 244—245
Algae—bacteria consortia, 13—14
  systems, 35—36
  wastewater treatment by, 32—38
    domestic wastewater, 32—35
    industrial wastewater, 35—38
Algae—bacteria treatment systems, 36—37
Algae—bacteria—based bioremediation, 13
AlgaeSim, 41—42
Algal biomass
  contaminants of emerging concern, 93
  contamination of, 92—95
  heavy metals, 94
  human pathogens, 94—95
Algal bioreactor configurations and operations,
  201—204
  alkalinity and temperature, 203
  influence of light, 203—204
  mono-and cocultivation of microalgae,
    204
Algal cultivation systems, 151—152
  economic analysis, 156—160
  economic performance of, 157t—159t
  life cycle impact assessment, 160—162
  liquid digestate
    algal cultivation systems, 151—152
    chemical analysis of, 151t
    chemical and EC treatment of, 150—151
    economic analysis, 153
    life cycle assessment, 153
    mass and energy balance analysis, 152—153
    mass and energy balance analysis, 153—156
Algal reactor products utilization and
    management, 207
Algal turf scrubber (ATS), 97

265

266 Index

Algal-based sewage treatment, 57–59
    mixotrophic sewage treatment and resource
        recovery system, 58
    STaRR system *vs.* high-rate algal ponds, 58–59
Alginates, 20–21
Alkalinity, 203
Ammonia (NH_3), 30, 115, 191–192
    inhibition, 131–132
Ammonium (NH_4), 114
    inhibition, 131–132
Anabaena, 18
Anaerobic bioprocesses, 3
Anaerobic biotechnology, 183
Anaerobic digesters, 168–170
Anaerobic digestion (AD), 2, 6, 61, 113, 149–150,
    165–166, 235
    digestates treatment by microalgae
      challenges and potential remedies for,
        131–135
      microalgal cultures, 114–129
      pilot scale plants, 135–140, 136t–138t
    microalgal growth in digestates, 129
Anaerobic Digestion Model No. 1 (ADM1), 41
Anaerobic fluidized membrane bioreactor
    (AFMBR), 56
Anaerobic membrane bioreactor (AnMBR), 56
Anaerobic sludge blanket reactors (UASB), 166
Anaerobically digested sewage sludge (ATSS), 125
Animal feed, 13
Aquaculture feed, 13
Aquaculture wastewater, 238–239
Aqueous phase (AP), 62
Arsenic (As), 126–127
Arthrospira, 18
Artificial intelligence (AI), 255–256
    enhancement of microalgae cultivation with,
      255–256
Ascaris lumbricoides, 94–95
Attached culture systems, 16–17
Attached microalgae cultivation, 248–249

B
Bacteria, 14, 85, 94–95
    removal, 66–69
      comparison of STaRR system with
        technologies, 67–69
Bicarbonate (HCO_3^-), 115
Biochemical oxygen demand (BOD), 51
Biocrude oil, recovery of, 69

Biodiesel, 240–241
    feedstock, 6–7
Biodiversity, 87
Bioeconomy, 13
Bioethanol, 241
Biofertilizers, 13
Biofilm systems, 97
Biofiltration systems, 236
Biofuels, 13, 38
    microalgae, 240–244
      biodiesel, 240–241
      bioethanol, 241
      hydrogen, 243–244
      methane, 241–243
Biogas, 2, 113, 115–116
    conditioning technologies, 187–194
      CO_2 removal, 187–192
      H_2S removal, 192–193
      novel microalgae, 194–199
      siloxane removal, 193–194
    constituents removal, 204–207
      CO_2 removal, 204–205
      energy density upgrading, 204–205
      H_2S removal, 205–206
      siloxanes removal, 206
      VOCs removal, 206–207
    production, 172–177
Biological filters, 193
Biological methods, 192
Biological oxygen demand (BOD), 14
Biological treatment of wastewater, 14–27
    biotransformation of organic micropollutants,
      22–27
    conventional biological treatment of wastewater,
      14–17
    metabolic interactions, 27–32
    nutrient removal, 17–21
      immobilized systems for microalgal wastewater
        treatment, 20–21
      suspended culture systems for microalgal
        wastewater treatment, 20
    removal of heavy metals, 21–22
Biological wastewater treatment (BWWT), 14
Biomass
    characteristics, 171–172
    production, 172
Biomethane (CH_4), 149–150
    analytical procedures, 171
    energy balances, 177–180

experimental design, 170—171
HRAP feed conditions, 173t
pilot plant description, 166—170
pilot plant performance, 171—176
   biogas production, 172—176
   biomass characteristics, 171—172
   biomass production, 172
   wastewater treatment, 171—172
production, 166
prototype performance, 176—177
   biogas production, 177
   wastewater treatment, 176—177
prototype plant description, 166—170
Biorefinery approaches, 149—150
Biorefining, 2
Biosequestration of CO_2 emissions by microalgal
   cultures, 4—5
Biosurfactants, 13
Biotransformation of organic micropollutants,
   22—27
comparison of micropollutant removal
   efficiencies in algal and bacterial treatment
   systems, 24t—25t
Boron (B), 114
Botryococcus, 18

C

Cadmium (Cd), 126—127
Calcium (Ca), 113—114
Capital expenditure (CapEx), 150, 153
Carbamazepine, 93
Carbon (C), 114—115
carbon—oxygen recycle, 29—30
constituents, 114—125
   inorganic carbon, 115—116
   organic carbon, 116—125
recovery, 56
sources for algae, 200—201
Carbon dioxide (CO_2), 2, 51—52, 114—115, 183
removal, 187—192, 194, 204—205
   biological methods, 192
   chemical absorption, 189
   cryogenic separation, 190—191
   membrane separation, 191—192
   organic solvent scrubbing, 189
   PSA, 189—190
   water scrubbing, 188, 188f
sequestration, 2
Carbon monoxide (CO), 115

Carbon-to-nitrogen ratio (C:N ratio), 6—7
Carbonate (CO_3^{2-}), 115
Carbonation station, construction of, 219—220
Carbonic acid (H_2CO_3), 115
Catalytic impregnation, 192—193
Cells, 85
Cellulose, 84
Centrifugation, 225, 254—255
Chemical absorption, 189
Chemical oxygen demand (COD), 32—33, 116,
   150—151, 228
Chemical treatment of liquid digestate, 150—151
Chinampas, 81—82
Chlamydomonas sp, 18—19, 27
   C. incerta, 18—19
   C. reinhardtii, 27
Chlorella sp., 4, 18, 22, 27, 88, 129
   C. vulgaris, 18
Chlorine (Cl), 114
Chromium (Cr), 126—127
Circular economy (CE), 1, 81, 235—236
potential role of microalgae in, 3—7
techno-economic feasibility, 7—8
waste valorization and, 1—3
Circular wastewater treatment, 81—84
principles of circularity in wastewater, 81—82
recoverable resources from wastewater, 82—84
Circularity
microalgae for, 84—95
principles of circularity in wastewater, 81—82
uncertainties and challenges microalgae-based
   technology for, 91—95
   contamination of algal biomass, 92—95
   technology application challenges, 91—92
Closed loop process, 3
Closed microalgae seed system, 7
Closed systems, 19
Cobalt (Co), 114, 126—127
Coccoid algae, 85
Combined heat and power technologies (CHP
   technologies), 184
Combustion, 61
Computational fluid dynamics (CFD), 246—247
Constructed wetlands (CWs), 227
Contaminants of emerging concern (CEC), 83, 93
Contaminants removal, 102
Continuously stirred tank reactor (CSTR),
   150—151

268 Index

Conventional activated sludge treatment process (CAS treatment process), 14
Conventional biological treatment of wastewater, 14—17
    attached culture systems, 16—17
    suspended culture systems, 14—16
Copper (Cu), 114, 126—127
Corn straw, 6—7
Cryogenic separation, 190—191
Cultivation, 167—168
    of microalgae, 235

D

Desmodesmus, 18
Digestates, 113—114, 125
    challenges and potential remedies for, 131—135
        ammonium/ammonia inhibition, 131—132
        carbon limitation, 133
        dominance of other communities, 133—134
        limitations regarding biogas, 134—135
        phosphorus limitation, 133
        turbidity, 131
    microalgal cultures, 114—129
    microalgal growth in, 129
Dinitrogen (N_2), 115
Dissolved air flotation (DAF), 165—166, 216, 224
Dissolved inorganic carbon (DIC), 201—203
Domestic wastewater, 32—35
Dry matter (DM), 113—114
Dry weight (DW), 114—115, 215
Drying process, 156
Dunaliella, 18
Dynamic algae—bacteria models for wastewater treatment, 39—42

E

Economic analysis, 150, 153, 156—160
Effluent of anaerobic digestion, 239—240
Electrocoagulation (EC), 150
    treatment of liquid digestate, 150—151
Energy
    balances, 177—180
    density upgrading, 204—205
    recovery, 69
        of biocrude oil, 69
        comparison of STaRR system with other technologies, 69
Energy dispersive X-ray analysis (EDX analysis), 63

Energy Return of Investment (EROI), 227
Enterobius vermicularis, 94—95
Enterococcus faecalis, 66—67
Escherichia coli, 66—67, 94—95
Ettlia, 18
European Environmental Agency (EEA), 1, 101
European Union (EU), 1—2, 90
Extracellular polymeric substances (EPSs), 84

F

Facultative ponds, 15
Fat oil and grease (FOG), 242
Feeds, microalgae-based, 245
Fermentation, 6
Fertilizers, microalgae, 244—245
Fertilizing Products Regulation (FPR), 231
Filtration, 225
Flaring gases, 4
Flat plate reactors, 89—90
Floating photobioreactors, 250
Flow deflectors, 222
Flue gases, 4, 13
Food security, 7—8
Fossil fuels, 13
Free ammonia (FA), 131—132
Fresh matter (FM), 113—114
Fresh weight (FW), 215
Fungi, 14

G

Galdieria sulphuraria, 39, 58, 69
γ-Proteobacteria, 31—32
Gas quality for heat and power equipment, 184—187
Gas-permeable membrane (GPM), 63
Gas-permeable membrane reactor (GPMR), 63
Generator sets (gen-sets), 183
Giardia spp., 94—95
Global warming potential (GWP), 153, 160—161
Glucose-based carbohydrates, 6
Glucosyl transferase, 25—26
Glutathione S-transferase, 25—26

H

Harvesting, 167—168
    by flotation, 169
    process, 223—225
        centrifugation, 225
        effluent conditioning, 224

filtration, 225
  optimal strategy to efficiently harvesting, 224
  preconcentration, 224
Heavy metals (HM), 87, 94
  phycoremediation of heavy metals by algae, 23f
  removal of, 21—22
Heavy metals and metalloids (HMMs), 126—127
Heterotrophic cultivation of microalgae, 97
High-rate algal ponds (HRAPs), 19, 57—59,
      127—129, 165, 216
Higher heating value (HHV), 69
Horizontal tube reactors, 89
Hydraulic retention time (HRT), 33, 151, 165,
      216, 241—242
Hydrochloric acid (HCl), 62—63
Hydrogen, 84, 243—244
Hydrogen sulfide (H_2S), 115, 183, 205—206
  removal, 192—193
    through adsorption, 192—193
    using biological filters, 193
Hydrothermal liquefaction (HTL), 38, 58
  of algal biomass, 61—62

I

Ibuprofen, 93
Immobilized systems for microalgal wastewater
      treatment, 20—21
Indole-3-acetic acid (IAA), 31
Inductively coupled plasma optical emission
      spectrometry (ICP-OES), 62
Industrial wastewater, 35—38
Inorganic carbon (IC), 114—116
Iron (Fe), 114, 254

K

k-Nearest Neighbor (k-NN), 255—256
Kitchen waste, 6—7

L

Large-scale raceway ponds construction, 217—222
  construction of carbonation station, 219—220
  construction of pond walls and bottom, 218—219
  covering or lining of ponds, 220
  mixing system, 220—222
    conventional HRAP with paddlewheel, 220
    flow deflectors, 222
    LEAR, 221—222
  settling sumps, 219—220
Lead (Pb), 126—127

Legal Readiness Level (LRL), 90
Length to width ratio (L/W ratio), 218—219
Lethal dose concentration of 50- (LD50),
      134—135
Life cycle assessment, 153
  analysis, 140—141
Life cycle impact assessment, 160—162
Light
  influence of, 203—204
  wavelength/photoperiod, 203—204
Light-emitting diode (LED), 203
Linear economy, 1
Lipids, 84
Liquid digestates
  algal cultivation systems, 151—152
  chemical analysis of, 151t
  chemical and EC treatment of, 150—151
  economic analysis, 153
  life cycle assessment, 153
  mass and energy balance analysis, 152—153
  N:P mole ratio of, 150
Liquid foam-bed photobioreactor, 97—98
Long—chain fatty acids, 5—6
Low Energy Algae Reactor (LEAR), 220—222

M

Macronutrients, 114
Magnesium (Mg), 113—114
Manganese (Mn), 114, 126—127
Mass and energy balance analysis, 152—156
Membrane bioreactors (MBRs), 17, 56—57
Membrane photobioreactor (MPBR), 20
Membrane separation, 191—192
Membrane-aerated bioreactors (MABRs), 17
Metagenomic dynamics, 13
Metals, 84
Methane (CH_4), 160—161, 166, 183, 241—243
Methanogens, 125
Methanol, 240—241
Methionine, 31
Metoprolol, 93
Michigan State University (MSU), 151—152
Microalgae, 165, 215
  advantageous physiology and biochemistry,
      84—87
  biofilm, 20—21
  biofuels, 240—244
    biodiesel, 240—241

Microalgae (*Continued*)
bioethanol, 241
hydrogen, 243—244
methane, 241—243
biogas upgrading, 202t
biotechnology approach, 88—90
flat plate reactors, 89—90
open ponds or raceways, 88
single-layer or horizontal tube reactors, 89
three-dimensional tubular reactors, 89
for circularity, 84—95
consortium, 223
cultivation systems, 246—250
attached microalgae cultivation, 248—249
floating photobioreactors, 250
open ponds systems, 246—247
photobioreactors, 247—248
ecosystem functioning approach, 87—88
fertilizers, 244—245
growth inhibition, 198—199
CO_2 affect, 198
H_2S toxicity, 198
toxicity issues, 198—199
microalgae-based feeds, 245
microalgae-based technologies
circular wastewater treatment, 81—84
microalgae for circularity, 84—95
for wastewater treatment, 95—102
role in circular economy, 3—7
biosequestration of CO_2 emissions by
microalgal cultures, 4—5
microalgal biorefineries, 5—7
wastewater treatment by microalgal cultures, 4
socio-economical approach, 90—91
uncertainties and challenges microalgae-based
technology for circularity, 91—95
Microalgae—bacteria consortium, 29—30
Microalgae—bacteria wastewater treatment
systems, 42
Microalgal biomass, 2, 113—114, 127
Microalgal biorefineries, 5—7
Microalgal carbohydrates, 5—6
Microalgal cultures, 4
digestates treatment by, 114—129, 117t—118t
carbon constituents, 114—125
heavy metal removals, 128t
HMMs, 126—127
microalgal treatment of digestates, 119t—124t
micropollutants, 127—129

nitrogen constituents, 125—126
phosphorus constituents, 126
Microalgal growth in digestates, 129
Microalgal—bacterial systems, 7
Microbial culture system, 15
Microbial diversity, 16
Microbial fuel cells (MFCs), 253—254
microalgae cultivation with, 253—254
Microelements, 83
Micronutrients, 114, 255—256
Microorganisms, 14
Microplastics, 83—84
Micropollutants, 22—23, 127—129
Million gallons (MG), 52
Mixotrophic cultivation, 58
Mixotrophic sewage treatment, 58
by *Galdieria sulphuraria*, 60—61
Modified Accelerated Cost Recovery System
(MACRS), 153
Molybdenum (Mo), 114
Mono-and cocultivation of microalgae, 204
Monod growth equation, 40
Monoraphidium arcuatum, 22
Multilayer photobioreactor, 19

N

Nannochloris sp, 20—21
Nannochloropsis, 18
Nanoparticles (NPs), 254—255
Nanotechnology for microalgae cultivation and
harvesting, 254—255
Nature-based solutions, 88
Nickel (Ni), 126—127
Nitrate (NO_3^-), 125
Nitric oxide (NO), 125
Nitrification, 132
Nitrification—denitrification process (ND
process), 52
Nitrite (NO_2^-), 125
Nitrogen (N), 114, 183—184, 196, 215—216
constituents, 125—126
N:P mole ratio, 150
nitrogen-fixing bacteria, 30
recovery, 56
recovery from aqueous product of HTL, 63
removal, 64, 126
Nitrous oxide (N_2O), 52, 160—161
Nonrenewable energy resources, 13
Nostoc pruniforme, 86

Index   271

Nutrients, 83
  recovery, 69–74
    aqueous phase characterization, 71
    biochar characterization, 71–72
    comparison of STaRR system with other
      technologies, 73–74
    recovery of N, 73
    recovery of P, 72–73
  sources for algae, 200–201

O

Olive wash water (OWW), 36
Open ponds systems, 246–247
Open raceway ponds (ORPs), 19, 246–247
Open systems, 19
Operational expenditure (OpEx), 150, 153
Organic carbon (OC), 114, 116–125
Organic micropollutants (OMPs), 127–129, 236
Organic rich wastes, 149–150
Organic solvents, 240–241
  scrubbing, 189
Organizational Readiness Level (ORL), 90
Orthophosphate, 126
Oxidoreductases, 25–26
Oxygen (O_2), 29–30, 86, 115
  evolving photosynthesis, 243–244
  production, 196–198
Oxygenic photosynthesis, 86

P

Palm oil mill effluent (POME), 18–19
Particle Tracking Velocity (PTV), 221–222
Passive immobilization systems, 20–21
Pathogen inactivation, 57
Personal care products (PCPs), 22–23
Phaeodactylum, 18
Pharmaceutical and personal care products
  (PPCPs), 93
Pharmaceutically active compounds (PhACs),
  22–23
Phormidium, 18
Phosphate (PO_4), 81, 114
  recovery from eluate of biochar, 63
  removal, 64–65
Phosphorus (P), 113–114, 196, 215–216
  constituents, 126
  P-rich products, 83–84
  recovery, 56

Photo-rotating biological contactor (PRBC),
  36–37
Photoautotrophy, 13
Photobioreactors (PBR), 18, 116, 247–248
Photosynthesis, 250
Photosynthetically Oxygenated Waste to Energy
  Recovery (POWER), 39
Phycoremediation, 39
Phytochelatin synthase (PCS), 22
Phytochelatins (PCs), 22
Pilot scale plants, 135–140, 136t–138t
Polyhydroxyalkanoates (PHAs), 84
Population equivalent (PE), 226
Potassium (K), 113–114
Pressure swing adsorption (PSA), 185, 189–190
Pretreatment method, 131
Proteins, 84
Protozoa, 14
Protozoans, 94–95
Publicly Owned Treatment Works (POTWs), 51
Pyrolysis, 61

Q

Quorum sensing, 87

R

Raw biogas, 204
  characteristics, 183–184
Reactive oxygen species (ROS), 86
Renewable resources, 1–2
Resource recovery, 81–82
  system, 58
River Water Quality Model No. 1 (RWQM-1),
  41
Root-Mean-Square-Error (RMSE), 255–256
Ruegeria pomeroyi, 31

S

Salmonella, 94–95
Scale up process, 168–170
Scanning electron microscope (SEM), 63
Scenedesmus sp., 18, 27, 129
  S. dimorphus, 18–19
SDGs. *See* United Nations sustainable
  development goals (SDGs)
Seawater, 216–217
Sequencing batch membrane photobioreactors
  (SB-MPBR), 19
Sequencing batch reactors (SBRs), 15

272   Index

Sewage
  farms, 81–82
  treatment, 63–66
    BOD removal, 65
    comparison of STaRR system with other technologies, 65–66
    nitrogen removal, 64
    phosphate removal, 64–65
Sewage treatment and resource recovery system (STaRR system), 60–63
  characterization of HTL byproducts, 62
  concerns about traditional POTWs, 51–53
  high-rate algal ponds *vs.*, 58–59
    comparison of salient features and performance characteristics of HRAP, 60t
  hydrothermal liquefaction of algal biomass, 61–62
  leaching of phosphates from biochar, 62–63
  mixotrophic sewage treatment by *Galdieria sulphuraria*, 60–61
  performance of, 63–74
    bacteria and virus removal, 66–69
    energy recovery, 69
    nutrient recovery, 69–74
    sewage treatment, 63–66
  recovery of nitrogen from aqueous product of HTL, 63
  recovery of phosphate from eluate of biochar, 63
  reinvention of POTWs, 53–54
    algal-based sewage treatment, 57–59
    emerging approaches for sewage treatment and resource recovery, 56–57
    potential for energy recovery from sewage, 54
    potential for nutrient recovery from sewage, 54
Shallow annular channels, 88
Silica gel, 193–194
Siloxanes, 185
  removal, 193–194, 206
Simultaneous saccharification and fermentation (SSF), 241
Single-layer reactors. *See* Horizontal tube reactors
Societal Readiness Level (SRL), 90
Socio-economical approach, 90–91
Sodium (Na), 114
Sodium hydroxide (NaOH), 62–63
Solar energy, 6
Spirulina, 88
Strongyloides stercoralis, 94–95
Sulfate reducing bacteria (SRB), 36–37

Sulfur (S), 113–116, 194–196
  in microalgae-based biogas upgrading system, 197t
Sulfur oxidizing bacteria (SOB), 193
Suspended culture systems, 14–16. *See also* Microalgal cultures
  algae–bacteria mutualistic relationship in facultative ponds, 15f
  for microalgal wastewater treatment, 20
Swine wastewater, 236–238
Switchgrass, 6–7
Switching functions, 40
Symbiosis mechanism between algae and bacteria, 27–32
Synechococcus, 18

T

Techno-economic analysis, 140–141
Techno-economic feasibility, 7–8
Technology Readiness Level (TRL), 90
Temperature, 203
Textile wastewater, 19
Thalassiosira pseudonana, 31
Thermo Scientific, 171
Three-dimensional tubular reactors, 89
Total nitrogen (TN), 62, 113–114, 176, 228
Total phosphorus (TP), 18, 62, 113–114, 126, 228
Total solids (TS), 150–151
Total suspended solids (TSS), 139, 171, 216
Trichomonas, 94–95
Trickling filters, 16–17
Turbidity, 131
Twin-layer photobioreactor, 19

U

United Nations sustainable development goals (SDGs), 90
Upflow anaerobic sludge blanket (UASB), 125
Urban wastewater, 19
US Environmental Protection Agency (US EPA), 153

V

Virus, 94–95
  removal, 66–69
    comparison of STaRR system with other technologies, 67–69
Volatile fatty acids (VFAs), 84, 116, 201

Index 273

Volatile organic chemicals (VOCs), 191—192
  removal, 206—207
Volatile solids (VS), 150—151, 242—243
Volatile suspended solids (VSS), 171
Volumetric removal rates (VRRs), 58—59
Volvox, 86

W

Waste
  paper, 6—7
  valorization, 1—3
Waste stabilization ponds (WSPs), 15
Wastewater (WW), 81, 216—217, 236
  principles of circularity in, 81—82
  recoverable resources from, 82—84
Wastewater resource recovery facility (WRRF),
      41—42
Wastewater treatment (WWT), 17—18, 82—83,
      92, 165, 171—172, 176—177, 215. *See also*
      Agricultural wastewater treatment
  using microalgae, 225—231

from commodities to higher value products,
      231
  energy consumption, 229—230
  process sustainability, 229—230
  wastewater performance, 228—229
microalgae-based technologies for, 95—102
  advantages of, 99—100
  currently technologies—large-scale application,
      96—97
  developing technologies—lab scale, 97—98
  futuristic perspective—drawing board, 98—99
  mitigation approaches for decreasing risks, 102
  risks involved in microalgae technology
      implementation, 100—101
  by microalgal cultures, 4
Wastewater treatment plants (WWTPs), 13
Water eutrophication potential, 161—162
Water scrubbing, 188

Z

Zinc (Zn), 81, 114, 126—127

Printed in the United States
by Baker & Taylor Publisher Services